Control of Power Electronic Converters and Systems

Control of Power Electronic Converters and Systems

Editor: Carolina Murray

MURPHY & MOORE
www.murphy-moorepublishing.com

www.murphy-moorepublishing.com

MURPHY & MOORE

Cataloging-in-Publication Data

Control of power electronic converters and systems / edited by Carolina Murray.
p. cm.
Includes bibliographical references and index.
ISBN 978-1-63987-744-7
1. Electric current converters. 2. Electric current converters--Control. 3. Electric current converters--Automatic control.
4. Electronic control. 5. Electric power supplies to apparatus. 6. Power electronics. I. Murray, Carolina.
TK7872.C8 C66 2023
621.381 532 2--dc23

Murphy & Moore Publishing
1 Rockefeller Plaza,
New York City,
NY 10020, USA

ISBN 978-1-63987-744-7

Contents

Preface

A power converter refers to an electrical circuit that converts electric energy from one form to another that is optimized for a particular load. Power electronic converters have been utilized in hybrid electric systems to achieve conversions from AC to DC, DC to AC, DC to DC, and AC to AC, which enable bidirectional power flow. They are primarily composed of solid-state switches including IGBT, power MOSFET, thyristors, power BJT, and lossless components such as capacitors and inductors. Precise timing control is needed for power electronic converters and systems. Power electronic systems are power systems that have an AC or DC source of electric power. This book unravels the recent studies on the control of power electronic converters and systems. It presents researches and studies performed by experts across the globe. The book will serve as a valuable source of reference for graduate and postgraduate students.

Various studies have approached the subject by analyzing it with a single perspective, but the present book provides diverse methodologies and techniques to address this field. This book contains theories and applications needed for understanding the subject from different perspectives. The aim is to keep the readers informed about the progresses in the field; therefore, the contributions were carefully examined to compile novel researches by specialists from across the globe.

Indeed, the job of the editor is the most crucial and challenging in compiling all chapters into a single book. In the end, I would extend my sincere thanks to the chapter authors for their profound work. I am also thankful for the support provided by my family and colleagues during the compilation of this book.

Editor

1

Comparison of Novel Approaches to the Predictive Control of a DC-DC Boost Converter, based on Heuristics

Robert Baždarić *, Danjel Vončina and Igor Škrjanc

Faculty of Electrical Engineering, University of Ljubljana, Tržaška 25, Ljubljana 1000, Slovenia; voncina@fe.uni-lj.si (D.V.); igor.skrjanc@fe.uni-lj.si (I.Š.)
* Correspondence: robert.bazdaric@fe.uni-lj.si

Abstract: This paper introduces novel approaches to the predictive control of a DC-DC boost converter and a comparison of the controllers built that consider all of the current objectives and minimize the complexity of the online processing. The primary concern is given to the applicability of the inclined methods for systems that are physically small but considered physically fast processes. Although the performed methodologies are simulated and applied to a DC-DC boost converter, they can have broader applicability for different switched affine systems as a subgroup of the hybrid systems. The introduced methods present an alternative way of building the process model based on the fuzzy identification that contributes to the final objective: the applicability of the predictive methods for fast processes.

Keywords: switched affine systems; hybrid systems; fuzzy identification; fuzzy modeling; two degrees of freedom; fuzzy model predictive control

---

1. Introduction

Predictive control has strict conjunction with the rendering of a proper mathematical model that will be used to determine a successor step of our control algorithm. Subsequently, the modeling paradigm must become the core issue in projecting the predictive control algorithm of the cyber-physical and the physical processes in general. Even though the process is purely natural, meaning it is a naturally continuous system, influenced by the control algorithm, it will undoubtedly evolve into the Hybrid System (HS). That is why this study points out the results from the aspects of HS controls in the field of Switched Affine Systems (SAS). Nowadays, the majority of power electronics uses a kind of Pulse Energy Conversion or Converters (PEC) in rendering the power sources or control signals. Consequently, these systems are in the group of SAS.

In this study, an alternative approach to modeling a DC-DC boost converter has been provided, and our SAS example is taken into consideration as a good representative of the aforementioned field. A renaissance of the HS theory occurred in 1990s. Simultaneously, the emerging of heuristic control principles happened: neural networks and fuzzy control. The findings of those two mostly separately developed fields will be integrated into the new approaches of the model-based predictive control of SAS. It is necessary to highlight that from the corner of the HS theory, the PEC is analytically definable and well posed [1]. From the aspects of applicability, we promote the transformation of HS to the purely continuous, although this principle is found to be incomplete by the HS theory [1]. This is mainly disproven by the inapplicability of the particular HS approaches to fast physical processes. Therein, the assumption that each particular physical process can be presented by Discrete Hybrid Automata (DHA) does not hold. In referring to the qualitative mathematical theory [2], and the need in DHA to synchronize all of the dynamic events, one faces the problem of multiple dynamics that cannot be unified simply. In previous work [3,4], the modeling was put into the focus and given possible solutions from the physical aspects of the system. First, in [3], the main problem is transferred to the idea of infinity horizon principle. It is comparable with the Model Predictive Control (MPC) infinite horizon, but it should not be mistaken for it. Second, in [4], the problem is grasped with an idea of two dynamics and unified into one by fuzzy logic (FL).

Some previous academic work have provided remarkable results in applying the advanced control algorithms for the relatively simple example of SAS: a DC-DC boost converter [5–13]. As a result, this article underlines the complexity of the controls of the DC-DC boost converter. This complexity has to be known in order to form a more solid basis for understanding the necessity for the advanced control algorithm. The assigned objectives in the control of DC-DC boost converters are the main complexity issue. Consequently, in this study, the standard objectives have been provided, as well as additionally extended with the objective of the applicability issue. The objectives are:

- Robustness to the wide system's parameter changes [PCR]
- Short response period and similar to all different operating points [SR]
- Optimality related to constraints [OPT]
- Stability [S]
- Minimization of the nonlinear phenomena for the selected robustness [NFM]
- Applicability to the systems with fast refresh rate. [A]

The above-mentioned has to determine the reason for finding a new and advanced approach that can integrate all of the objectives, even though established and notable approaches already exist [5,7,11]. Those methodologies are successful in newer releases [14,15], but are still a continuation in a way similarly improving the main and present issues, the processing complexity, and the system discontinuity. The originators of the modeling principles in their releases [5,6,16] direct the modeling to the direction of identification, because of the compelling interest in minimizing the processing complexity. We see that there is still a possibility to improve the well-articulated problems of system discontinuity and processing complexity by reducing the prediction horizon [17], which is the fundamental stability pillar of the MPC in general. Also, generally for the HS, it is essential to bear in mind that distributing or decoupling the modeling complexity into the several and less complex HS models might be the solution, i.e., current, voltage and feedback; however, then the synchronization of those models to the one and overall DHA should not be endangered [1]. Considering that the main fundament of this study is the presentation of Identification-Based (IDB) modeling and a comparison of those modeling-based control algorithms, Figure 1 presents it in a broader view. In contrast to typical and basic model representatives in HS [1], i.e., Piece Wise Affine (PWA) models, Linear Complementarity (LC) models, Extended Linear Complementarity models (ELC), Mixed Logical Dynamical (MLD), and Max-Min-Plus-Scaling (MMPS), the Identification-Based (IDB) modeling represents a new group of models rendered by different identification approaches and algorithms that partly identify the uncertainties in HS models or completely, as in the example of this study. Additionally, and to achieve the selected objectives, the new approach will contain the model-based predictive control strategy reformulated to accommodate the above-mentioned nonlinear system's dynamics problem. As a continuation of previous work [3,4], this study will concentrate more on the comparison of controllers that are rendered by the novel methodologies in seeking more advanced control solutions. Additionally, the subsequent discussion has to contribute to a better understanding of the novelty in the predictive controls that can be widely exploited. Originally, the main goal was to provide a more compact and applicable solution for the DC-DC boost converter that can be integrated into different, more complex, and practical problems. The main complexity is here grasped in the system's model structure and performed offline. The lifted modeling accuracy that is gained by the identification of a constructed process on the site can help in distinguishing and diagnosing other process problems, as discussed in [18]. In contrast, it can block the nonlinearities produced by the semiconductors realistically, and not be injected by the trivial nonlinear switching functions, but still be compact in forming the overall model for the more complex examples in [19,20]. It will be all based on assumptions that the system states are measurable and the switching period T_s is fixed. Subsequently, Section 2 provides a discussion of the distinctive problems in the modeling of SAS that contributes to the complexity and places the novel solution into the methodology frames that are of crucial importance for the understanding of latter compared controllers. Section 3 presents the successfully applied methods and comparison of the achieved results, and Section 4 contains the short conclusion.

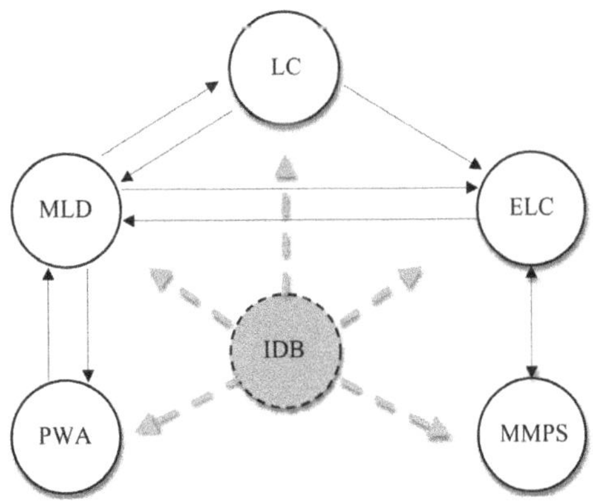

Figure 1. Established types of hybrid models and novel modeling in a Hybrid System (HS).

2. Novel Modeling Paradigm and Predictive Control of a DC-DC Boost Converter, Minimization of Online Processing

Predictive control must be the main strategy in achieving the robustness and optimality of the final control algorithm of a DC-DC boost converter. Optimality in the controls of the DC-DC boost converter must be seen through the system's constraints. The main constraints are the voltage source $E(t)$, the inductor current $i_L(t)$ and duty cycle $d(t)$. In Figure 2, a principal electronic circuit of a DC-DC boost converter is presented, and assigning the mentioned circuit constraints additionally including the converter's voltage set point s, the output voltage v_o, the load current i_R and the capacitor's voltage v_c.

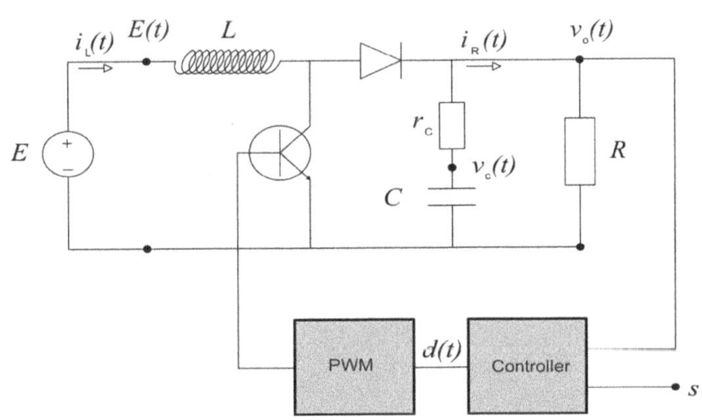

Figure 2. DC-DC boost converter circuit diagram.

To predict a successor control signal of a new control algorithm, a process model has to be known, as well as measured or observed antecedent data. In our case the process model is a mathematical model of the system presented in Figure 2. Mostly, and as already mentioned, in the broader literature of DC-DC boost converters, several different models are known. Based on the authors' opinions and knowledge about this particular system, the following distinctive and established methodologies are available:

- Small and large signal modeling [12,13]

- Poincaré mapping in discontinuous conduction mode (DCM) [7]
- HS based modeling; Mixed Logical Dynamical and Linear Complementarity [11,21,22].

Those modeling principles address most of the modeling problems of a DC-DC boost converter, but from the control theory point of view and applicability, there is room for improvement. It is unlikely that the answer or a predefined block that precisely models our natural process could be found in the modern computer toolboxes. Looking from the simulation side and finally from the control side, different natural processes can be similar. Furthermore, the modeling formalisms are prone to upgrades. A good example is HS formalisms and their development [1]. In the high refresh-rate systems, and before developing the control algorithm, the awareness about proven anomalies that can produce a significant impact on our control algorithm must be present. In the example of a DC-DC boost converter, we highlight the three most distinctive ones:

- Anomalies caused by the system nature [1,2]
- Anomalies caused by the modeling formalisms [23,24]
- Anomalies caused by the control algorithm itself [2,7].

A DC-DC boost converter is a cyber-physical process, which by its nature has a controllable switching process that automatically groups it in SAS as a subgroup of HS. If the switching event would be a time-wise insignificantly short in comparison with the discretization, or if the switching event is just occasionally appearing, then the importance might be minimized or simply distinguished by one bit of information. Otherwise, the significance of that event has to be reconsidered. The pulse width modulation (PWM) of our example is exactly the core issue in the control of a DC-DC boost converter, and subsequently, it has to be viewed through the mentioned existence of the possible anomalies.

The switching event over the semiconductors is burdened with the sticking and grazing effects over the switching manifolds [2]. For the selected physical system, these anomalies are considered natural.

In an application in which we use the HS formalisms, we must be aware of the typical anomalies as "Live lock" and "Zeno" effect [1,2,23,24].

Conclusively, when the control algorithm is in operation, the awareness of the nonlinearity phenomena recognized in bifurcations and chaos come into focus and must be treated by the control algorithm itself.

An elevation of the grade of modeling accuracy, in order to avoid previously mentioned anomalies, raises the complexity of the model and consequently the control algorithm, but it is of great importance in the stability of predictive control. Additionally, at the same time, it minimizes the nonlinearity phenomena. The mentioned complex view on modeling and the contemporary minimization of the limited time of processing bring this discussion to an alternative solution that integrates and conciliates two from the first sight disputable extremes. In our previous work [3,4], as the answer on demanding tasks, the fuzzy logic-based control algorithms successfully provide the possible solution for expressed complexity, and these will be compared here, including the two additional MPC controllers that are not yet presented, but are based on the same modeling principle. The fuzzy logic is not used traditionally [9,10,25] in a way to formulate the nonlinear gain interpolation, but differently to approximate the complex system model and answer positively to all of the objectives that have been collected together. The modeling benefits that are basically derived from the fuzzy system identification are threefold.

First, apart from the HS structure, in this way the system can be interpreted as a purely continuous system; in the worst-case scenario, it maintains the same grade of accuracy as the HS modeling approach.

Second, it provides the final model that is a more compact solution and minimizes the online processing burden of the control algorithm.

Third, in practical solutions, and because it is based on identification, this modeling principle provides the physical characteristics of the electronic circuits in real operation, which overcomes all possible theoretical miscalculations at once.

To understand the novelty of subsequently compared control algorithms, and to frame the modeling methodologies to the unique approach in studies of a DC-DC boost converter, we highlight the main idea. The most distinct characteristic of the new control algorithms is found in the non-analytical but heuristically-based modeling of a DC-DC boost converter. The source models are Takagi-Sugeno (T-S) fuzzy models derived from the mathematically transferred and normed space of measured system states [3,4]. Thus, naturally, the process of measuring, or in the simulation environment the numerical integration, transfers the originally and analytically two-dimensional state space problem of the DC-DC boost converter states $x_a = [v_c\, i_L]^T$ in the six-dimensional $\overline{x} = [\overline{v}_o\, \overline{i}_L\, \overline{E}\, \overline{R}\, d\, t]^T$ normed vector space $(V, \|\cdot\|)$. The newly formed and augmented pseudo-Banach space is derived from the Lebesgue 2 norming of the process variables $\|\cdot\|$, which is expressed by

$$\|x_i\|_{L^p} = \frac{1}{t}\left(\int |x_i|^p dt\right)^{\frac{1}{p}} = \overline{x}_i,\ for\ p = 2\ and\ i \in [1,4]. \tag{1}$$

In the above-developed space, the new methodologies perform a different grey box fuzzy identification process to derive the final fuzzy model of a DC-DC boost converter.

The physical process constraints due to the electrical components that were used in the electric circuit presented in Figure 2 will be the main experimental foundation in forming the Fuzzy Universes of Discourses $\mathbf{X}_i$ and $\mathbf{Y}$, of regression vectors and the scaled duty cycle for each particular identification process subsequently explained. In Expression (2), the universes are formed from the excited DHA simulation model of a DC-DC boost converter to provide the necessary robustness, including the discontinuous conduction mode (DCM) and the continuous conduction mode (CCM) of a DC-DC boost converter:

$$\begin{aligned} \overline{v}_o &= x_1 \in \mathbf{X}_1 = [0\ V,\ 700\ V] \\ \overline{i}_L &= x_2 \in \mathbf{X}_2 = [0\ A,\ 1030\ A] \\ E &= x_3 \in \mathbf{X}_3 = [10\ V,\ 16\ V] \\ R &= \tfrac{\overline{v}_o}{\overline{i}_R} = x_4 \in \mathbf{X}_4 = [10\ \Omega,\ 32\ \Omega] \\ d_u &= y \in \mathbf{Y} = [0.015 \cdot d,\ 0.985 \cdot d] = [0.65,\ 0.0.1]. \end{aligned} \tag{2}$$

2.1. Fuzzy Model Structure of the Two Degrees of Freedom Controller, the Prediction and Processing Minimization

Baždarić et al. in [3] demonstrated the FL identification powerful tool in reconstructing all of the system's possible steady states. It is done in the wide range of operating points that are defined by altering different process parameters in combinations and providing the expected real and extreme system conditions.

In the five-dimensional normed space (excluding the time) $(V^{(-1)}, \|\cdot\|)$, the modeling provides a fuzzy model, which is expressed by Equation (3):

$$\begin{aligned} \boldsymbol{x}_{d_4} &= [\overline{v}_o\ \overline{i}_L\ E\ R]^T \Leftrightarrow \boldsymbol{x} \\ f(\boldsymbol{x}|\boldsymbol{\theta}) &= \frac{\sum_{j=1}^{b} (a_{j,0}+a_{j,1}x_1+\ldots+a_{j,4}x_4)\mu_{Hj}(\boldsymbol{x})}{\sum_{j=1}^{b} \mu_{Hj}(\boldsymbol{x})} \\ d_{FEM} &= \mathbf{F}(\boldsymbol{x}_{d_4}) = f(\boldsymbol{x}|\boldsymbol{\theta}) \end{aligned} \tag{3}$$

that relates the system states and process parameters. In Equation (3), the $a_{j,0}$ is the member (real number) of the fuzzy model parameter matrix $\boldsymbol{\theta}$ in the jth row and the first column, while the μ_{Hj} denotes specially j^{th} c-clustering defined membership function from the b number of clusters [3]. Four measured variables of the regression vector $\boldsymbol{x}$ will give information about the process parameters' change. Instantly and from the previous offline identification process, the fuzzy model relates a linear contribution of inputs, as the T-S consequence functions (3), to the infinity horizon steady duty cycle d_{FEM}.

The model in Equation (3) is a basis for the predictive control of a DC-DC boost converter. As a derived model, Equation (3) is not a dynamic system model, it is used in [3] in the feed-forward line of the optimized PI controller. The resulting control law is

$$\begin{aligned} d_u(t) &= \tfrac{1}{T_a}\int d_{FEM}(kT_s)dt + d_{PI}(t) \quad T_a - \textit{current constraint based tuning parameter} \\ d_{FEM}(k) &= \mathbf{F}(\boldsymbol{x}_{\boldsymbol{d_4}}(k)) = \mathbf{F}(s(k), \mathrm{I}_L, E(k), \tfrac{\overline{v}_o(k)}{\overline{i}_R(k)}) \end{aligned} \tag{4}$$

where d_{PI} denotes the output from the PI controller. The control signal d_u is a scaled duty cycle $d(t)$ for the used PWM. The I_L in Equation (4) is the infinity horizon prediction of the steady-state inductor current $\overline{i}_L$, corresponding to the load current $\overline{i}_R$ and the model based reconstructed efficiency ratio.

The control law will guarantee as fast as possible response of a DC-DC boost converter in finding the steady state for a new possible operating point, mostly limited only by the maximum inductor current. As the new predicted steady state is just an approximation with the projected maximum process/model error, the optimized PI controller will be in function to compensate the steady state error. Here, the optimized PI controller's main function is to regulate the process operation near the operating point, where this type of linear controller was found to be the most effective. The posed control solution agrees with all the previously mentioned objectives and forms the Two Degrees of Freedom controller. Thus, the controller predicts the new steady duty cycle instantly and based on the measured process parameters. The fuzzy model $d_{FEM}(k) = \mathbf{F}(\boldsymbol{x}_{\boldsymbol{d_4}}(k))$ is an explicit model as the solution of Equation (3) and does not need calculation of the inverse matrix to fulfill the controller's prediction. It is a Fuzzy Model-Based predictive control that can be compared with the MPC in a sense of infinity horizon prediction [3]. This feature drastically minimizes the controller's processing time.

2.2. *Fuzzy Model Structure of the Fuzzy Model Predictive Control (FMPC) Controllers and their Processing Minimization*

Furthermore, Baždarić et al. in [4] demonstrated the FL identification powerful tool in reconstructing the system's dynamic model. In the similarly developed normed space $(V, \|\cdot\|)$ as the one in [3], but now 6-dimensional, including the time, the FL identification and modeling will provide the alternative to the analytically derived model of a DC-DC converter. Slightly different than in Equations (2) and (3), the derived fuzzy model will be constructed from the aspects that $\overline{v}_o$ is the output and d_u is an input with the corresponding universes. Moved from the HS theory, but also from the traditional "averaged switched model", this fuzzy model renders all benefits of the alternative modeling as mentioned earlier.

As stated in [4], the fuzzy model $\boldsymbol{F}(\boldsymbol{\varphi}_k)$ is expressed by

$$\begin{aligned} y_m(k+1) &= \boldsymbol{F}(\boldsymbol{\varphi}_k) = \boldsymbol{\beta}(\boldsymbol{\varphi}_2(k))\boldsymbol{\theta}_1\boldsymbol{\varphi}_1(k)^T \\ \boldsymbol{\varphi}_1(k) &= [\overline{v}_o(k) \quad \overline{v}_o(k-1) \quad \overline{i}_L(k) \quad d_u(k) \quad 1] \\ \boldsymbol{\varphi}_2(k) &= [\overline{E}(k) \quad \overline{v}_o(k)/\overline{i}_R(k) \quad \overline{i}_L(k)]. \end{aligned} \tag{5}$$

In the Equations (5) the $\boldsymbol{\varphi}_1(k)$ denotes a first-grade and $\boldsymbol{\varphi}_2(k)$ a second-grade regression vector. Furthermore, $\boldsymbol{\theta}_1$ denotes the parameter matrix of the 1st-grade identification, and $\boldsymbol{\beta}(\boldsymbol{\varphi}_2(k)) \in \mathrm{R}^p$ (p-number of fuzzy rules) is the vector of normalized degrees of fulfillment [4]. The mentioned vector is the compact expression of the T-S Fuzzy System that consists of Gaussian membership function in the process of fuzzification, the product as the conjunction function in the premise and the center-average defuzzification. The new type of Fuzzy Model that is now expressing the DC-DC boost converter is a T-S fuzzy nonlinear interpolation of p-linear models identified in the first-grade of identification and further forming the T-S consequence functions. The novelty is that the linear models in the different operating points are not analytically driven, and those are devised by the Least-Squares

(LS) identification process in the wider neighborhood of the operating point. Furthermore, this type of approach minimizes the number of rules in the fuzzy rule base and additionally smoothes the nonlinear interpolation [4]. Equation (6) presents the final state space expression of the DC-DC boost converter, where the matrixes $\boldsymbol{A}_{m_i}$, $\boldsymbol{B}_{m_i}$, $\boldsymbol{R}_{m_i}$ are the LS identified matrixes in the neighborhood of the i-th operating point and $w(k)$ is the unit step function that is used to integrate the identification process error.

$$\begin{aligned} \boldsymbol{x_m}(k+1) &= \sum_{i=1}^{p} [\boldsymbol{A}_{m_i}\boldsymbol{x}_m(k) + \boldsymbol{B}_{m_i}d_u(k) + \boldsymbol{R}_{m_i}w(k)]\beta_i(\boldsymbol{\varphi}_2(k)) \\ \beta_i(\boldsymbol{\varphi}_2(k) \in [0,1] & \quad \sum_{i=1}^{p} \beta_i(\boldsymbol{\varphi}_2(k) = 1 \quad i = 1,\ldots,p \quad \boldsymbol{x_m} = [\overline{v}_o(k)\ \overline{v}_o(k-1)]^T \end{aligned} \tag{6}$$

As the typical and compared MPC controllers have the time variable inner models (kT_s), the state space form evolves in a more compact form, as provided in Model (7):

$$\begin{aligned} \boldsymbol{A}_{m_k} &= \begin{bmatrix} a_{m1_k} & a_{m2_k} \\ 1 & 0 \end{bmatrix} \quad \boldsymbol{B}_{m_k} = \begin{bmatrix} a_{m4_k} \\ 0 \end{bmatrix} \\ \boldsymbol{R}_{m_k} &= \begin{bmatrix} a_{m3_k} i_L(k) + a_{m5_k} \\ 0 \end{bmatrix} \\ \boldsymbol{C}_{m_k} &= [1 \quad 0] \quad \boldsymbol{D}_{m_k} = 0 \\ \mathbf{x}_m(k+1) &= \boldsymbol{A}_{m_k}\mathbf{x}_m(k) + \boldsymbol{B}_{m_k}d_u(k) + \boldsymbol{R}_{m_k}w(k) \\ y_m(k) &= \boldsymbol{C}_{m_k}\mathbf{x}_m(k) + \boldsymbol{D}_{m_k}y_m(k) = \boldsymbol{C}_{m_k}\mathbf{x}_m(k). \end{aligned} \tag{7}$$

Herein, $\boldsymbol{a}_m(k) = [a_{m1}(k)a_{m2}(k)a_{m3}(k)a_{m4}(k)a_{m5}(k)] = \boldsymbol{\beta}(\boldsymbol{\varphi}_2(k))\boldsymbol{\theta}_1$ denotes the compact vector of the fuzzy model coefficients at the time k. Those coefficients are profiling our novel FMPC algorithms in the controls of a DC-DC boost converter. As Section 3 compared FMPCs that, after the modeling part, are built typically (reference model, preceding horizon principle and filtering) [26,27], the control laws will be omitted. Here, we present the cost functions $J(k)$ and the suppression factor λ that have been used to form the compared MPC algorithms:

- Fuzzy Dynamic Matrix Control (FDMC)

$$J(k) = \sum_{j=1}^{N_u} q_j(y_m(k+j|k) - r(k+j|k))^2 + \sum_{j=0}^{N_u-1} \lambda_{j+1}(k)\Delta u(k+j|k)^2. \tag{8}$$

- Fuzzy Generalized Predictive Control (FGPC)

$$\begin{aligned} \boldsymbol{J}(k) &= \sum_{j=1}^{N_u} \boldsymbol{x_J}^T(k+j|k)\boldsymbol{\Gamma}(j)\boldsymbol{x_J}(k+j|k) \\ \boldsymbol{x_J}(k) &= \begin{bmatrix} \boldsymbol{x_{J1}}(k) \\ \boldsymbol{x_{J2}}(k) \end{bmatrix} = \begin{bmatrix} y_p(k+1) - r(k+1) \\ \lambda\Delta u(k) \end{bmatrix} \end{aligned} \tag{9}$$

- Fuzzy Predictive Functional Control (FPFC)

$$J(\boldsymbol{u},k) = \sum_{j=1}^{n_H} (r(k+H_j|k) - y_m(k+H_j|k))^2 + \sum_{j=1}^{n_H} \lambda_j{}^2(k)\boldsymbol{u}(k+j-1|k)^T\boldsymbol{u}(k+j-1|k). \tag{10}$$

During the control horizon N_u, the FDMC cost function (8) minimizes the error between the model-based prediction y_m and first-order reference model r, which is additionally weighted by the state weighting factor q. The control signal u (8) is suppressed by the suppression factor λ. In Equation (9), we see the augmented model state vector $\boldsymbol{x_J}$ which contains the filtered output from the second order filter $\boldsymbol{y}_p$ [22] and no state weighing factor. The FPFC cost function in Equation (10) is different and additionally minimizes the computation burden, as it calculates for n_H a number of coincidence points H during the control horizon and minimizes the vector of the control signals for those points $\boldsymbol{u}$ [4,21,22]. It is important to add that the suppression factor λ, when comparing FMPCs, is the inner model-based adaptive and time variable factor $\lambda(k) = a_{SUPP} \cdot y_m(k+1) \cdot u(k)^{-1}$, where a_{SUPP} denotes the tuning coefficient [4].

To demonstrate a minimization of the online processing done in [4] by the state space dimension reduction, the model state space matrixes that are presented in the MLD model in [28] are provided:

$$
\begin{gathered}
A = \begin{bmatrix} A_{d_3} & 0_{2x1} & 0_{2x2} \\ 0_{2x2} & 0_{2x1} & 0_{2x2} \\ 0_{1x2} & 1 & 0_{1x2} \end{bmatrix} \\
B_2 = \begin{bmatrix} 0_{2x1} & B_{d_2} - B_{d_3} & B_{d_1} - B_{d_3} & B_{d_4} - B_{d_3} \\ 0 & 1 & 0 & 0 \\ 0 & 0 & 1 & 0 \\ 0 & 0 & 0 & 0 \end{bmatrix} \\
B_3 = \begin{bmatrix} I_{2x2} & I_{2x2} & I_{2x2} & 0_{2x1} \\ 0_{3x2} & 0_{3x2} & 0_{3x2} & 0_{3x1} \end{bmatrix} \quad B_1 = 0_{5x1} \\
C = \begin{bmatrix} 1 & 0 & 0 & 0 & 0 \end{bmatrix} \\
D_1 = 0 \quad D_2 = 0_{1x4} \quad D_3 = 0_{1x7} \\
x(k+1) = Ax(k) + B_1u(k) + B_2\delta(k) + B_3z(k) \\
y(k) = Cx(k) + D_1u(k) + D_2\delta(k) + D_3z(k) \\
E_2\delta(k) + E_3z(k) \leq E_1u(k) + E_4x(k) + E_5.
\end{gathered} \tag{11}
$$

The state-space equations in Model (11) present the final solution in the MLD approach, which consists of the characteristic state space matrixes A, B, C, D, where the A, B are derived from the intermediate matrixes A_{d_i}, B_{d_i} as the result of discretization processes during the T_s time. Thus, in Model (11), the state vector $\boldsymbol{x}$ consists of the inductor current and output voltage, the input is provided by $\boldsymbol{\delta}$ as the five-dimensional vector of logic variables, and $\boldsymbol{z}$ is the MLD seven-dimensional auxiliary variables' vector. The matrixes E_i form the set of selected inequalities as the subject to the optimization [28].

It is obvious that in the examples in which we use the Model Predictive Control (MPC), a dimension of the state space model matrixes (A in Model (11) and $\boldsymbol{A}_{m_k}$ in Model (7)) has the extreme impact on the online processing time. Thus, $\boldsymbol{A}_{m_k} \in R^{2x2}$ in comparison with the $A \in R^{5x5}$ simplifies the final model predictive state matrix. For the n horizon prediction, the model predictive state matrix is $A_{m_p} \in R^{n \cdot n_r \ x \ n_c}$, if the n_r and n_c are the number of rows and columns, respectively, of the original model state matrix.

3. Simulation Results of the Novel Control Methods and Their Comparison

Although in this study a Discrete Hybrid Automata (DHA) and the equivalent model of a DC-DC boost converter are considered to be the online processing time complex, it is a recommended basis for forming the simulation model of the MATLAB/SIMULINK platform [29]. For the novel control methods, it is a source in exploring the identification methodologies and later application to the physical processes. Thus, the developed methods will be simulated and applied to the DHA-developed simulation model. A minimum discretization or the sampling time in modeling the PWM, and switching events of the semiconductors, is set to 1 μs while a DC-DC boost converter model sampling time T_s is 333 μs. The simulation process was not done in real time; we left enough time to process the DHA simulation model itself together with the developed control algorithms. To provide a meaningful comparison, we use the test parameters or system events that are applied in [3]. The selection of the operating points and the process parameter change have to prove the robustness of the control algorithms driven from the minimum borders of a DCM to the maximum borders of the CCM of a DC-DC boost converter. In this comparison, we will also present the MPC-developed methods that were not found to be sufficiently compact, and their processing time was short enough to be applied in the physical experiment of [4]. Subsequently, the Fuzzy Generalized Predictive Control (FGPC) algorithm, the Fuzzy Dynamic Matrix Control (FDMC), the Fuzzy Predictive Functional Control (FPFC), the optimized PI control, and finally the Two Degrees of Freedom (TDOF) control will be tested for the same combination of the system parameters' change, and their dynamics and steady states will be subsequently compared.

The simulation block diagrams of discussing control algorithms can be found in [3,4]. The FMPC methods are based on the same internal model (7) as described in Section 2.2 and [4]. Furthermore, a TDOF control methodology is described in Section 2.1 and [3], and an optimized PI controller as the reference to all of the comparisons is described in both articles. The latter control solution is intentionally chosen to simplify any of the following comparisons, as its construction is based purely on the standard toolbox of the used simulation platform [29] and optimized by the Integral of the Time Weighted Absolute Error (ITAE) criterion.

In Figure 3a, the response of the most favored predictive control solution FPFC is presented. It comes from the online processing time span and aggressiveness in the transition time. The online processing time is again proportionally related to the number of the prediction points, but in the nature of this MPC method, it is the reduction of coinciding points by the implementation of basis functions [26]. Furthermore, in the figure, there is the curve of the inductor current function that demonstrates a controller's controlled manner of the most critical constraint of the DC-DC boost converter. Separately from Figure 3a, Figure 3b presents the manipulated or controlling parameter that is a duty cycle of a DC-DC boost converter.

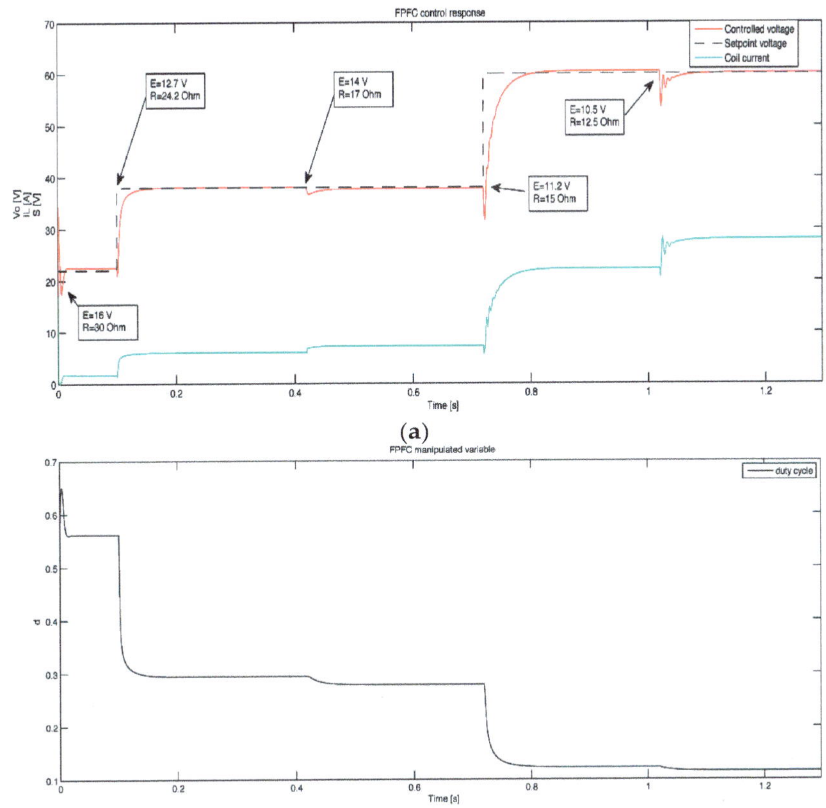

(b)

Figure 3. Simulation results: (**a**) Controlled variable by the Fuzzy Predictive Functional Control (FPFC) controller v_o ($MSRE = 0.0023$ for 129601 samples related to the reference) and coil current i_L; (**b**) Manipulated variable from the FPFC controller (duty cycle) $d_u = [0.01,\ 0.66]$ *range* $98.5 \div 1.5\%$.

In the sequel, the result of the simulation process for the FGPC, FDMC, the optimized PI, and the TDOF method is presented in Figures 4–7 respectively.

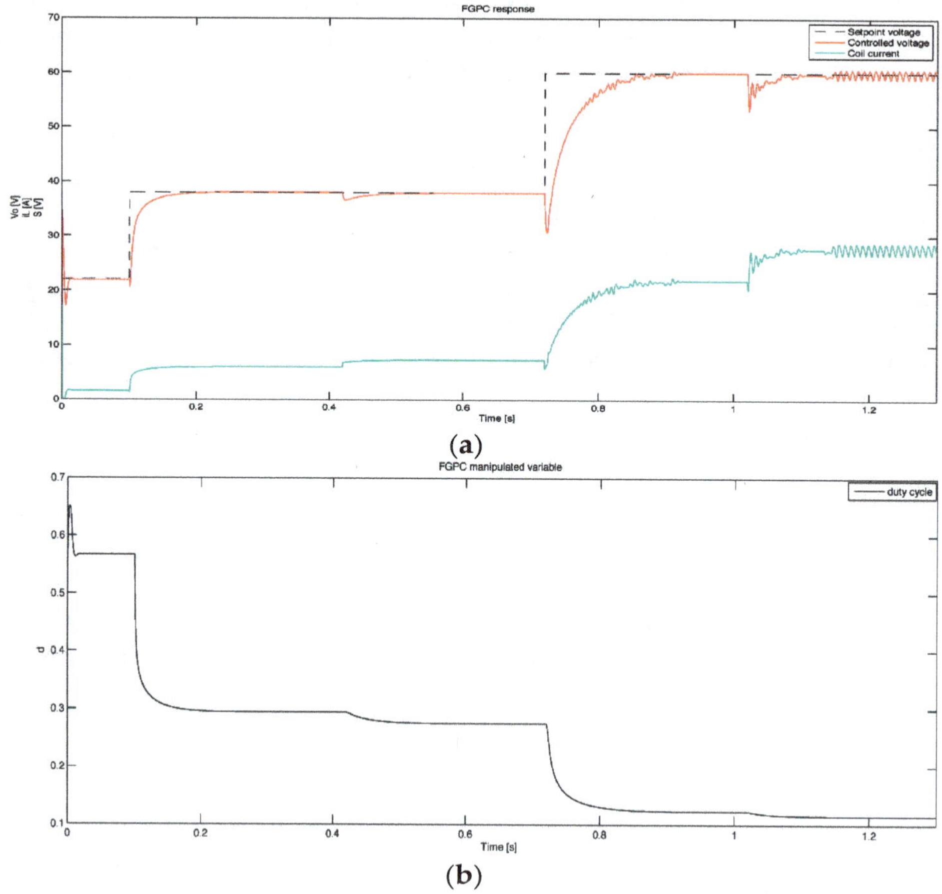

Figure 4. Simulation results: (**a**) Controlled variable by the Fuzzy Generalized Predictive Control (FGPC) controller v_o ($MSRE = 0.0044$ for 129601 samples related to the reference) and coil current i_L; (**b**) Manipulated variable from the FGPC controller (duty cycle) $d_u = [0.01,\ 0.66]\ range\ 98.5 \div 1.5\%$.

In Figure 4, the FGPC responses are significantly slower and have a slightly worse mean square relative error (MSRE) in comparison with the FPFC. We see that even the second-order filter integrated into the prediction model and filtering the model output did not significantly improve the steady state stability, but it reduces the aggressiveness in the response. The small, steady-state chattering in the last two changes of system parameters assign the border of the controller's robustness. It is well-suppressed by the adaptive controller's control signal suppression factor. Similarly, the FDMC simulation in Figure 6 presents the slightly faster response in the transitions and a worse MSRE, but in Section 2, it is explained that it has less of a processing time burden due to the absence of the model output filter.

From the results, the similarities related to the aggressiveness in the transitions and the accuracy in the steady-state of the applied FMPC methods are noticeable. The best performance is experienced with the FPFC in Figure 3, and other methods can be compared to that reference. Closer to the performance of the FPFC is the FDMC. The FGPC has been found to be a bit coarser in the steady state, but on some occasions, such as a system start, it is faster in the response. As expected, the optimized PI controller in Figure 6 is slower in transitions, especially for the operating points far from the center of the performed optimization for the time periods in which the lower system gains have been experienced.

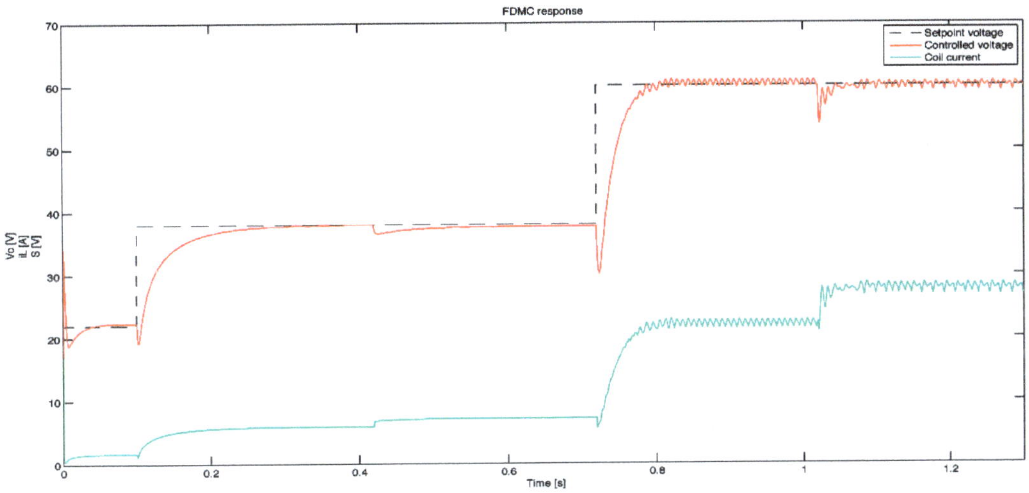

(a)

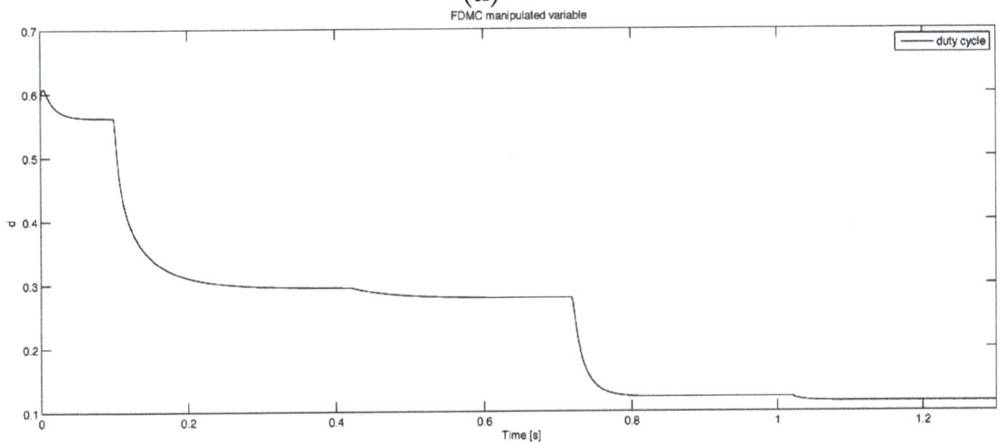

(b)

Figure 5. Simulation results: (**a**) Controlled variable by the Fuzzy Dynamic Matrix Control (FDMC) controller v_o ($MSRE = 0.0062$ for 129601 samples related to the reference) and coil current i_L; (**b**) Manipulated variable from the FDMC controller (duty cycle) $d_u = [0.01,\ 0.66]$ *range* $98.5 \div 1.5\%$.

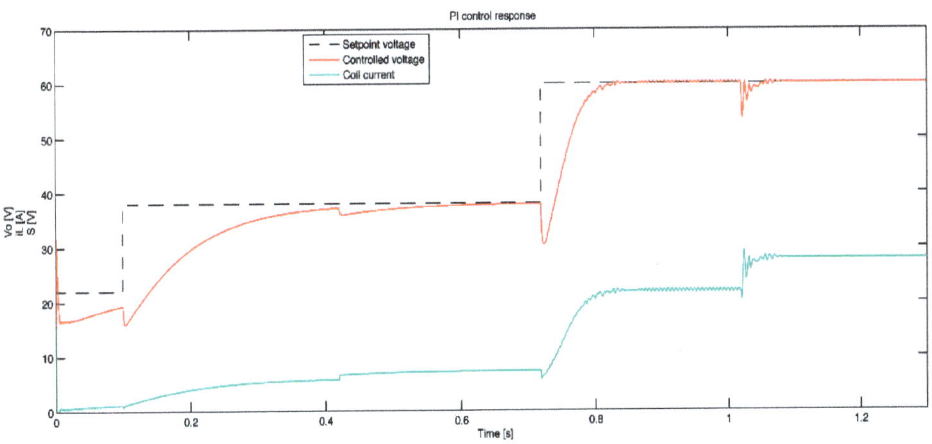

(a)

Figure 6. *Cont.*

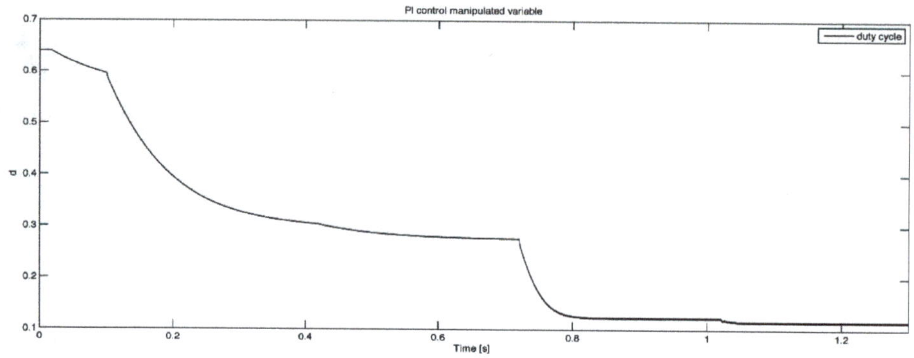

(b)

Figure 6. Simulation results: (**a**) Controlled variable by the Optimized PI controller v_o ($MSRE = 0.0220$ for 129601 samples related to the reference) and coil current i_L; (**b**) Manipulated variable from the optimized PI controller (duty cycle) $d_u = [0.01,\ 0.66]\ range$ 98.5 ÷ 1.5%.

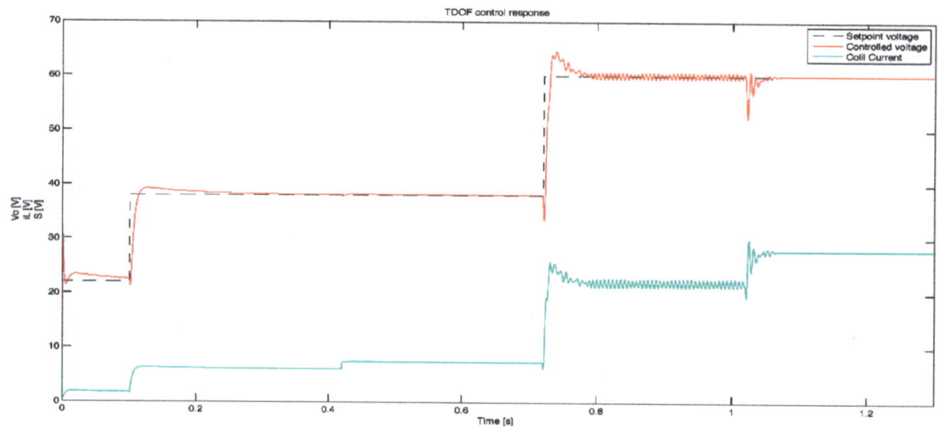

(a)

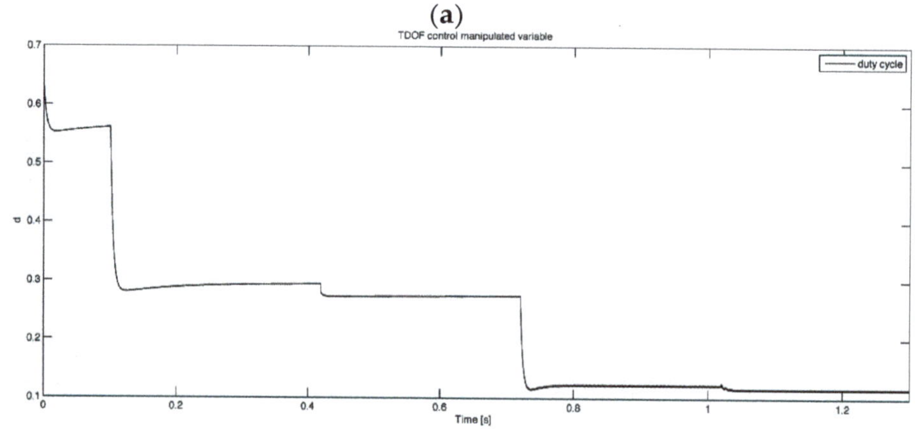

(b)

Figure 7. Simulation results: (**a**) Controlled variable by the Two Degrees of Freedom (TDOF) controller v_o ($MSRE = 0.0015$ for 129601 samples related to the reference) and coil current i_L; (**b**) Manipulated variable from the TDOF controller (duty cycle) $d_u = [0.01,\ 0.66]\ range$ 98.5 ÷ 1.5%.

Different from the aforementioned, the TDOF methodology that is presented in Figure 7 performs as the fastest in transitions, but during that particular time, the control algorithm cannot attenuate the gradient of change, and on some occasions, it violates the reference point. This scenario is initiated by the error of the identified steady state, which is filtered out by the receding horizon strategy in other FMPCs. In contrast, the TDOF algorithm will certainly demand significantly less processor time of execution. From that point of view, the optimized PI controller will use the shortest time of execution, but it is not the robust solution, and in some occasions, it cannot reach the reference value before another is selected (see Figure 6).

Figure 8 presents all methods together and additionally simplifies the comparison.

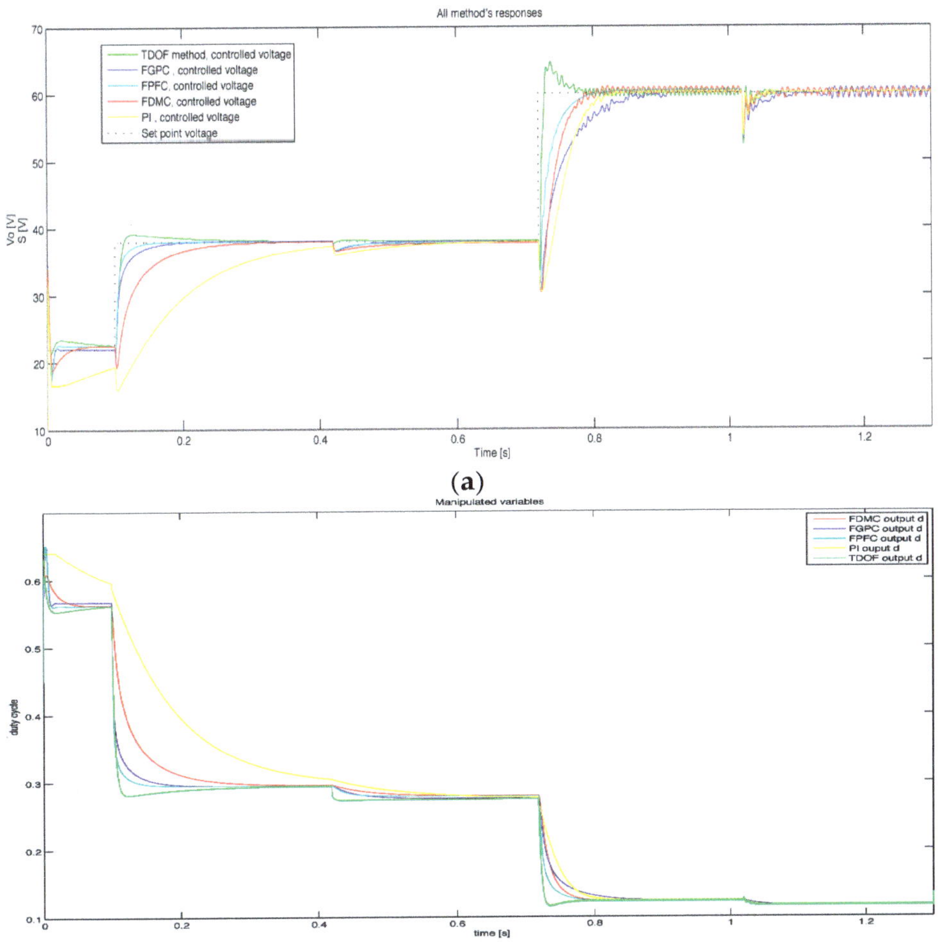

(a)

(b)

Figure 8. Simulation results: (**a**) Controlled variable v_o compared for all methods; (**b**) Manipulated variable d_u compared for all methods.

The overall comparison is quantified by the time of response and the MSRE that can be seen in Table 1.

Table 1. Comparison of the performances for the applied controllers based on 129601 samples of simulation for the 1.3 s of the simulation duration.

Controller	MSRE	$_rt_a$ [s] [1]	t_{st} [s] [2]
FPFC	0.0023	0.0038	0.0110
TDOF	0.0015	0.0099	0.0728
FDMC	0.0062	0.0207	0.0256
FGPC	0.0044	0.0331	0.0103
Opt_PI	0.0220	0.0429 [3]	NA [4]

[1] $_rt_a$ is an averaged response time while the controller reaches 0.03 of the relative error to the reference, excluding the start time; [2] t_{st} is the controllers' start time; [3] the time valid only for two comparable responses, [4] NA- non-applicable.

The above successful results led to the selection of the FPFC method as the most favored and comprehensive solution to be applied in the experimental test of the real DC-DC boost converter. Table 2 presents the controller's scores related to the assigned objectives in Section 1.

Table 2. The table of achieved objectives stated in Section 1.

Controller	PCR	SR	OPT	S	NFM	A
FPFC	Y	Y	SO	ES	Y	A
TDOF	Y	Y	ITAEO	ES	Y	A
FDMC	Y	M	SO	ES	Y	M
FGPC	Y	M	SO	ES	Y	M
Opt_PI	N	NA	ITAEO	ES	N	A

Y-yes, N-no, NA-non-applicable, M-moderate, SO-suboptimal, ITAEO-ITAE criterion optimality, ES-experimental stability, A-applicable.

Referring [4], our experimental results prove the assigned objectives mentioned in the Introduction and in Section 2 for the FPFC as the controller with the highest score.

4. Conclusions

The primary objective of this article was the new control methodology in the direction of implementing the most recently developed strategies for systems that are physically small, but complex in their behavior. That task required the fundamental decomposition of the problem and the learning of the most basic system's behavior. Differently from here, most of the known methods performed on this particular problem employ the analytical methodology in forming the model-based solution [5]. In this study, the model is identified with fuzzy logic. Additionally to widely known objectives, the developed modeling allows the forming of control solutions that conciliate the method's complexity with applicability to the physical processes.

The novel controllers presented are developed with the intention of reducing the online processing complexity. This intention guides the development of two different control methodologies. Both of the control methodologies utilize model-based predictive control. In the case of TDOF control, the problem was decomposed in a way to compensate for the main drawback of the numerically optimized linear controller, which is recognized for its robustness for a wider range of the process parameters' change.

The transition period in TDOF methodology is fast, but it sometimes overshoots the reference point, depending on the positivity of the system/model error at that operating point. To synthesize the solution that is considered to be a dynamic model-based prediction, and to rectify the overshoots, a fuzzy model-based MPC solution is derived. In this study, it is shown that for the reasonable longer processing time, the FMPC successfully follows the aggressiveness of the TDOF controller and the contemporary reference model. Furthermore, the modern predictive control methodology here presented has suboptimal control over the complete transition period and the smooth, steady state of the system. Therefore, in general, the presented FMPCs provide the MPC methodologies that apply to the fast response systems.

Further research will go in the direction of implementing these methodologies into the more complex SAS, as well as continuing to implement the qualitative mathematical theory in designing and exploring the new control methodologies.

Author Contributions: Investigation, R.B.; Methodology, R.B.; Software, R.B.; Supervision, I.Š.; Validation, D.V.; Writing—original draft, R.B.

References

1. Lunze, J.; Lamnabhi-Lagarrigue, F. *Handbook of Hybrid Systems Control: Theory, Tools, Applications*; Cambridge University Press: New York, NY, USA, 2009.
2. Di Bernardo, M.; Budd, C.J.; Champneys, A.R.; Kowalczyk, P. *Piecewise-Smooth Dynamical Systems—Theory and Applications*; Springer: London, UK, 2008.

3. Baždarić, R.; Škrjanc, I.; Matko, D. Two Degrees of Freedom in the Control of a DC-DC Boost Converter, Fuzzy Identified Explicit Model in Feed-forward Line. *J. Intell. Robot. Syst.* **2016**, *82*, 479–483. [CrossRef]
4. Baždarić, R.; Matko, D.; Leban, A.; Vončina, D.; Škrjanc, I. Fuzzy model predictive control of a DC-DC boost converter based on non-linear model Identification. *Math. Comput. Model. Dyn. Syst.* **2017**, *23*, 116–134. [CrossRef]
5. Mariéthoz, S.; Almér, S.; Bâja, M.; Beccuti, A.G.; Patino, D.; Wernrud, A.; Buisson, J.; Cormerais, H.; Geyer, T.; Fujioka, H.; et al. Comparison of Hybrid Control Techniques for Buck and Boost DC-DC Converters. *IEEE Trans. Control Syst. Technol.* **2010**, *18*, 1126–1145. [CrossRef]
6. Beccuti, A.G.; Papafotiou, G.; Frasca, R.; Morari, M. Explicit Hybrid Model Predictive Control of the dc-dc Boost Converter. In Proceedings of the 2007 IEEE Power Electronics Specialists Conference (PECS), Orlando, FL, USA, 17–21 June 2007; pp. 2503–2509.
7. Tse, C.K. Flip Bifurcations and Chaos in Three-State Boost Switching Regulators. *IEEE Trans. Circuits Syst. I Fundam. Theory Appl.* **1994**, *41*, 16–23. [CrossRef]
8. Chen, Y.; Tse, C.K.; Wong, S.C.; Qiu, S.S. Interaction of Fast-Scale and Slow-Scale Bifurcation in Current-Mode Controlled DC/DC Converters. *Int. J. Bifurc. Chaos* **2007**, *17*, 1609–1622. [CrossRef]
9. Guesmi, K.; Essounbouli, N.; Hamzaoui, A.; Zaytoon, J.; Manamanni, N. *Shifting Nonlinear Phenomena in a DC-DC Converter Using Fuzzy Logic Controller*; IMACS, Elsevier: Amsterdam, The Netherlands, 2007.
10. Mehran, K.; Giaouris, D.; Zahawi, B. Stability Analysis and Control of Nonlinear Phenomena in Boost Converters Using Model-Based Takagi–Sugeno Fuzzy Approach. *IEEE Trans. Circuits Syst. I Regul. Pap.* **2010**, *57*, 200–212. [CrossRef]
11. Vasca, F.; Iannelli, L.; Kanat Camlibel, M.; Frasca, R. A New Perspective for Modeling Power Electronic Converters: Complementarity Framework. *IEEE Trans. Power Electron.* **2009**, *24*, 456–468. [CrossRef]
12. Middlebrook, R.D. Small-Signal Modeling of Pulse-Width Modulated Switched-Mode Power Converters. *Proc. IEEE* **1988**, *76*, 343–354. [CrossRef]
13. Erickson, R.W.; Ćuk, S.; Middlebrook, R.D. Large-Signal Modelling and Analysis of Switching Regulators. In Proceedings of the 1982 IEEE Power Electronics Specialists Conference (PESC), Cambridge, MA, USA, 14–17 June 1982; pp. 240–250.
14. Seok-Kyoon, K.; Chang, R.P.; Jung-Su, K.; Young Il, L. A Stabilizing Model Predictive Controller for Voltage Regulation of a DC/DC Boost Converter. *IEEE Trans. Control Syst. Technol.* **2014**, *22*, 2016–2023.
15. Errouissi, R.; Al-Durra, A.; Muyeen, S.M. A Robust Continuous-Time MPC of a DC–DC Boost Converter Interfaced With a Grid-Connected Photovoltaic System. *IEEE J. Photovolt.* **2016**, *6*, 1619–1629. [CrossRef]
16. Alessio, A.; Bemporad, A. A Survey on Explicit Model Predictive Control. In *Nonlinear Model Predictive Control*; Lecture Notes in Control and Information Sciences Book Series; Springer: Berlin/Heidelberg, Germany, 2009; Volume 384, pp. 345–369.
17. Cheng, L.; Acuna, P.; Aguilera, R.P.; Jiang, J.; Wei, S.; Fletcher, J.E.; Lu, D.D.C. Model Predictive Control for DC–DC Boost Converters With Reduced-Prediction Horizon and Constant Switching Frequency. *IEEE Trans. Power Electron.* **2018**, *33*, 9064–9075. [CrossRef]
18. Li, J.; Zhang, Z.; Li, B. Sensor Fault Detection and System Reconfiguration for DC-DC Boost Converter. *Sensors* **2018**, *18*, 1375. [CrossRef] [PubMed]
19. García-Rodríguez, V.H.; Silva-Ortigoza, R.; Hernández-Márquez, E.; García-Sánchez, J.R.; Taud, H. DC/DC Boost Converter–Inverter as Driver for a DC Motor: Modeling and Experimental Verification. *Energies* **2018**, *11*, 2044. [CrossRef]
20. Hernández-Márquez, E.; Avila-Rea, C.A.; García-Sánchez, J.R.; Silva-Ortigoza, R.; Silva-Ortigoza, G.; Taud, H.; Marcelino-Aranda, M. Robust Tracking Controller for a DC/DC Buck-Boost Converter–Inverter–DC Motor System. *Energies* **2018**, *11*, 2500. [CrossRef]
21. Bemporad, A.; Morari, M. Control of systems integrating logic, dynamics, and constraints. *Automatica* **1999**, *35*, 407–427. [CrossRef]
22. Geyer, T.; Papafotiou, G.; Morari, M. On the Optimal Control of Switch-Mode DC-DC Converters. In Proceedings of the 7th International Workshop on Hybrid Systems: Computation and Control, Philadelphia, PA, USA, 25–27 March 2004; pp. 342–356.

23. Lygeros, J. *Lecture Notes on Hybrid Systems*; GR-26500; Department of Electrical and Computer Engineering University of Patras: Patras, Greece, 2004.
24. Lygeros, J.; Johansson, K.H.; Simić, S.N.; Zhang, J.; Shankar Sastry, S. Dynamical Properties of Hybrid Automata. *IEEE Trans. Autom. Control* **2003**, *48*, 2–17. [CrossRef]
25. El Beid, S.; Doubabi, S. DSP-Based Implementation of Fuzzy Output Tracking Control for a Boost Converter. *IEEE Trans. Ind. Electron.* **2014**, *61*, 196–209. [CrossRef]
26. Maciejowski, J.M. *Predictive Control*; Prentice-Hall: Englewood Cliffs, NJ, USA, 2002.
27. Stoorvogel, A.A.; van den Boom, T.J.J. *Model Predictive Control*; DISC, Course Lecture Notes: Delft, The Netherlands, 2010.

2

Design of Current Programmed Switching Converters using Sliding-Mode Control Theory

Javier Calvente, Abdelali El Aroudi *, Roberto Giral, Angel Cid-Pastor, Enric Vidal-Idiarte and Luis Martínez-Salamero

Departament d'Enginyeria Electrònica, Elèctrica i Automàtica, Universitat Rovira i Virgili, 43007 Tarragona, Spain; javier.calvente@urv.cat (J.C.); roberto.giral@urv.cat (R.G.); angel.cid@urv.cat (A.C.-P.); enric.vidal@urv.cat (E.V.-I.); luis.martinez@urv.cat (L.M.-S.)

* Correspondence: abdelali.elaroudi@urv.cat

Abstract: This paper presents a comprehensive approach to analyze and design the voltage and current loops of switching DC-DC converters by using sliding-mode control theory. The approach is interchangeably applied to switching converters under current-programmed control with both fixed and variable frequency modulation. An ideal sliding-mode dynamics model is then obtained together with its circuit schematic representation that can be used for designing the output voltage compensator, as well as to predict the large signal behavior such as during start-up and under large disturbances. Simulations and experimental measurements illustrate the theoretical approach for two different examples of switching converters.

Keywords: switching converters; sliding-mode control; current-mode control; hysteresis control

1. Introduction

Switched mode power converters are widely used for power processing in different applications such as in portable devices, solid-state lighting drivers and technologies for renewable energy production. With the aim of either regulating an output variable to a desired reference or balancing the values of different variables, a control strategy is needed [1–3].

One of the key factors affecting the response of these systems is the control mode used. Conventional Voltage Mode Control (VMC) is a simple single feedback loop with only the output voltage as a control variable. However, its first generation version features poor response against load changes [2]. Current Mode Control (CMC), also known in the literature as current-programmed control [4], utilizes the inductor current as an additional control variable, which improves the transient response [2] at the expense of extra cost and more complex controller design. CMC of power converters is nowadays an extensive design practice due to its intrinsic advantages and the existence of an important number of dedicated commercial chips that facilitate its implementation. CMC offers superior performances over conventional VMC in both parallel operation of power converters [5] and voltage regulation of non-minimum phase switching conversion structures [6]. In the first case, current sharing among the different converters of the parallel ensemble is a relatively simple task for both static and dynamic sharing [5], while in the second case, CMC is the best solution for an indirect output voltage regulation [7].

A usual way of controlling switching converters consist of imposing the duration during which a switch is maintained closed or open. In CMC, the control strategy must impose switching instants for the switches by comparing the inductor current with a desired reference. Under fixed switching frequency operation, there are three main types of CMC strategies, namely peak CMC, valley CMC and average CMC. The peak and valley cases are dual, i.e., the internal control loop forces respectively the maximum and the minimum value of the current to track the reference given by the voltage control loop. Average control, in turn, uses the current average value in a switching cycle to track the reference supplied by the outer loop. Although most commercial ICs of CMC are based on the peak-current control principle due to its simplicity of implementation, the existence of sub-harmonic oscillations for duty cycles bigger than 50% makes necessary the use of a compensating ramp, which constitutes the main drawback of this technique.

From the seminal paper of Deisch [8] until now, more than three decades have elapsed during which a great number of researchers have contributed to building a solid representation of the dynamic behavior of the internal loop by means of either continuous or discrete-time models, or a combination of both. Dozens of papers were reported on this topic during the 1980s, the most representative being [4,9]. Some important contributions to this topic were published in the early 1990s [10–12] until the subject reached enough maturity to be treated as a chapter in some recent text books [2].

In the last decade, a renewed interest has emerged in variable frequency CMC strategies such as constant ON-time CMC [13], which, at the expense of a variable switching frequency operation, offers high efficiency under light load operation and precludes the existence of sub-harmonic oscillations in most cases. A detailed analysis of the dynamic behavior of this control technique can be found in [14], where the main difference with respect to peak CMC is investigated in the buck converter by means of an equivalent circuit consisting of the output load and the LC output filter, which is supplied by two current sources with a resistor and a capacitor in parallel.

Variable switching frequency would also result from the insertion of a hysteretic comparator instead of the Pulse Width Modulator (PWM) with constant switching frequency in the case of current programming, or instead of the PWM with constant ON-time duration in the case of constant ON-time CMC.

The voltage regulation of DC-DC switching converters using hysteretic control, also called free running or bang-bang controlled converters, goes back in time to the early years of modern power electronics when conditions for stable limit cycles in a buck switching regulator were first established [15,16].

Recently, hysteretic controllers were used in Voltage Regulator Module (VRM) applications that require current control with fast response [17–19], and some commercial chips are already available [20].

Traditionally, hysteresis-based control systems have been analyzed by means of frequency domain techniques like the descriptive function or Tsypkin's method [21–23].

On the other hand, Sliding-Mode Control (SMC) theory is a time-domain analytical technique that predicts with high precision the dynamic behavior of a Variable Structure System (VSS) [24,25]. Power converters can be classified as VSSs, and SMC is the natural way to regulate their outputs and deal with their dynamical behavior because the generation of chattering, which is intrinsic to the use of SMC, is also inherent to the nature of power converters operation. In other words, the sliding chattering becomes converter ripple, which is a physical manifestation of the way a converter absorbs energy from the input source, storing it usually in an inductor and then transferring it to an output load.

There are two main results in the SMC theory applicable to switching converters. The first one is the fact that for single input systems, a suitable Lyapunov function $V(x) = \frac{1}{2}\sigma^2(x)$ exists, where $\sigma(x)$ is the switching function. If the switching feedback gains are chosen so that $\dot{V} = \sigma\dot{\sigma} < 0$ in the domain of attraction, then the state trajectory converges to the surface $\sigma = 0$ and is restricted to it for all subsequent time. The second one provides the ideal sliding-mode dynamics on the switching surface. In this case, Filippov's method [26] and the equivalent control approach yield identical sliding equations when applied to systems that are linear with respect to the control input [27].

These techniques have been used in [28] considering feedback switching conditions of the form $\sigma = x_j - K = 0$, expressed in terms of a suitable state variable x_j that is desired to be regulated at a suitable level K. They have been employed in output voltage regulation of DC-DC switching converters in early works such as in [29–33] and also in some recent contributions to this field such as [34–38]. In [39], SMC theory has been applied to control paralleled inverters of the buck type and in [40,41]. In [42,43], it has been used in the analysis of interleaved boost converters with hysteretic control in a ring configuration. In [44], sliding mode control theory was used to synthesize canonical elements for power processing such as in impedance matching in PV systems and in power factor correction. These techniques were also combined with other robust control methods such as fuzzy logic [45] and H_∞ [46] for designing these systems.

In the field of switching converters, two different kinds of studies using SMC theory exist in the literature. The one that uses this theory to design the switching decision and to analyze the ideal sliding dynamics and that ultimately uses a hysteretic loop to limit the switching frequency [7]. Other works use this theory to derive the equivalent control, and this is used finally to implement a PWM control [32,33]. Actually, with this strategy, all the advantages of SMC are not maintained. Small-signal analysis of sliding mode-controlled switching converters, based on hysteresis modulation, were reported in [47], where Bode plot and root locus analysis were used to reveal the effect of the parameters on the system behavior.

In this paper, on the basis of the equivalent control method, a procedure to design the control loop of DC-DC power converters operating in CMC with both variable and fixed frequency modulation strategies is proposed. The controller employs one switching function of the form $\sigma = i_r - i_L$, where i_L is a converter current and i_r is the output of an outer compensating network that processes the output voltage error. The paper comprehensively explains the dynamics of switching converters regardless of their modulation strategies. In particular, it will be demonstrated that under sliding mode conditions, the behavior of the switching converters under all the control strategies tend to the same dynamics known as the ideal sliding mode dynamics when the switching period tends to zero. The advantages, limitations and drawbacks of each strategy in a practical implementation are also discussed. It is worth noting that in this paper, it is not pretended to control switching converters using sliding mode control that ultimately end-up in a variable frequency implementation of the modulator, but the dynamics of these systems under both constant and variable frequency modulation schemes are explained by using sliding mode control theory, while providing circuit equivalent models that can be used for designing these systems by utilizing conventional frequency-domain methods.

The rest of the paper is organized as follows. The description of the behavior of current-programmed DC-DC converters is presented in Section 2 in the light of sliding mode control theory. A small-signal model is used to design the voltage loop by means of linear techniques in the frequency domain, as well as simulation results on two practical examples of switching converters in Section 3. The examples consist of a boost converter under fixed and variable frequency operation and a buck converter with current-mode hysteretic control and high bandwidth voltage regulation. An experimental validation of the results corresponding to the buck converter is given in Section 4. The application of the approach to single-loop ripple-based voltage mode control is discussed in Section 5. Finally, the conclusions of this work are summarized in Section 6.

2. Sliding-Mode Control of Switching Converters Based on Two-Loop Current Mode Control

In CCM operation, a switching converter is described by a piecewise linear state-space model that can be written in the following form:

$$\dot{x}(t) = \begin{cases} A_1x(t) + B_1w(t) & u = 1 \\ A_2x(t) + B_2w(t) & u = 0 \end{cases} \quad (1)$$

where u is a binary signal that can take the values of zero or one. It is assumed that the switching is instantaneous. $x(t)$ is the vector of state variables, and $w(t)$ is the vector of independent sources. Equation (1) can be expressed as follows:

$$\dot{x}(t) = [A_1x(t) + B_1w(t)]u + [A_2x(t) + B_2w(t)](1-u). \quad (2)$$

Conventionally, the output variable y is a linear combination of the state variables. However, the approach is also applicable to the case with a nonlinear switching surface as in high-order boundary control schemes [48]. Hereinafter, let the linear combination of state variables $y(t) = c^{\mathsf{T}} x(t)$ be the output to be controlled, where c is a suitable vector to select the output $y(t)$ from the state variables $x(t)$. The derivative of the controlled output $y(t)$ is of the form:

$$\dot{y} = (m_1(x,t) + m_2(x,t))u - m_2(x,t) \tag{3}$$

where $m_1(x,t)$ and $m_2(x,t)$ are the rising and falling slopes of the signal $y(t)$ that can be expressed as follows:

$$m_1(x,t) = \quad c^{\mathsf{T}}[A_1 x(t) + B_1 w(t)] \tag{4}$$

$$m_2(x,t) = \quad -c^{\mathsf{T}}[A_2 x(t) + B_2 w(t)]. \tag{5}$$

Our interest is in situations where y can be controlled in sliding mode, so that it follows the reference smooth function $r(t)$ (continuous and differentiable). The switching function can be expressed as follows:

$$\sigma(x,t) = r(t) - y \tag{6}$$

along with the following control law (Figure 1):

$$u = \begin{cases} 1 & \text{if} \quad \sigma > 0 \\ 0 & \text{if} \quad \sigma < 0. \end{cases} \tag{7}$$

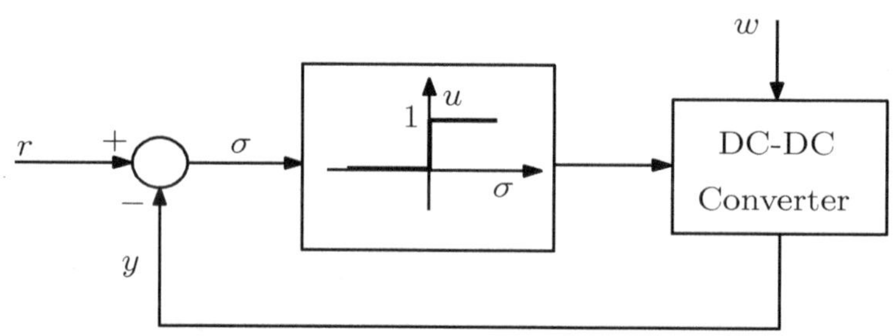

Figure 1. Simplified diagram of a DC-DC converter under Current Mode Control (CMC).

For the existence of a sliding regime on the time-variant surface Σ defined by $\{x|\sigma(x,t) = 0\}$, the sliding condition $\sigma\dot{\sigma} < 0$ must be fulfilled, which is equivalent to:

$$\dot{\sigma}_{\sigma>0} < 0 \tag{8}$$

$$\dot{\sigma}_{\sigma<0} > 0. \tag{9}$$

According to (3), the derivative $\dot{\sigma}$ of the switching function σ can be expressed as follows:

$$\dot{\sigma}(x,t) = \dot{r}(t) - ((m_1(x,t) + m_2(x,t))u - m_2(x,t)). \tag{10}$$

By using (7), the sliding conditions (8) and (9) are:

$$\dot{\sigma}_{\sigma>0} = \dot{r}(t) - m_1(x,t) < 0 \tag{11}$$

$$\dot{\sigma}_{\sigma<0} = \dot{r}(t) + m_2(x,t) > 0 \tag{12}$$

or in compact form:

$$-m_2(x,t) < \dot{r}(t) < m_1(x,t) \tag{13}$$

Equation (13) implies that the rate of change of the signal r must be lower than the absolute values of the slopes $m_1(x,t)$ and $m_2(x,t)$ of the control signal y. This is easily met if the bandwidth of the outer loop is much lower than the switching frequency, as is the case in any practical design. Therefore, under this condition, the trajectory of the system (2) with the control decision (7) will reach the sliding surface Σ in finite time and stay on it in sliding mode [27].

To maintain the trajectory in this regime, it is necessary to switch continuously at infinite frequency in the ideal case. This is not an acceptable behavior for DC-DC switching converters, which are designed to operate in a specific switching frequency range. Operating at higher frequencies, the converter efficiency decreases, and the instantaneous switching model (2) is no longer valid. This problem is solved by introducing a boundary layer around Σ and replacing the switching decision (7) by:

$$u = \begin{cases} 1 & \text{if} \quad \sigma > +\Delta \\ 0 & \text{if} \quad \sigma < -\Delta \end{cases} \tag{14}$$

where $\Delta > 0$. With this control law, the switching frequency in sliding-mode will be finite. In turn, the sliding motion will not occur strictly on the surface Σ, but in a neighborhood that meets the condition $|\sigma| \leq \Delta$. This means that a certain switching ripple will exist in the output to be controlled and other state variables as a penalty of a bounded switching frequency. This is a natural way of exchanging the energy between the reactive components of the converter.

The ideal sliding-mode dynamics is the behavior of the system (2) in the sliding regime with a control of the form (14) and in the limit case when Δ tends to zero. In order to find the ideal sliding-mode equations, the equivalent control method [27] is used. A necessary condition for the existence of an equivalent control u_{eq} and therefore to apply this method is that u appears explicitly on the right side of (10), i.e.,

$$m_1(x,t) + m_2(x,t) \neq 0. \tag{15}$$

Remark 1. *The previous condition is largely related to the fact that the relative degree of the converter u-to-σ loop is equal to one. The fast response characterizing CMC is due to this fact. The relative degree of a switched system is the number of differentiations one needs to perform on the switching function σ to make the input explicitly appear in its derivative, i.e., the minimum n that satisfies the following equation:*

$$\frac{\partial}{\partial u}\frac{\mathrm{d}^n\sigma(x,t)}{\mathrm{d}t^n} \neq 0 \tag{16}$$

Remark 2. *At high frequencies, if the relative degree is one, the system behaves as an integrator, which converts the square-wave signal u to a triangular signal σ.*

Remark 3. *It should be noted that apart from the widely-used hysteresis comparison, any other strategy to limit the switching frequency can be used without invading the sliding motion. Other possible strategies to limit the switching frequency are time delays, filtering and clocked switching.*

In all the cases, if (15) is accomplished, the equation $\dot{\sigma} = 0$ can be solved with respect to u to obtain the equivalent leading to the following expression:

$$u_{eq}(x,t) = \frac{\dot{r}(t) + m_2(x,t)}{m_1(x,t) + m_2(x,t)}. \tag{17}$$

The equivalent control caused by any modulation scheme such as the variable frequency hysteretic modulation and constant switching-frequency peak/valley control methods are the same since the equivalent control corresponds to the ideal sliding dynamics with theoretically infinite switching frequency or to the averaged dynamics of the converter under the switching constraints' imposed previous modulation strategies. The order of the averaged dynamics or the ideal sliding dynamics is reduced due to the relationship imposed between the state variables by the switching constraint. From (17), if $m_1(x,t) + m_2(x,t) \neq 0$, the equivalent control u_{eq} will exist, and according to (13), its value will be comprised between zero and one. The ideal sliding-mode dynamics is obtained by substituting u in (2) by u_{eq} given by (17) and introducing the order reduction due to the constraint $\sigma = 0$. In DC-DC converters, the ideal sliding-mode dynamics should have an asymptotically-stable equilibrium point.

Note that u is undetermined when $|\sigma| \leq \Delta$; therefore, there is a plurality of controls compatible with (14) [49]. For example, in the hysteretic control (Figure 2a), from the moment that the condition $|\sigma| < \Delta$ is fulfilled, a switching occurs whenever $|\sigma| = \Delta$. Other examples are constant-switching-frequency controls [50], such as peak (Figure 2b) and valley (Figure 2c) current, in which time-driven switching occurs periodically, while event-driven switching occurs whenever $|\sigma| = \Delta$. DC-DC switching converters typically have a periodic behavior in steady-state with two switchings per period. In the limit as Δ tends to zero, the periodic solution tends to the equilibrium point of the ideal sliding-mode dynamics.

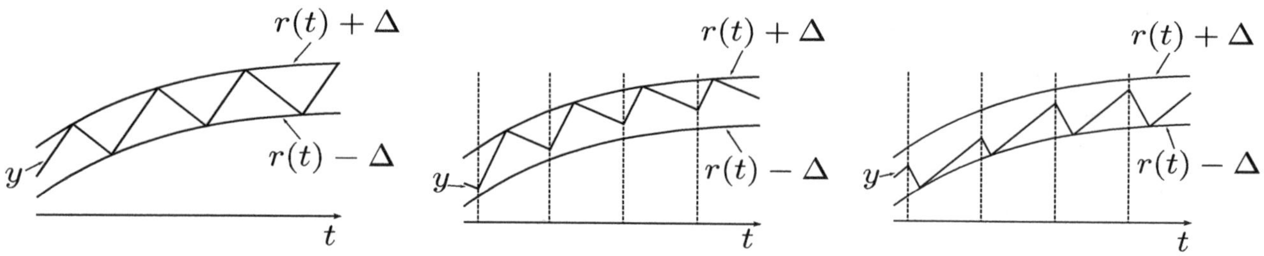

Figure 2. Evolution of the controlled output and its reference in sliding-mode for three types of control implementing (14). (**a**) Hysteresis control. (**b**) Peak control at constant switching frequency. (**c**) Valley control at constant switching frequency.

The DC-DC converter with sliding-mode current control is a system, the control input of which is the current reference $r(t)$. To regulate the output voltage of the converter, a second slower control loop that dictates $r(t)$ from the voltage error must be added. The examples considered in the next section show how to design such a controller using the ideal sliding-mode dynamics model.

3. Practical Examples

3.1. Example 1: A Boost Converter under Fixed and Variable Frequency CMC

Let us consider the boost converter depicted in Figure 3a. The state variables are the inductor current i_L and the capacitor voltage v_o. The independent sources correspond to the input voltage $v_g(t)$ and the current $i_d(t)$, which models load variations. For this example, (2) becomes as follows:

$$\frac{di_L}{dt} = \frac{v_g(t)}{L} - \frac{v_o(1-u)}{L} \tag{18}$$

$$\frac{dv_o}{dt} = \frac{i_L(1-u)}{C} - \frac{v_o}{RC} + \frac{i_d(t)}{C}. \tag{19}$$

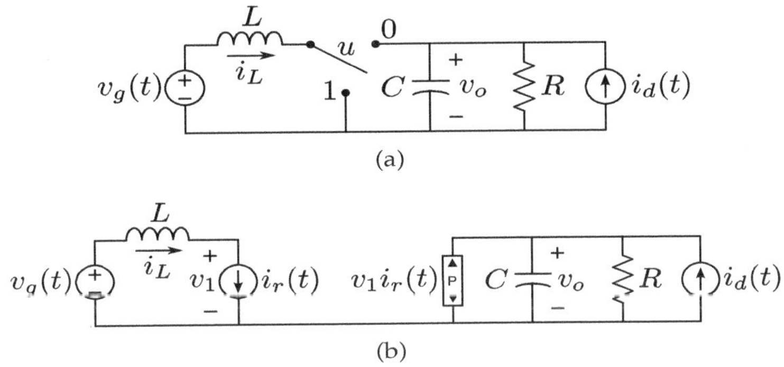

Figure 3. Schematic circuit diagram of (**a**) a Boost converter and (**b**) its ideal sliding-mode dynamics model under CMC.

Let $i_r(t)$ be the current reference. Using the switching function $\sigma = i_r(t) - i_L$ together with the control law (7), the sliding condition (13) leads to the following inequality:

$$\frac{v_g(t) - v_o}{L} < \frac{di_r(t)}{dt} < \frac{v_g(t)}{L}. \tag{20}$$

When $i_r(t)$ is constant, (20) becomes $0 < v_g(t) < v_o$. In this case, starting from zero initial conditions, as soon as v_o reaches $v_g(t)$, the sliding condition is fulfilled, and after a finite time, the system trajectory is constrained in the sliding manifold defined by the constraint $i_r - i_L = 0$. When $i_r(t)$ is time varying or state-dependent, according to (17), the equivalent control for this example becomes as follows:

$$u_{eq} = \frac{L\frac{di_r(t)}{dt} + v_o - v_g(t)}{v_o}. \tag{21}$$

After substituting u by u_{eq} in (18) and (19) and using $\sigma = 0$ ($i_L = i_r(t)$), the system order is reduced, and the following ideal sliding-mode dynamics is obtained:

$$i_L = i_r(t) \tag{22}$$

$$\frac{dv_o}{dt} = \frac{i_r(t)\left[v_g(t) - L\frac{di_r(t)}{dt}\right]}{Cv_o} - \frac{v_o}{RC} + \frac{i_d(t)}{C}. \tag{23}$$

Expression (23) also establishes a power balance between the input port and the output port of the boost converter provided that $i_L = i_r$. These equations can be represented by means of an equivalent circuit using a power source [51] to model the nonlinear term (Figure 3b). The ideal sliding-mode dynamics (23), with the following constant inputs:

$$i_r(t) = I_r, \;\; v_g(t) = V_g, \;\; i_d(t) = 0, \tag{24}$$

has the following equilibrium point:

$$v_o{}^* = V_r = \sqrt{I_r R V_g} \tag{25}$$

which is asymptotically stable, as can be demonstrated by means of the Lyapunov function $V(v_o) = (1/2)(v_o - V_r)^2$ [52].

Note that $v_o{}^*$ depends on both the line voltage and the supplied load resistance. That is why a regulation of the voltage v_o is needed when disturbances in $i_d(t)$ and $v_g(t)$ take place. The regulation can be accomplished by adding another external voltage loop and making the current reference i_r be the output of this loop.

Figure 4 shows a double-loop control scheme, which is proposed for output voltage regulation. The voltage controller consists of three cascaded stages. Namely, a Proportional-Integral (PI) block to process the error, a limiter to avoid the current reference overpassing an admissible level and a low-pass filter to ensure the smoothness of $i_r(t)$. The current controller is of the type (14) with $\Delta = I_\Delta$. The transfer function of the voltage controller in the linear region has the following form:

$$G_c(s) = K_p\left(1+\frac{\omega_I}{s}\right)\frac{1}{1+s/\omega_h}. \tag{26}$$

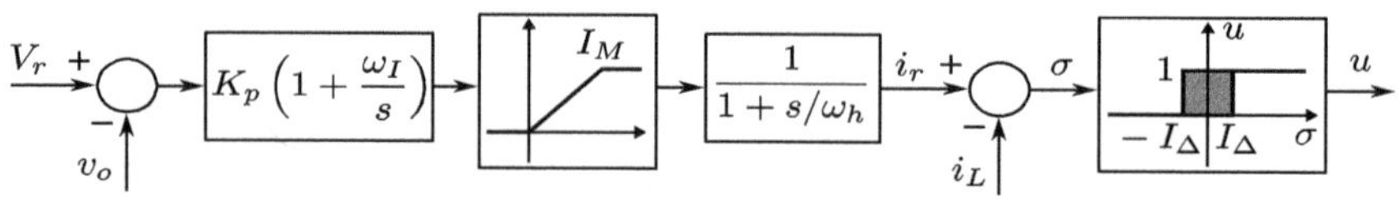

Figure 4. Control scheme with a double loop voltage regulation.

To design $G_c(s)$, a small-signal model of the ideal sliding-mode dynamics will be used. Linearizing (23) around (24) and (25) leads to the following equation for the small-signal variables represented by a "hat" (^) over the letter:

$$\frac{d\hat{v}_o}{dt} = -\frac{2}{CR}\hat{v}_o + \frac{V_g}{CV_r}\hat{i}_r(t) - \frac{LV_r}{CRV_g}\frac{d\hat{i}_r(t)}{dt} + \frac{V_r}{CRV_g}\hat{v}_g(t) + \frac{1}{C}\hat{i}_d(t). \tag{27}$$

Applying the Laplace transform to (27) yields the following model in terms of transfer functions:

$$\hat{V}_o(s) = G_{vir}(s)\hat{I}_r(s) + G_{vvg}(s)\hat{V}_g(s) + G_{vid}(s)\hat{I}_d(s) \tag{28}$$

where:

$$G_{vir}(s) = \frac{\hat{V}(s)}{\hat{I}_r(s)} = \frac{RV_g}{2V_r}\frac{1-s/\omega_z}{1+s/\omega_p} \tag{29}$$

$$G_{vvg}(s) = \frac{\hat{V}V(s)}{\hat{V}_g V(s)} = \frac{V_r}{2V_g}\frac{1}{1+s/\omega_p} \tag{30}$$

$$G_{vid}(s) = \frac{\hat{V}V(s)}{\hat{I}_d V(s)} = \frac{R}{2}\frac{1}{1+s/\omega_p} \tag{31}$$

$$\omega_z = \frac{R{V_g}^2}{L{V_r}^2};\ \omega_p = \frac{2}{RC}. \tag{32}$$

The loop gain is $T(s) = G_c(s)G_{vir}(s)$, and the closed-loop output impedance is given by $Z_o(s) = G_{vid}(s)/(1+T(s))$. Besides, the closed-loop input to output transfer function is expressed as $G_{vvg}(s)/(1+T(s))$. The coefficients of $G_c(s)$ are chosen to ensure that the frequency response of the total loop gain exhibits a high value of the modulus and a sufficient stability margin. The higher the value of $|T(j\omega)|$, the better will be the disturbance rejection, but for a poor relative stability, there can be a range of frequencies very sensitive to disturbances. In the boost converter, the right half-plane zero of $G_{vir}(s)$ (29) imposes an upper bound on the achievable bandwidth [53].

Let us consider an example of a boost converter with the following set of parameter values: $L = 30\ \mu$H, $C = 100\ \mu$F, $R = 10\ \Omega$, $V_g = 10$ V and $V_r = 30$ V. Two cases of CMC will be considered, i.e., valley CMC at constant switching frequency (Figure 2c) with $I_\Delta = 2.5$ A and $f_s = 50$ kHz and hysteresis CMC (Figure 2a) with $I_\Delta = 2.22$ A, which in the steady-state exhibits the same switching frequency of 50 kHz. In Figure 5, the frequency response from the theoretical expression of $G_{vir}(j\omega)$ is compared to the simulated corresponding frequency response of the two mentioned CMC controllers. The simulated frequency response was obtained using the power electronics simulator software

PSIM©, which has a specific feature to get different types of frequency responses, basically input, output and closed loop gains. The AC sweep module manages the frequency sweep (amplitude, initial frequency, final frequency, number of points) as in a frequency response analyzer.

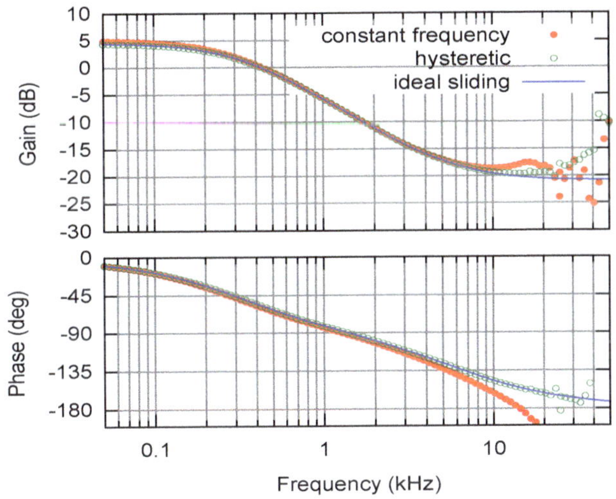

Figure 5. Theoretical and simulated frequency responses from the current reference to output voltage in a boost converter.

A good agreement among the three frequency responses can be clearly observed at low frequencies until approximately 5 kHz ($f_s/10$). The right half-plane zero is located at 6 kHz. Therefore, the control bandwidth will be placed within the region wherein the three responses almost coincide, and the design based on the ideal sliding-mode dynamics model will be valid for the other two cases.

It can be verified in a Bode diagram of $T(j\omega)$ that choosing $\omega_h = \omega_z = 37\,\text{krad/s}$, $\omega_I = 1.2\,\text{krad/s}$ and $K_p = 3.7\,\text{A/V}$, the crossover frequency at 0 dB is $f_c = 2\,\text{kHz}$, the phase margin is 57° and the gain margin is 10 dB at 6 kHz. The Bode diagrams obtained by simulation for the two cases with finite switching frequency show similar results to the ideal case, not only in the loop gain (Figure 6), but also in the output impedance (Figure 7). The closed-loop converter response to a step change in $i_d(t)$ is shown in Figure 8, where the simulated ideal sliding-mode dynamics based on the circuit of Figure 3b is compared to the two previous cases of finite switching frequency. The concordance of the three responses is in agreement with the frequency results depicted in Figure 7.

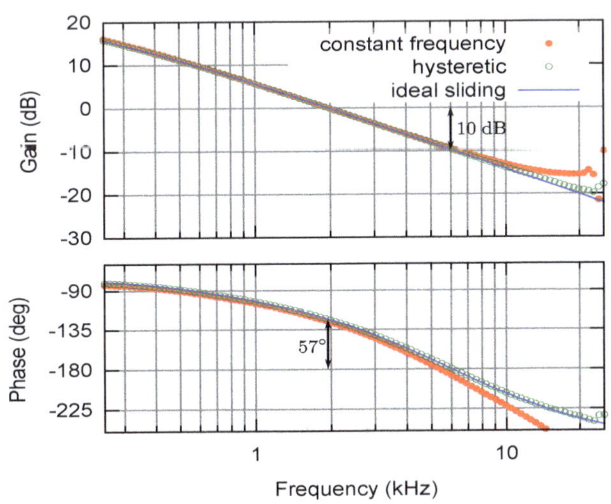

Figure 6. Theoretical and simulated loop gain Bode plots in a boost converter.

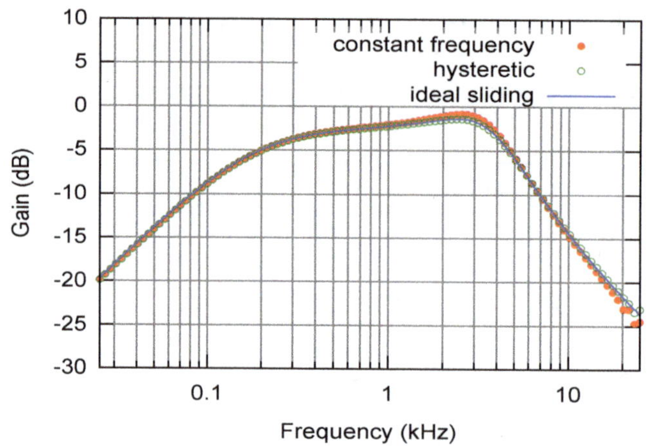

Figure 7. Theoretical and simulated closed-loop output impedance in a boost converter.

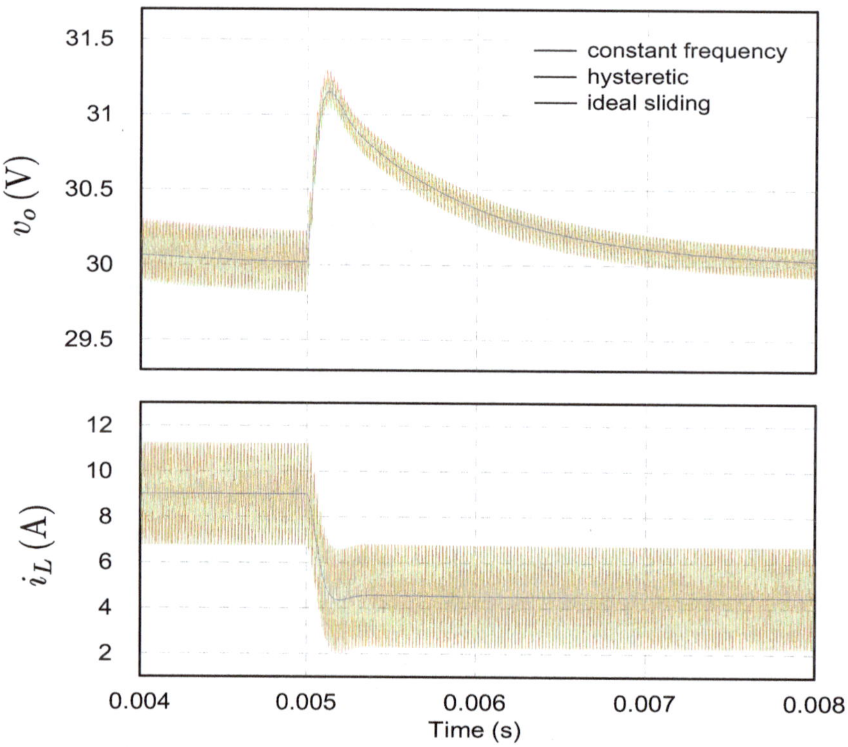

Figure 8. Response of the state variables v_o and i_L of the boost converter in closed-loop to a step change of 1.5 A in $i_d(t)$. Comparison of the ideal sliding-mode dynamics with that corresponding to two different types of control methods at finite switching frequency.

Let us analyze now the time-domain closed-loop converter response from zero initial conditions and without the presence of any disturbance. Figure 9 depicts the case of valley CMC at constant switching frequency. The case of hysteresis control is very similar. It is worth mentioning that the inductor current must be limited in the transient-state. The current control ensures that the inequality $|i_r(t) - i_L| \leq I_\Delta$ is satisfied in sliding mode. The limiter of the voltage controller guarantees that $i_r(t) < I_M$, so that, in the sliding regime, the inductor current satisfies at any instant the inequality $i_L < I_{max}$ where $I_{max} = I_M + I_\Delta$. In this example, the inductor current must reach at least the value $V_r{}^2/(RV_g) = 9$ A, which corresponds to the equilibrium point. The maximum current was fixed to $I_{max} = 15$ A, so that $I_M = 12.78$ A in the case of hysteresis CMC and $I_M = 12.5$ A in the case of valley

CMC. In both cases, the same regions in the response can be observed. Initially $v_o = 0$ V, so that the sliding condition (20) is not accomplished, and the trajectory goes away from the switching manifold defined by $\sigma = 0$ while i_L is increasing. Since the voltage error is positive, $i_r(t)$ also increases until its saturation. This ensures that in some instant, the inequality $i_L > i_r(t) + I_\Delta$ will be satisfied so that $u = 0$ and v_o will be also increasing. When v_o overpasses V_g, the sliding condition is fulfilled, the trajectory alters course and later enters in sliding mode as predicted by the theory. In this regime and with $i_r(t)$ saturated, v_o tends asymptotically towards an equilibrium point in which $v_o{}^* > V_r$. Once v_o surpasses V_r, the voltage controller enters in the linear region, and the trajectory tends towards the desired equilibrium point. An anti-windup system in the integrator can improve this start-up transient by reducing the voltage overshooting, although in this example, it has not been included. Moreover, the converter structure can be modified to avoid the initial current peak [7].

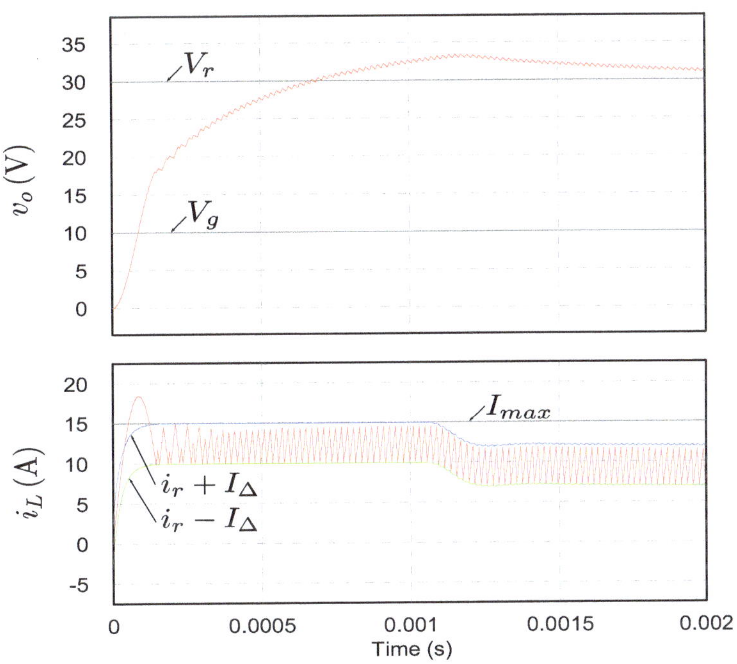

Figure 9. Boost converter response from zero initial conditions using a valley current control at constant switching frequency and voltage regulation.

3.2. Example 2: Buck Converter under Fixed and Variable Frequency CMC

Let us consider the buck converter depicted in Figure 10a, the state equations of which are:

$$\frac{di_L}{dt} = \frac{v_g(t)u - v_o}{L} \tag{33}$$

$$\frac{dv_o}{dt} = \frac{i_L}{C} - \frac{v_o}{RC} + \frac{i_d(t)}{C} \tag{34}$$

$$i_g = i_L u \tag{35}$$

$$i_o = \frac{v_o}{R} - i_d(t). \tag{36}$$

Using the switching function $\sigma = i_r(t) - i_L$ and the control law (7), the sliding condition (13) for this converter becomes:

$$-\frac{v_o}{L} < \frac{di_r(t)}{dt} < \frac{v_g(t) - v_o}{L}. \tag{37}$$

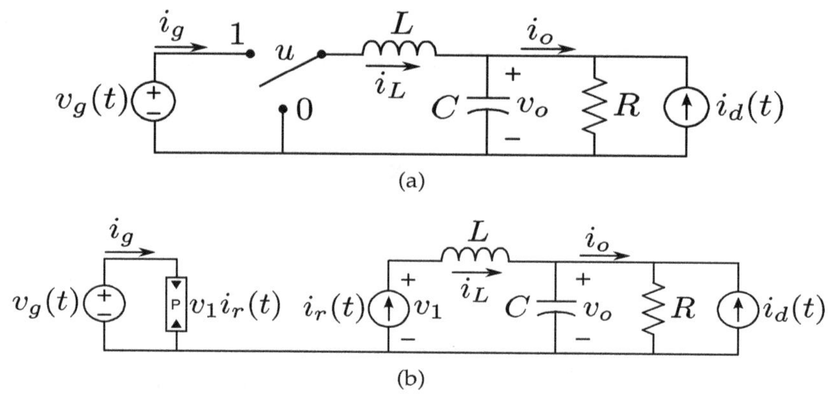

Figure 10. Schematic circuit diagram of (**a**) a Buck converter and (**b**) its ideal sliding-mode dynamics model under CMC.

Inequalities (37) are fulfilled for $i_r(t)$ constant provided that $0 < v_o < v_g(t)$, which are the normal operating conditions of a buck converter. For this converter, the sliding motion is guaranteed from the start-up even with zero initial conditions. The ideal sliding-mode dynamics, obtained by means of the equivalent control method, is the following:

$$i_L = i_r(t) \tag{38}$$

$$\frac{\mathrm{d}v_o}{\mathrm{d}t} = \frac{i_r(t)}{C} - \frac{v_o}{RC} + \frac{i_d(t)}{C} \tag{39}$$

$$i_g = \frac{i_r(t)\left[v_o + L\frac{\mathrm{d}i_r(t)}{\mathrm{d}t}\right]}{v_g(t)} \tag{40}$$

$$i_o = \frac{v_o}{R} - i_d(t) \tag{41}$$

and it can be represented by the circuit illustrated in Figure 10b. The equilibrium point for constant inputs (24), $v_o{}^* = V_r = I_r R$ is asymptotically stable. Linearizing the ideal sliding-mode dynamics around the equilibrium point yields a small-signal model, which is used to design the voltage controller or to analyze the system stability when an input filter is added.

The voltage loop consists of the current reference $\hat{I}_r(s)$ to the output voltage $\hat{V}_o(s)$ transfer function, $G_{vir}(s)$, i.e.,

$$G_{vir}(s) = \frac{R}{1 + sRC}, \tag{42}$$

and the same controller (26) employed in the previous example. In this converter, $G_{vir}(s)$ is a minimum phase transfer function. In the ideal case, the loop gain and the control bandwidth can be made boundlessly high without affecting the system stability. For instance, if a particular value of ω_c is chosen and the coefficients of $G_c(s)$ are calculated according to the following criteria:

$$\omega_c \gg 1/(RC); \;\; K_p = C\omega_c; \;\; \omega_I = \omega_c/4; \;\; \omega_h = 4\omega_c, \tag{43}$$

then ω_c will be the crossover frequency at 0 dB of the loop gain, the phase margin will be higher than 60° and the gain margin will be infinite.

However, the sliding condition (37) is a constraint that must be respected. For example, let us assume that the system is in equilibrium and that a step disturbance of amplitude I_d in current $i_d(t)$ is applied at $t = 0$. If the controller has been designed according to (43), the slope of $i_r(t)$ during the transient-state will be maximal at $t = 0.55/\omega_c$ approximately, reaching a value of:

$$\left.\frac{\mathrm{d}i_r(t)}{\mathrm{d}t}\right|_{t=0.55/\omega_c} \approx -0.8\,\omega_c I_d \tag{44}$$

and therefore, the conditions to keep the sliding mode with this disturbance are:

$$-V_r < -0.8\,\omega_c I_d L < V_g - V_r. \tag{45}$$

Hence, if the bandwidth is too large, small disturbances could provoke very fast variations of $i_r(t)$, violating (37). Moreover, in the case of finite switching frequency, the distortion of the frequency response near the switching frequency degrades the phase margin and limits in turn the control bandwidth. As was observed in Figure 5, the phase decrease is higher in the case of constant switching frequency than in the case of hysteresis, and therefore, a larger bandwidth can be achieved by means of the hysteretic control.

4. Experimental Results

A prototype of the buck converter with hysteretic CMC has been constructed for the set of parameter values $L = 3.3\,\mu\text{H}$, $C = 350\,\mu\text{F}$, $R = 1\,\Omega$, $V_g = 15\,\text{V}$ and $V_r = 5\,\text{V}$. The switching frequency has been tuned to 100 kHz in the equilibrium point. The gain of the current sensor was $R_s = 44\,\text{mV/A}$, and the coefficients of the voltage controller have been calculated according to (43) with $\omega_c = 2\pi \cdot 40\,\text{krad/s}$. A picture of the the experimental setup is shown in Figure 11, and the schematic diagram is depicted in Figure 12. While in the numerical simulations, we used the scheme depicted in Figure 4 in which the saturation block is inserted between the PI compensator and the low-pass filter, our experimental measurements were obtained from a system under the control scheme of Figure 12 in which the saturation block acted after the low-pass filter. From an implementation point of view, it is much simpler to use Figure 12 rather than Figure 4. It is clear that if no saturation takes place, both schemes are equivalent, and this is the case of all the small signal responses (time and frequency domains) presented in the paper. A small difference may arise however under large signal responses such as for instance during the transient. Moreover, it is worth noting that this saturation block was added to limit the current reference during start-up, and this can be accomplished by the scheme of Figure 4, as well as by the one shown at the bottom of Figure 12.

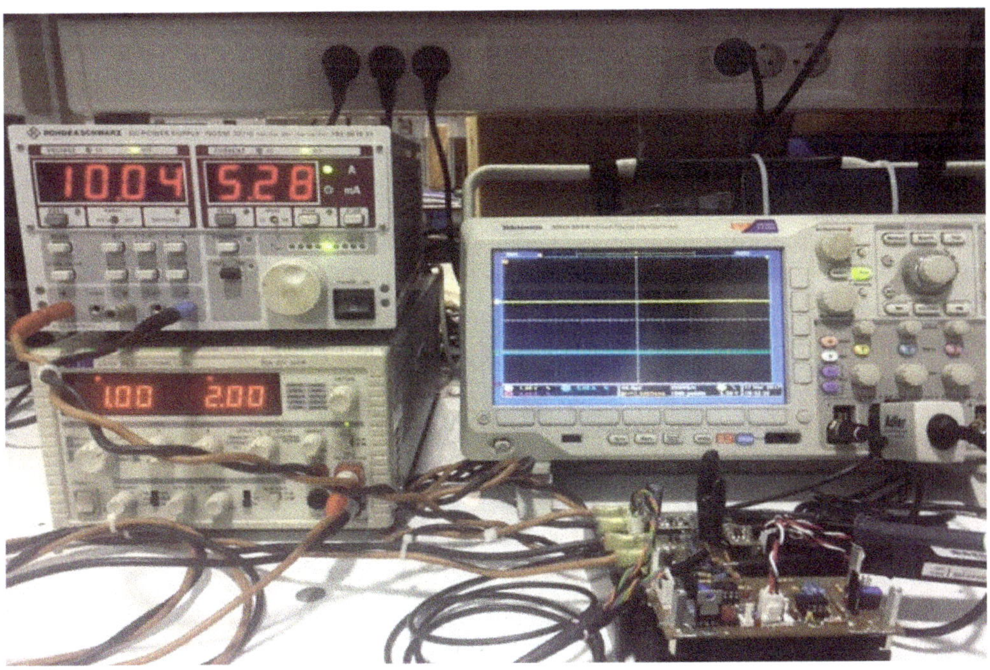

Figure 11. The experimental setup used to validate the theoretical and the simulation results corresponding to the buck converter under two-loop hysteretic CMC.

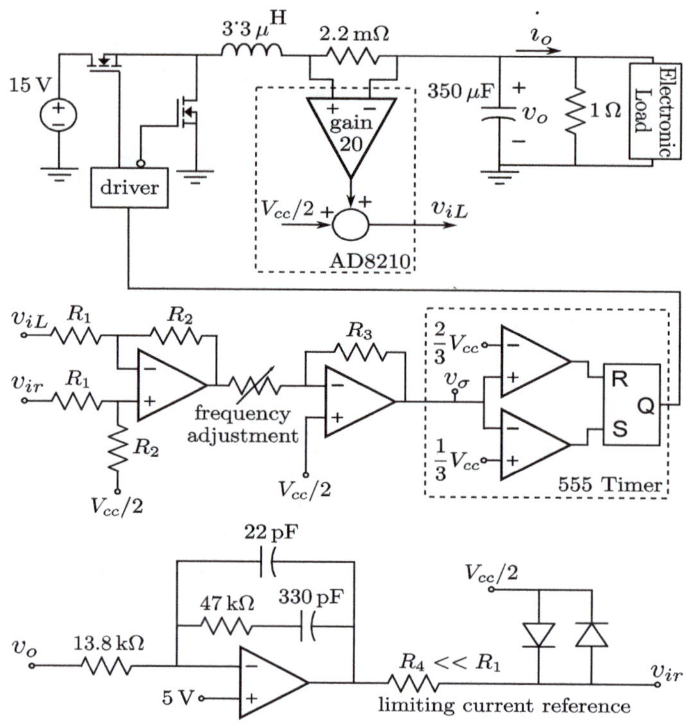

Figure 12. Schematic circuit diagram of the buck converter with hysteretic CMC and output voltage regulation.

The frequency response of the system being the input $\hat{v}_{ir}$ and the output $\hat{v}_o$, in the case of an open voltage loop, has been obtained experimentally, and it is depicted in Figure 13 with the Bode diagram of $G_{vir}(s)/R_s$. The experimental frequency response in Figure 13 was obtained by using the Venable 3120 FRA. In this particular case, the analog control circuit is the one in Figure 12; with the voltage loop opened, the FRA provides the voltage reference v_{ir}, which is composed of a sinusoidal variable frequency reference and a DC component to polarize the system around the desired steady-state operating point. Closed loop gains in Figure 14 were obtained in closed loop using also the Venable 3120 FRA; in this case, a wideband medium frequency injection transformer model Bode Box 200-002 and two probes to obtain differential measurements are connected to the R4-100 Ω resistor.

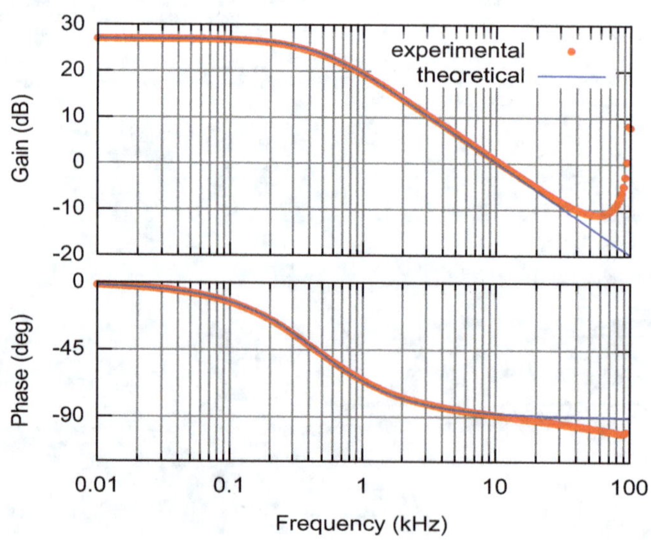

Figure 13. Theoretical and experimental frequency responses from current reference $\hat{v}_{ir}$ to output voltage $\hat{v}_o$ in the buck converter of Figure 12.

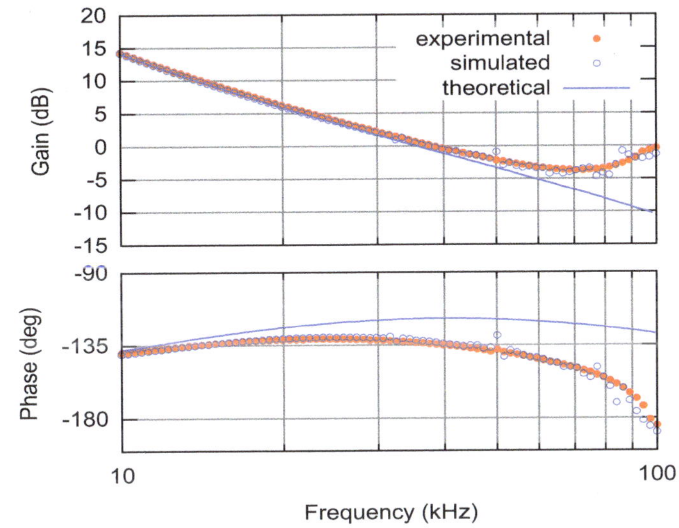

Figure 14. Theoretical, simulated and experimental Bode diagrams of the voltage loop gain in the buck converter.

It can be observed that the model based on the ideal sliding-mode dynamics is valid up to almost one half of the switching frequency. The loop gain has been also measured, and it is shown in Figure 14 with the Bode diagram of $G_c(s)G_{vir}(s)/R_s$ and with the frequency response obtained by numerical simulation of the switched model. It can be observed that the experimental closed-loop bandwidth (ω_c) hardly deviated from the theoretical one, but the phase margin was reduced from 60°–45° due to non-ideal high frequency effects. To match the simulation with the experimental results at high frequency, it was necessary to include in the simulation a dynamic model of the current sensor (a low-pass second-order filter with a cutoff frequency of 450 kHz and a damping ratio of 0.7) and pure delays for both control logic circuits and switching elements (300 ns in total). Figures 15 and 16 show respectively the experimental and simulated response of the converter to a load variation of step type that satisfies the condition (45) by including these non-ideal effects. Note that signal v_σ representing the switching function is not totally within the hysteresis band due to propagation delays and to the dynamics of the current sensor.

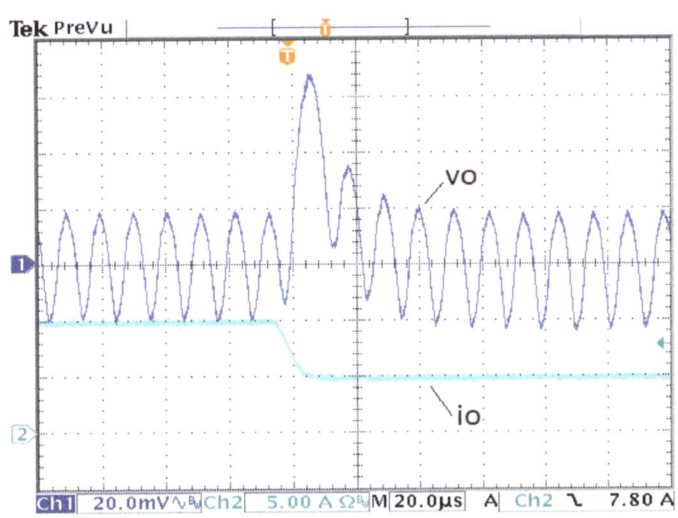

Figure 15. Experimental output voltage response to a load current step in the buck converter from 10 A–5 A.

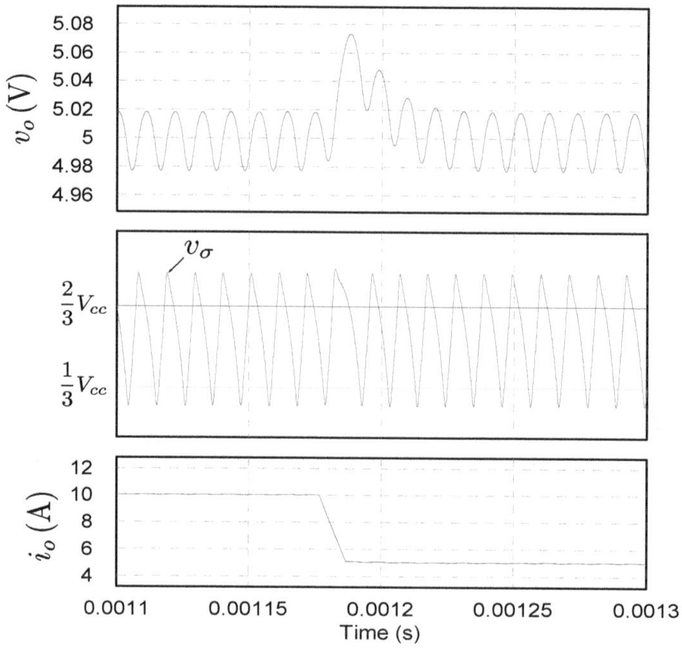

Figure 16. Simulated response to a load current step in the buck converter from 10 A–5 A corresponding to Figure 15.

Remark 4. *The analysis of the same converter with fixed frequency V^2 control and enhanced V^2I_L control [54–56], the V^2I_C [57] and the V^1 control concept [58] can be done following the same procedures with a little effort.*

5. Discussion: Extension to Single-Loop Ripple-Based VMC Strategies

Although the CMC of power converters is nowadays an extensive design practice due to its intrinsic advantages and the existence of an important number of dedicated commercial chips that facilitate its implementation, different ripple-based VMC schemes, that use the parasitic output voltage ripple instead of the inductor current as an additional feedback signal have been also proposed to improve the load transient response of switching converters. For instance, conventional hysteretic voltage mode controllers use the output voltage ripple as the control signal, and a hysteretic comparator is used to generate the square wave for the switch drivers. This is the case of hysteretic voltage regulator modules [59], the V^2 and the enhanced V^2I_L control strategies [54–56] and also the case of the V^2I_C [57] and its equivalent scheme known as the V^1 control concept [58]. Except the V^2I_L control scheme, in all these control strategies, the inductor current is used indirectly in the feedback, and the system works as desired only with a non-ideal output capacitor characterized by a high Equivalent Series Resistance (ESR). Although it is conventionally claimed that hysteretic controllers show faster response than traditional fixed frequency PWM controllers, this happens only with relatively large ESR, and similar responses can be obtained with both modulations if a similar ESR is used. In fact, this ESR introduces indirectly inductor current feedback, and this is the main reason for making the response faster. The enhanced V^2I_L ripple-based control [56] and the I^2 control [60] directly introduce this current feedback. Other approaches for introducing current feedback indirectly in the controller are the so-called raster control surfaces in [61], the use of the capacitor current as in [62] and the introduction of the derivative term of the error signal as in [63].

Although the interest for hysteretic controllers seems a recent practice, due to its immediate application in Voltage Regulation Modules (VRMs) [17,19] and the existence of some commercial chips [20], its use in the voltage regulation of DC-DC switching converters goes back in time to the early years of modern power electronics when conditions for stable limit cycles in a buck switching regulator were first established [15,16]. Other control schemes that can be placed into the same category are boundary control [48,64] and synergetic control [65] strategies.

All the previous strategies can be implemented either with the variable frequency modulation strategy or a fixed frequency modulation scheme. Variable frequency would result from the use of a hysteretic comparator. It would be also the case for constant ON-time and constant OFF-time strategies. Fixed frequency operation will result from peak CMC (or VMC), valley CMC (or VMC) and average CMC (or VMC) using a latch and a clock signal in the modulator. The same theory can be applied interchangeably to all the previous strategies.

6. Conclusions

A two-loop control design technique for DC-DC switching converters has been presented. The double loop consists of an inner current loop in sliding-mode and an outer voltage regulation loop that includes a limiter of the current reference.

When the sliding condition is accomplished, the converter enters in sliding mode after a finite time, and the current tracks its reference. The ideal sliding-mode dynamics describes the system behavior, and a small-signal model around the equilibrium point is used to design the voltage regulation loop by means of the frequency response method.

It has to be pointed out that sliding-mode control theory does not specify the nature of the switching law when $|\sigma| < \Delta$. Hence, hysteresis control and constant switching frequency control have been compared for a boost converter. The frequency response in both cases almost coincides with the theoretical prediction of the ideal sliding-mode dynamics at low frequencies, the hysteresis control response being much more similar to the ideal case near the switching frequency region. However, in the example reported here for the boost converter, the most limiting factor in the regulation bandwidth is the right half-plane zero, and for that reason, the closed-loop response in both cases is very similar to the ideal one.

In the example of the buck converter, an expression of the sliding condition in terms of both voltage regulation bandwidth and load disturbance amplitude has been derived. A bandwidth of the voltage loop near one half the switching frequency has been obtained using a hysteretic current mode controller. In the same example, it has been found that the actual dynamic behavior of the current sensor and the delay of both switching and control logic circuits can have an appreciable influence upon the phase margin of the voltage loop.

Author Contributions: Conceptualization, J.C., A.E.A. and L.M.-S; All the authors contributed equally to the other parts of the work.

References

1. El Fadil, H.; Giri, F. Robust nonlinear adaptive control of multiphase synchronous buck power converters. *Control Eng. Pract.* **2009**, *17*, 1245–1254. [CrossRef]
2. Erickson, R.W.; Maksimovic, D. *Fundamentals of Power Electronics*; Kluwer: Alphen aan den Rijn, The Netherlands, 2001.
3. Yfoulis, C.; Giaouris, D.; Stergiopoulos, F.; Ziogou, C.; Voutetakis, S.; Papadopoulou, S. Robust constrained stabilization of boost DC-DC converters through bifurcation analysis. *Control Eng. Pract.* **2015**, *35*, 67–82. [CrossRef]
4. Middlebrook, R. Topics in multiple-loop regulators and current-mode programming. *IEEE Trans. Power Electron.* **1987**, *PE-2*, 109–124. [CrossRef]
5. Ramos, R.; Biel, D.; Fossas, E.; Gri, R. Sliding mode controlled multiphase buck converter with interleaving and current equalization. *Control Eng. Pract.* **2013**, *21*, 737–746. [CrossRef]

6. Liu, Y.-F.; Sen, P.C. Large-signal modeling of hysteretic current-programmed converters. *IEEE Trans. Power Electron.* **1996**, *11*, 423–430.
7. Martinez-Salamero, L.; Garcia, G.; Orellana, M.; Lahore, C.; Estibals, B. Start-up control and voltage regulation in a boost converter under sliding-mode operation. *IEEE Trans. Ind. Electron.* **2013**, *60*, 4637–4646. [CrossRef]
8. Deisch, C.W. Simple switching control method changes power converter into a current source. In Proceedings of the PESC'78, Power Electronics Specialists Conference, Syracuse, NY, USA, 13–15 June 1978; pp. 300–306.
9. Redl, R.; Sokal, N. Current-mode control, five different types, used with the three basic classes of power converters-small-signal ac and large-signal dc characterization, stability requirements, and implementation of practical circuits. In Proceedings of the PESC'85 Record, 16th Annual IEEE, Toulouse, France, 24–28 June 1985; pp. 771–785.
10. Perreault, D.; Verghese, G.C. Time-varying effects and averaging issues in models for current-mode control. *IEEE Trans. Power Electron.* **1997**, *12*, 453–461. [CrossRef]
11. Ridley, R. A new, continuous-time model for current-mode contro. *IEEE Trans. Power Electron.* **1991**, *6*, 271–280. [CrossRef]
12. Tan, F.; Middlebrook, R. A unified model for current-programmed converters. *IEEE Trans. Power Electron.* **1995**, *10*, 397–408. [CrossRef]
13. Yan, Y.; Lee, F.; Mattavelli, P. Comparison of small signal characteristics in current mode control schemes for point-of-load buck converter applications. *IEEE Trans. Power Electron.* **2013**, *28*, 3405–3414. [CrossRef]
14. Li, J.; Lee, F. New modeling approach and equivalent circuit representation for current-mode control. *IEEE Trans. Power Electron.* **2010**, *25*, 1218–1230.
15. Babaa, I.; Wilson, T.; Yu, Y. Analytic solutions of limit cycles in a feedback-regulated converter system with hysteresis. *IEEE Trans. Autom. Control* **1968**, *13*, 524–531. [CrossRef]
16. Schweitzer, B.P.; Rosenstein, A.B. Free Running-Switching Mode Power Regulator: Analysis and Design. *IEEE Trans. Aerosp.* **1964**, *2*, 1171–1180. [CrossRef]
17. Huerta, S.; Soto, A.; Alou, P.; Oliver, J.; Garcia, O.; Cobos, J. Advanced control for very fast DC-DC converters based on hysteresis of the C_{out} current. *IEEE Trans. Circuits Syst. I Regular Pap.* **2013**, *60*, 1052–1061. [CrossRef]
18. Ki, W.-H.; Lai, K.-M.; Zhan, C. Charge balance analysis and state transition analysis of hysteretic voltage mode switching converters. *IEEE Trans. Circuits Power Syst.* **2011**, *58*, 1142–1152. [CrossRef]
19. Wen, Y.; Trescases, O. Analysis and comparison of frequency stabilization loops in self-oscillating current mode dc-dc converters. *IEEE Trans. Power Electron.* **2013**, *28*, 4753–4766. [CrossRef]
20. Castilla, M.; Guerrero, J.; Matas, J.; Miret, J.; Sosa, J. Comparative study of hysteretic controllers for single-phase voltage regulators. *IET Power Electron.* **2008**, *1*, 132–143. [CrossRef]
21. Atre, S.R.; Lele, V.P. Analytic application of Tsypkin's method relay with hysteresis and dead zone. *Proc. Inst. Electr. Eng.* **1977**, *124*, 1241–1242. [CrossRef]
22. Barrado, J.A.; El Aroudi, A.; Valderrama-Blavi, H.; Calvente, J.; Martinez-Salamero, L. Analysis of a self-oscillating bidirectional DC-DC converter in battery energy storage applications. *IEEE Trans. Power Deliv.* **2012**, *27*, 1292–1300. [CrossRef]
23. Tsypkin, Y.Z. *Relay Control Systems*; Cambridge Univ. Press: Cambridge, UK, 1984.
24. Boiko, I. Variable-structure PID controller for level process. *Control Eng. Pract.* **2013**, *21*, 700–707. [CrossRef]
25. Utkin, V.; Guldner, J.; Shi, J. *Sliding Mode Control in Electro-Mechanical Systems, (Automation and Control Engineering)*; CRC Press: Boca Raton, FL, USA, 2009.
26. Filippov, A.F. Differential equations with discontinuous right-hand side. *Am. Math. Soc. Transl.* **1964**, *42*, 199–231.
27. Utkin, V.I. *Sliding Modes in Control and Optimization*; Springer-Verlag: Berlin, Germany, 1992.
28. Sira-Ramirez, H. Sliding motions in bilinear switched networks. *IEEE Trans. Circuits Syst.* **1987**, *34*, 919–933. [CrossRef]
29. Bilalovic, F.; Music, O.; Sabanovic, A. Buck converter regulator operating in the sliding mode. In Proceedings of the 7th Internatinal PCI Conference, Orlando, FL, USA, 19–21 April 1983; pp. 331–340.
30. Malesani, L.; Rossetto, L.; Spiazzi, G.; Tenti, P. Performance optimization of Cuk converters by sliding-mode control. *IEEE Trans. Power Electron.* **1995**, *10*, 302–309. [CrossRef]

31. Spiazzi, G.; Mattavelli, P. Sliding-mode control of switched-mode power supplies. In *The Power Electronics Handbook*; Chapter 8; Lappeenranta University of Technology: Lappeenranta, Finland, 2002.
32. Tan, S.-C.; Lai, Y.; Cheung, M.; Tse, C. On the practical design of a sliding mode voltage controlled buck converter. *IEEE Trans. Power Electron.* **2005**, *20*, 425–437. [CrossRef]
33. Tan, S.-C.; Lai, Y.; Tse, C.; Cheung, M. Adaptive feedforward and feedback control schemes for sliding mode controlled power converters. *IEEE Trans. Power Electron.* **2006**, *21*, 182–192.
34. Chen, Z.; Gao, W.; Hu, J.; Ye, X. Closed-loop analysis and cascade control of a nonminimum phase boost converter. *IEEE Trans. Power Electron.* **2011**, *26*, 1237–1252. [CrossRef]
35. Chen, Z. PI and sliding mode control of a Ćuk converter. *IEEE Trans. Power Electron.* **2012**, *27*, 3695–3703. [CrossRef]
36. Komurcugil, H. Non-singular terminal sliding-mode control of DC-DC buck converters. *Control Eng. Pract.* **2013**, *21*, 321–332. [CrossRef]
37. Singh, P.; Hote, Y.; Garg, M. Comments on PI and sliding mode control of a Cuk converter. *IEEE Trans. Power Electron.* **2014**, *29*, 1551–1552. [CrossRef]
38. Sira-Ramirez, H.; Luviano-Juarez, A.; Romero, J.C. Robust input-output sliding mode control of the buck converter. *Control Eng. Pract.* **2013**, *21*, 671–678. [CrossRef]
39. Ramos, R.; Biel, D.; Fossas, E.; Guinjoan, F. Interleaving quasi-sliding-mode control of parallel-connected buck-based inverters. *IEEE Trans. Ind. Electron.* **2008**, *55*, 3865–3873. [CrossRef]
40. Cid-Pastor, A.; Martinez-Salamero, L.; Alonso, C.; Estibals, B.; Alzieu, J.; Schweitz, G.; Shmilovitz, D. Analysis and design of power gyrators in sliding-mode operation. *IEE Proc. Electr. Power Appl.* **2005**, *152*, 821–826. [CrossRef]
41. Haroun, R.; Cid-Pastor, A.; El Aroudi, A.; Martinez-Salamero, L. Synthesis of canonical elements for power processing in dc distribution systems using cascaded converters and sliding-mode control. *IEEE Trans. Power Electron.* **2014**, *29*, 1366–1381. [CrossRef]
42. Cid-Pastor, A.; Giral, R.; Calvente, J.; Utkin, V.; Martinez-Salamero, L. Interleaved converters based on sliding-mode control in a ring configuration. *IEEE Trans. Circuits Syst. I Regular Pap.* **2011**, *58*, 2566–2577. [CrossRef]
43. Utkin, V. Sliding mode control of dc/dc converters. *J. Frank. Inst.* **2013**, *350*, 2146–2165. [CrossRef]
44. Cid-Pastor, A.; Martinez-Salamero, L.; El Aroudi, A.; Giral, R.; Calvente, J.; Leyva, R. Synthesis of loss-free resistors based on sliding-mode control and its applications in power processing. *Control Eng. Pract.* **2013**, *21*, 689–699. [CrossRef]
45. Vidal-Idiarte, E.; Martinez-Salamero, L.; Guinjoan, F.; Calvente, J.; Gomariz, S. Sliding and fuzzy control of a boost converter using an 8-bit microcontroller. *IET Electr. Power Appl.* **2004**, *151*, 5–11. [CrossRef]
46. Vidal-Idiarte, E.; Martínez-Salamero, L.; Calvente, J.; Romero, A. An H_∞ control strategy for switching converters in sliding-mode current control. *IEEE Trans. Power Electron.* **2006**, *21*, 553–556. [CrossRef]
47. Mattavelli, P.; Rossetto, L.; Spiazzi, G. Small-signal analysis of DC-DC converters with sliding mode control. *IEEE Trans. Power Electron.* **1997**, *12*, 96–102. [CrossRef]
48. Chan, M.P.; Mok, P.K.T. A monolithic 2nd-order boundary controller for buck converter with fast transient response. In Proceedings of the 2010 IEEE Asia Pacific Conference on Circuits and Systems, Kuala Lumpur, Malaysia, 6–9 December 2010.
49. Biel, D.; Fossas, E. SMC applications in power electronics. *Var. Struct. Syst. Princ. Implementation* **2004**, *66*, 265–293.
50. Silva, J.; Paulo, S. Fixed frequency sliding mode modulator for current mode PWM inverters. In Proceedings of the 24th Annual IEEE Power Electronics Specialist Conference—PESC '93, Seattle, WA, USA, 20–24 June 1993; pp. 623–629.
51. Singer, S.; Erickson, R. Power-source element and its properties. *IEE Proc. Circuits Devices Syst.* **1994**, *141*, 220–226. [CrossRef]
52. Sira-Ramirez, H.; Silva-Ortigoza, R. *Control Design Techniques in Power Electronics Devices, Power Systems Series*; Springer: London, UK, 2006.
53. Doyle, J.; Francis, B.; Tannenbaum, A. *Feedback Control Theory*; Macmillan Publishing Co.: Basingstoke, UK, 1990.
54. Goder, D.; Pelletier, W.R. *V2 Architecture Provides Ultra-Fast Transient Response in Switch Power Supplies*; Intertect International: Las Vegas, NV, USA, 1996; pp. 19–23.

55. Redl, R.; Sun, J. Ripple-based control of switching regulators: an overview. *IEEE Trans. Power Electron.* **2009**, *24*, 2669–2680. [CrossRef]
56. Sun, J. Characterization and performance comparison of ripple-based control for voltage regulator modules. *IEEE Trans. Power Electron.* **2006**, *21*, 346–353.
57. Del Viejo, M.; Alou, P.; Oliver, J.A.; García, O.; Cobos, J.A. V^2IC control: A novel control technique with very fast response under load and voltage steps. In Proceedings of the 2011 Twenty-Sixth Annual IEEE Applied Power Electronics Conference and Exposition (APEC), Fort Worth, TX, USA, 6–11 March 2011.
58. Cortes, J.; Svikovic, V.; Alou, P.; Oliver, J.; Cobos, J.A. V^1 concept: Designing a voltage mode control as current mode with near time-optimal response for Buck-type converters. *IEEE Trans. Power Electron.* **2015**, *30*, 5829–5841. [CrossRef]
59. Chen, M.; Zhao, M.; Wu, X. A fixed-frequency fast transient response DC-DC controller for VRMs. *J. Semiconduct.* **2013**, *34*, 125002. [CrossRef]
60. Yan, Y.; Lee, F.C.; Mattavelli, P.; Liu, P.H. I^2 average current mode control for switching converters. *IEEE Trans. Power Electron.* **2014**, *29*, 2027–2036. [CrossRef]
61. Pitel, G.E.; Krein, P.T. Minimum-time transient recovery for DC-DC converters using raster control surfaces. *IEEE Trans. Power Electron.* **2009**, *12*, 2692–2703. [CrossRef]
62. Kapat, S.; Krein, P.T. Improved time optimal control of a buck converter based on capacitor current. *IEEE Trans. Power Electron.* **2012**, *27*, 1444–1454. [CrossRef]
63. Kapat, S.; Krein, P.T. Formulation of PID control for DC-DC converters based on capacitor current: A geometric context. *IEEE Trans. Power Electron.* **2012**, *27*, 1424–1432. [CrossRef]
64. Onwuchekwa, C.N.; Kwasinski, A. Analysis of boundary control for buck converters with instantaneous constant-power loads. *IEEE Trans. Power Electron.* **2010**, *2*, 2018–2032. [CrossRef]
65. Santi, E.; Monti, A.; Li, D.; Proddutur, K.; Dougal, R.A. Synergetic control for DC-DC boost converter: Implementation options. *IEEE Trans. Ind. Appl.* **2003**, *39*, 1803–1813. [CrossRef]

3

Novel Step-Down DC–DC Converters based on the Inductor–Diode and Inductor–Capacitor–Diode Structures in a Two-Stage Buck Converter

Mauricio Dalla Vecchia [1,2,*], Giel Van den Broeck [1,2], Simon Ravyts [1,2] and Johan Driesen [1,2]

1 Departement Elektrotechniek, KU Leuven, Kasteelpark Arenberg 10, 3001 Leuven, Belgium; giel.vandenbroeck@esat.kuleuven.be (G.V.d.B.); simon.ravyts@kuleuven.be (S.R.); johan.driesen@kuleuven.be (J.D.)
2 EnergyVille, Thor Park 8310, 3600 Genk, Belgium
* Correspondence: mauricio.dallavecchia@kuleuven.be

Abstract: This paper explores and presents the application of the Inductor–Diode and Inductor-Capacitor-Diode structures in a DC–DC step-down configuration for systems that require voltage adjustments. DC micro/picogrids are becoming more popular nowadays and the study of power electronics converters to supply the load demand in different voltage levels is required. Multiple strategies to step-down voltages are proposed based on different approaches, e.g., high-frequency transformer and voltage multiplier/divider cells. The key question that motivates the research is the investigation of the aforementioned Inductor–Diode and Inductor–Capacitor–Diode, current multiplier/divider cells, in a step-down application. The two-stage buck converter is used as a study case to achieve the output voltage required. To extend the intermediate voltage level flexibility in the two-stage buck converter, a second switch was implemented replacing a diode, which gives an extra degree-of-freedom for the topology. Based on this modification, three regions of operation are theoretically defined, depending on the operational duty cycles δ_2 and δ_1 of switches S_2 and S_1. The intermediate and output voltage levels are defined based on the choice of the region of operation and are mapped herein, summarizing the possible voltage levels achieved by each configuration. The paper presents the theoretical analysis, simulation, implementation and experimental validation of a converter with the following specifications; 48 V/12 V input-to-output voltage, different intermediate voltage levels, 100 W power rating, and switching frequency of 300 kHz. Comparisons between mathematical, simulation, and experimental results are made with the objective of validating the statements herein introduced.

Keywords: DC–DC converter; experimental verification; Inductor–Diode; Inductor–Capacitor–Diode; nonisolated; step-down; two-stage buck converter; voltage regulation

---

1. Introduction

In the last century, AC systems predominated as the most widespread method for energy generation, transmission, and distribution. For regular AC generation, large power plants, e.g., coal, hydro, and nuclear, have been used to centrally generate energy. Alternatives, such as wind and solar generation, are more frequently being installed to decentralize the energy production and lower the CO_2 emitted by the energy production [1,2].

Wind farms (AC generation) are more common in remote regions (on and off-shore) where the wind is conducive for generation. Solar (DC generation), on the other hand, started to become viable for application in high populated areas where the sun irradiation is high enough to achieve maximum power generation. DC systems started to be developed and implemented in pilot projects and real applications because they have proved to be more efficient than AC systems in terms of transformation steps [3], costs [4], and power transfer capability [5]. Applications of DC system such as in micro- and nanogrids [6,7], datacenters [8,9], and aircrafts, where redundancy and backup is mandatory, are currently on the rise. DC system technology can also be implemented in isolated/remote/emerging regions [10]. In these systems, redundancy normally is indispensable in order to keep the system operational even if a component and/or a converter fails. Furthermore, DC loads are common especially in low voltage/low power applications, for instance, in computer and battery chargers, USB and USB-C ports, and LED lights. These type of loads are commonly found in offices and residences.

Discussions among the voltage levels that are implemented in these systems are extensively presented in the literature [11], for instance, with a voltage between 350 and 400 V for distribution in unipolar or bipolar [6,12,13] DC configurations to supply high power equipments. The 48 V level is [14,15] normally adopted as an intermediate bus stage to supply the low voltage/low power loads aforementioned. The 48 V level is also getting more common in battery systems for Electric Vehicles (EV), and it is considered for residential applications due to its user safety approach.

A DC picogrid, based on state-of-the-art architectures [3,6,16], is depicted in Figure 1, which emulates a building and/or residence with a DC grid implementation. In this grid, a bipolar DC network is realized (P0N poles), which has the possibility of injecting the energy generated by the panels in the AC grid or, alternatively, directly supply DC loads bypassing the DC–AC–DC transformation.

A lot of effort has been spent to propose new topologies and systems to interconnect PV generation in AC or DC grids with, respectively, microinverters [17–19] and microconverters (DC power optimizers) [20–24] which, in essence, are step-up converters since the PV panel output voltage is lower than the AC grid or DC bus voltages. The control aspects of this system and MPPT strategies are common subjects in the literature [20,25–27].

Different strategies to step-up the voltage have been studied and proposed in the literature, such as: isolated step-up topologies [17,21,28], coupled-inductor, nonisolated topologies [29–35], nonisolated topologies with Capacitor–Diode/switch structure [19,36–38], nonisolated topologies with Inductor–Diode/switch [39,40] and nonisolated topologies with Capacitor–Diode/switch and Inductor–Diode/switch structures [41,42]. These structures are known, respectively, as switched-capacitors and switched-inductors. An adaptation of the Inductor–Diode structure is proposed in [43], where a capacitor replaces one diode and changes the gain of the structure for step-up applications. Each of the above-mentioned strategies to step-up voltages has its advantages and disadvantages. For instance, isolated topologies and coupled inductor topologies have the flexibility of, based on the turns ratio of the magnetic device, select the suitable output voltage requirement, although it is known that these magnetic components are, together with heatsinks, the bulkiest components in a converter. Switched capacitor and switched inductors do not bring this flexibility once the gain is defined based on the operational duty cycle of the active switches. Inductors are more reliable components than capacitors, giving an advantage in terms of reliability, while capacitors are more compact and more suited for high power density applications.

In step-down conversion ratio converters, the Capacitor–Diode/switch structures are commonly used to improve the step-down ratio of the converters [44–46]. On the other hand, the implementation of an Inductor–Diode/switch and Inductor–Capacitor–Diode structures in the step-down conversion ratio is not extensively discussed in the literature, with few exceptions that explore both step-up and step-down structures [39]. The structures evaluated in this paper are shown in Figure 2. The Inductor–Diode structure shown in Figure 2 can offer a redundant path if one diode of the structure fails in open-circuit, as proposed in [47]. In steady-state, the inductors can be considered as current sources and, due to the structure operability discussed in Section 4, mitigate the implementation of current sensors for control purposes, since the current balance between the inductors is naturally achieved without any control strategy (switched-inductor characteristic). The only sensor that might be consider for implementation is a voltage sensor, if the application requires a regulated and constant output voltage level for the entire load range of operation. For instance, a contraction technique was proposed to control a regular step-down converter [48].

With these considerations, the objective of this paper is to investigate the influence of the Inductor–Diode (ID) and Inductor–Capacitor–Diode (ICD) structures in the step-down conversion stage from 48 V to 12 V input-to-output voltage. The main contributions of this paper are (1) an analysis of the influence of the ID and ICD structures in a step-down conversion system; (2) the presentation of a full theoretical analysis of the converters generated; (3) applying the ID and ICD structures in a quadratic buck converter; (4) analyzing the intermediate voltage level behavior for different operational points, showing the intermediate voltage flexibility of the circuit, which is beneficial for a voltage regulation point of view; (5) creating a flexible test bench to be able to validate the proposed converters and; (6) experimentally validatation with waveforms of the topologies studied in the paper.

The paper is organized as follows; Section 2 presents the analysis of the ID and ICD structure in a regular single-stage buck converter and its impact on the gain of the converter. Section 3 presents the two-stage buck converter, known as quadratic buck converter [49], to enable the converter applying the ID and ICD structures to achieve the required output voltage levels. The operational principle of the topologies generated by the proposed intra-exchangeability is explained in Section 4. Section 5 explores the simulation results of the proposed topologies, showing the intermediate and output voltage levels that are obtained for each possible configuration. Section 6 describes the experimental test bench built for further validation, showing experimental waveforms that validate the theoretical analysis presented in Sections 3 and 4. Section 7 presents the main conclusions and future work suggestions.

2. Impact of Inductor–Diode and Inductor–Capacitor–Diode Structures in a Step-Down Converter

This section has the objective of investigating the influence of the ID and ICD structure in a step-down buck converter. Inductor I, shown in Figure 3, is replaced by ID and ICD structures and the PWM modes are described herein. The main difference between the ID and ICD structures is that the capacitor of the ICD is also charged during the magnetization of the inductors. When discharging, the capacitor also discharges, providing energy to the load. The capacitor stores part of the energy that is delivered to the load, which helps decrease the size of the inductors, since less energy is stored in their magnetic field. Additionally, the capacitor changes the voltage applied across the inductors during demagnetization, which changes the gain of the structures, as described herein.

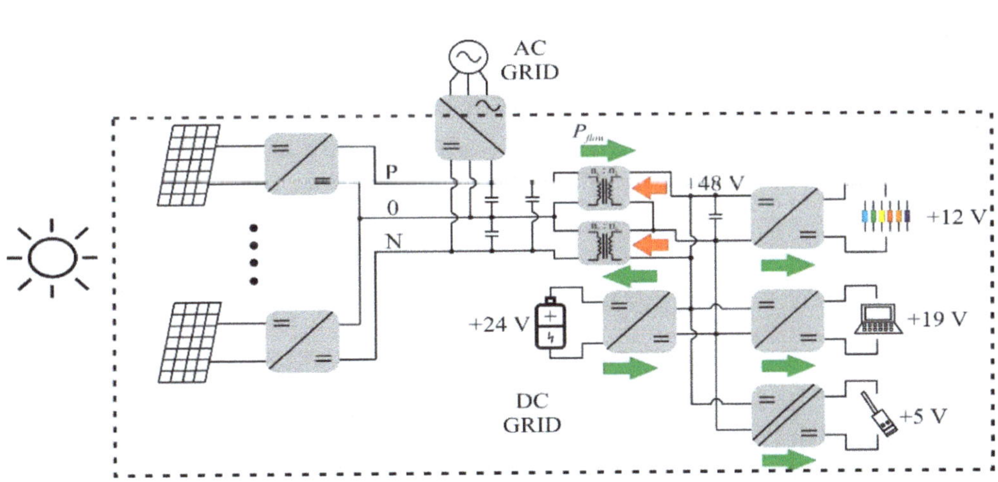

Figure 1. A basic DC picogrid, highlighting the PV generation and DC step-up conversion; the bipolar DC network (P0N); the battery energy storage system (BESS); and the DC loads supplied by DC–DC step-down converters.

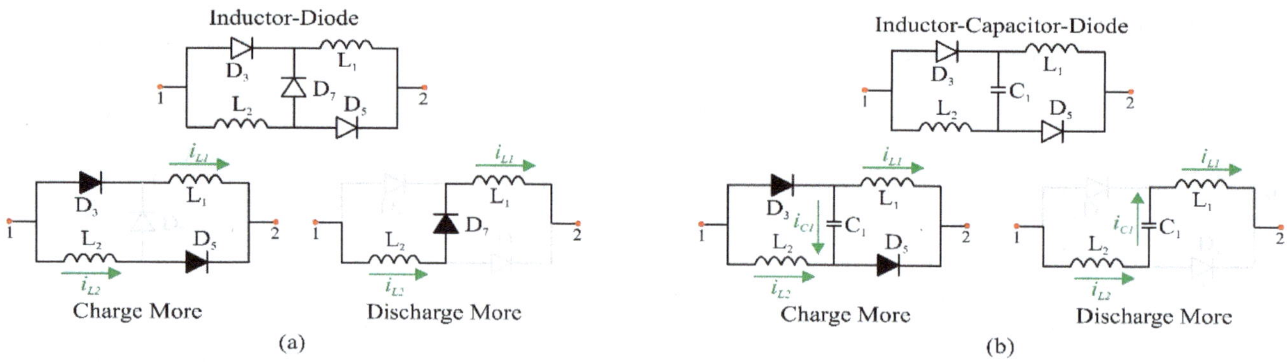

Figure 2. (**a**) Inductor–Diode (ID) structure and its charge and discharge modes and (**b**) Inductor–Capacitor–Diode (ICD) structure and its charge and discharge modes.

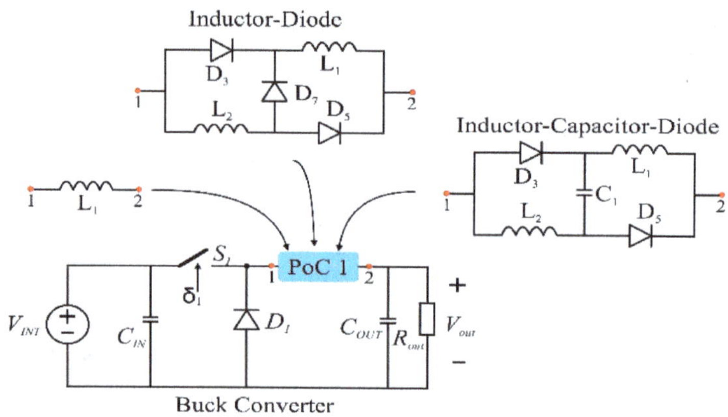

Figure 3. DC–DC Buck Converter with Point-of-Connection 1 (PoC1) to interchange inductor (I), Inductor–Diode (ID) and Inductor–Capacitor–Diode (ICD) structures.

2.1. PWM Mode 1

In this PWM mode, the switch S_1 commutates from 'open' to 'close', changing the equivalent circuit of the converter. The equivalent circuit for this mode is shown in Figure 4a,b for the ID and ICD applications in PoC1, respectively. During this mode, the voltage applied in the inductor terminals V_{L1} and V_{L2} is positive, due to the step-down mechanism, and energy is magnetically stored in the inductors during the process. The voltage applied across the inductors terminals has the same level (internally in ID and ICD structures), which leads to similar current level in the inductors of the structures. In the ID structure, the inductors store energy during this PWM Mode while in the ICD structure the inductors and the capacitor of the structure store energy.

2.2. PWM Mode 2

In this PWM mode, the switch S_1 commutates from 'close' to 'open'. The equivalent circuit for this mode is shown in Figure 4c,d for the ID and ICD structures, respectively. During this PWM mode, the voltage applied across the inductors terminals V_{L1} and V_{L2} is negative, since diode D_1 is forward biased, leading to a demagnetization of these inductors. The current across both inductors is equal as these components are in series, which guarantees the same negative voltage level across the inductors of the structures during the demagnetization. The demagnetization mechanism for both structures is similar, since in both structures the inductors are connected in series and the same current flows through the passive components.

From the above PWM mode analysis, the theoretical voltage profile for all the inductors in both structures are presented in (1). These equations are applied in (2) to obtain the conversion gain of the structures, in steady-state.

$$\underset{ID}{\begin{cases} V_{L1} = V_{L2} = V_{IN} - V_{OUT} \ during \ \delta_1 \\ V_{L1} = V_{L2} = -\frac{V_{OUT}}{2} \ during \ (1-\delta_1) \end{cases}} \qquad \underset{ICD}{\begin{cases} V_{L1} = V_{L2} = V_{C1} = V_{IN} - V_{OUT} \ during \ \delta_1 \\ V_{L1} = V_{L2} = \frac{V_{IN}}{2} - V_{OUT} \ during \ (1-\delta_1) \end{cases}} \tag{1}$$

$$\overline{V_L} = \int_0^{T_S} v_L dt = 0 \tag{2}$$

2.3. Theoretical Gain in Continuous Conduction Mode

To obtain the theoretical voltage gain profile of the topologies, operation in steady-state is considered, which allows the application of (1) in (2) to obtain the ideal gain of the structures. After some mathematical manipulation, the theoretical expression that represents the gain of the Inductor–Diode (ID) and Inductor–Capacitor–Diode (ICD) structures, respectively, is graphically represented in Figure 5. From Figure 5, it is noticed that the operational voltage range tackled in this paper cannot be realized with the ICD structure, since its gain range is, ideally, between [0.5 ; 1]. To overcome this, the quadratic buck converter—a nonlinear step-down converter generated by two buck converters connected in series with each other—as depicted in Figure 6, is the base topology used to generate the converters proposed in this paper. This converter presents an intermediate stage, giving an extra degree of freedom for implementation, since two voltage levels are generated in the same converter and both can be used to supply loads.

Only one switch is implemented on the power stage of the conventional quadratic buck, which has the benefit of control strategy simplification; and the intermediate voltage level is in the function of the duty cycle applied to switch S_1. If a second switch is introduced in the power stage, the degree of freedom for voltage regulation in the intermediate stage is guaranteed by the duty cycle of the second switch S_2, although the control strategy might increase in complexity, if applicable. Since the objective of the topologies is to improve the flexibility of the conversion ratio and the overall system, the quadratic buck with two switches is the converter that will be extensively explored in the next section, together with the intra-exchangeable PoC structures presented in Figure 3.

3. Double-Switch Two-Stage Quadratic Buck Converter for Voltage Regulation

As previously mentioned, the two-stage buck converter, most known as quadratic buck converter, is the base topology used for the intra-exchangeable methodology used herein for testing multiple configurations, based on the ID and ICD structures, applied for voltage regulation. Since a second stage is now implemented in the power circuit, the intra-exchangeability increases the amount of topologies that can be generated. More specifically, with three structures and two stages, the number of topologies generated is $3^2 = 9$, where only the conventional quadratic buck is already explored in the literature. Figure 6 exemplifies the possible configurations that can be obtained from the proposed intra-exchangeability between a regular inductor (I) and the ID and ICD structures. When an I structure is connected to PoC1 and PoC2, the regular quadratic buck is obtained. In total, nine topologies are generated from this method and, since two switches are implemented in the power stage with independent duty cycles, three regions of operation can be explored and will be briefly discussed herein. The regions of operation are highlighted in Figure 6, named as region 1, when duty cycle δ_2 is higher than duty cycle δ_1; region 2, when δ_1 is higher than δ_2; and region 3, when both duty cycles are equal. The switch realization is based on a MOSFET with an antiparallel diode, as shown in Figure 6.

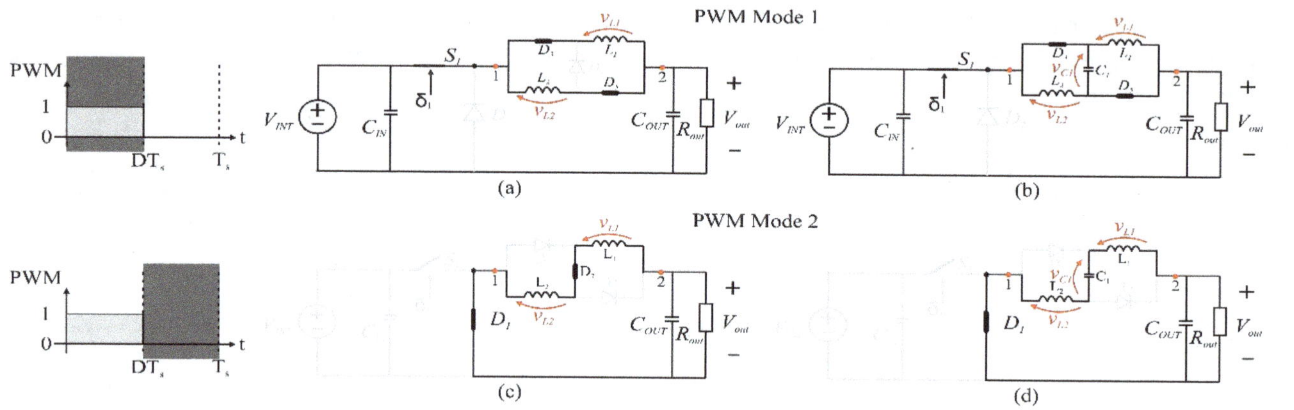

Figure 4. (**a**) PWM mode 1 for ID structure; (**b**) PWM mode 1 for ICD structure; (**c**) PWM mode 2 for ID structure, and (**d**) PWM mode 2 for ICD structure.

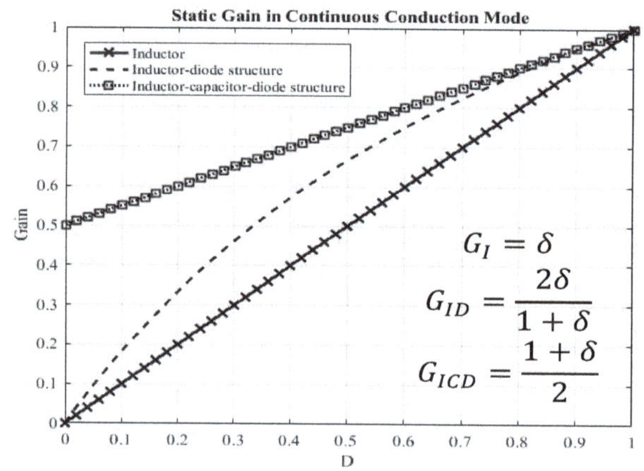

Figure 5. Continuous conduction mode graphical gain representation.

3.1. Region 1 ($\delta_2 > \delta_1$)

Operating in Region 1, the two stages of the quadratic buck structure operate independently, leading to intermediate and output voltages, depending on the duty cycles δ_2 and δ_1. The intermediate voltage across capacitor C_{inter} depends on the structure connected in PoC2 and the duty cycle δ_2 to command switch S_2. The output voltage, on the other hand, depends on both structures connected in PoC2 and PoC1, since the intermediate voltage level is the input voltage of the first stage. The equations presented in Figure 5 are reused to find the intermediate and output voltages generated by this integration. Table 1 summarize the analysis, showing the theoretical gain profile of the intermediate stage and output stage in function of the input voltage of the circuit and the duty cycle of switches S_1 and S_2. For instance, if the ID structure is applied in PoC1 and the I structure in PoC2, the intermediate and output voltages of the circuit are, respectively, related to equations δ_2 and $\delta_2\frac{2\delta_1}{1+\delta_1}$, , which means that the intermediate voltage is regulated by the nominal set point of the duty cycle applied in S_2 and the output voltage depends on the duty cycles applied to S_1 and S_2. The analysis considers that the duty cycles are independent, and with no phase-shift between each other.

3.2. Region 2 ($\delta_2 < \delta_1$)

Operating in this region, the second stage of the quadratic buck depends on the first stage, since the duty cycle of switch S_1 is higher than the duty cycle of switch S_2. Consequently, the intermediate capacitor continues to discharge even if the switch S_2 is in 'off'-state (due to the antiparallel diode between drain-source points of S_2). In this operational region, the duty cycle δ_1 defines the intermediate voltage level and the output voltage level. Table 2 summarizes the voltage levels in this region of operation.

3.3. *Region 3 ($\delta_2 = \delta_1 = \delta$)*

Region 3 is the region when both switches operate with the same duty cycle ($\delta_2 = \delta_1 = \delta$) and the duty cycles are in phase with each other. For this case, the intermediate and output voltages are subject to the operational duty cycle of the circuit. The theoretical analysis, shown in Section 4, is related to operation in Region 3. Table 3 summarize the voltage gain of the topologies generated by this intra-exchangeability.

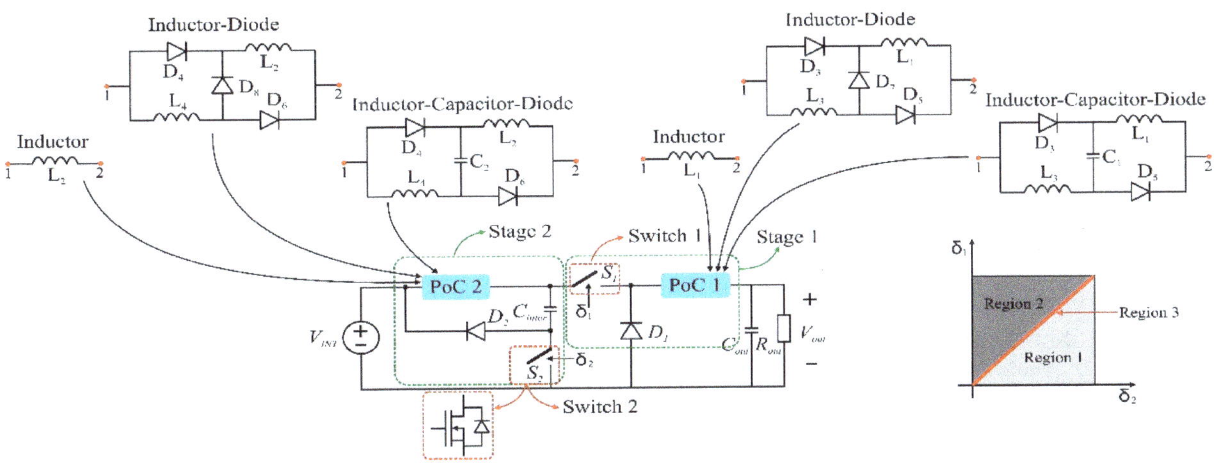

Figure 6. Quadratic Buck with two Point-of-Conncetions (PoCs) implemented and possible configurations achieved by the intra-exchangeability between I, ID, and ICD structures. Additionally, the operational regions for different duty cycle possibilities and switch realization.

Table 1. Theoretical Gain Analysis for all possible combinations between I, ID, and ICD structures in the quadratic buck configuration operating in Region 1. Intermediate and output voltage profiles in function of the input voltage of the circuit.

PoC1 / PoC2	I		ID		ICD	
	$\frac{V_{inter}}{V_{IN}}$	$\frac{V_{OUT}}{V_{IN}}$	$\frac{V_{inter}}{V_{IN}}$	$\frac{V_{OUT}}{V_{IN}}$	$\frac{V_{inter}}{V_{IN}}$	$\frac{V_{OUT}}{V_{IN}}$
I	δ_2	$\delta_2\delta_1$	δ_2	$\delta_2\frac{2\delta_1}{1+\delta_1}$	δ_2	$\delta_2\frac{1+\delta_1}{2}$
ID	$\frac{2\delta_2}{1+\delta_2}$	$\frac{2\delta_2}{1+\delta_2}\delta_1$	$\frac{2\delta_2}{1+\delta_2}$	$\frac{2\delta_2}{1+\delta_2}\frac{2\delta_1}{1+\delta_1}$	$\frac{2\delta_2}{1+\delta_2}$	$\frac{2\delta_2}{1+\delta_2}\frac{1+\delta_1}{2}$
ICD	$\frac{1+\delta_2}{2}$	$\frac{1+\delta_2}{2}\delta_1$	$\frac{1+\delta_2}{2}$	$\frac{1+\delta_2}{2}\frac{2\delta_1}{1+\delta_1}$	$\frac{1+\delta_2}{2}$	$\frac{1+\delta_2}{2}\frac{1+\delta_1}{2}$

where δ_2 and δ_1 are the duty cycle of switches S_2 and S_1, respectively.

Table 2. Theoretical Gain Analysis for all possible combinations between I, ID, and ICD structures in the quadratic buck configuration operating in Region 2. Intermediate and output voltage profiles in function of the input voltage of the circuit.

PoC1 / PoC2	I		ID		ICD	
	$\frac{V_{inter}}{V_{IN}}$	$\frac{V_{OUT}}{V_{IN}}$	$\frac{V_{inter}}{V_{IN}}$	$\frac{V_{OUT}}{V_{IN}}$	$\frac{V_{inter}}{V_{IN}}$	$\frac{V_{OUT}}{V_{IN}}$
I	δ_1	δ_1^2	δ_1	$\frac{2\delta_1^2}{1+\delta_1}$	δ_1	$\delta_1\frac{1+\delta_1}{2}$
ID	$\frac{2\delta_1}{1+\delta_1}$	$\frac{2\delta_1^2}{1+\delta_1}$	$\frac{2\delta_1}{1+\delta_1}$	$\left(\frac{2\delta_1}{1+\delta_1}\right)^2$	$\frac{2\delta_1}{1+\delta_1}$	δ_1
ICD	$\frac{1+\delta_1}{2}$	$\frac{1+\delta_1}{2}\delta_1$	$\frac{1+\delta_1}{2}$	δ_1	$\frac{1+\delta_1}{2}$	$\left(\frac{1+\delta_1}{2}\right)^2$

Table 3. Theoretical Gain Analysis for all possible combinations between I, ID, and ICD structures in the quadratic buck configuration operating in Region 3. Intermediate and output voltage profiles in function of the input voltage of the circuit.

PoC2 \ PoC1	I		ID		ICD	
	$\frac{V_{inter}}{V_{IN}}$	$\frac{V_{OUT}}{V_{IN}}$	$\frac{V_{inter}}{V_{IN}}$	$\frac{V_{OUT}}{V_{IN}}$	$\frac{V_{inter}}{V_{IN}}$	$\frac{V_{OUT}}{V_{IN}}$
I	δ	δ^2	δ	$\frac{2\delta}{1+\delta}\delta$	δ	$\delta\frac{1+\delta}{2}$
ID	$\frac{2\delta}{1+\delta}$	$\frac{2\delta}{1+\delta}\delta$	$\frac{2\delta}{1+\delta}$	$\left(\frac{2\delta}{1+\delta}\right)^2$	$\frac{2\delta}{1+\delta}$	$\frac{2\delta}{1+\delta}\frac{1+\delta}{2}$
ICD	$\frac{1+\delta}{2}$	$\frac{1+\delta}{2}\delta$	$\frac{1+\delta}{2}$	$\frac{1+\delta}{2}\frac{2\delta}{1+\delta}$	$\frac{1+\delta}{2}$	$\left(\frac{1+\delta}{2}\right)^2$

Although nine different topologies are realized in this analysis, only six different output voltage profiles are obtained, since the topologies I-ID/ID-I, I-ICD/ICD-I and ID-ICD/ICD-ID provide the same output voltage profile, while different intermediate voltage profiles are observed. Figure 7 shows the graphical representation of the output voltage profiles presented in Table 3. It is noticed that, due to the equations of the voltage profile shown in Figure 5, the highly nonlinear behavior of the quadratic buck converter is attenuated when the ID or ICD structures are used in one of the PoCs. The gain of the topologies, mathematically expressed in Table 3 and graphically in Figure 7, is less nonlinear if compared with the conventional quadratic converter, which is interesting for control purposes. These converters, however, are not suitable for applications that require high step-down capability, since the ID and ICD structures decrease the step-down capability of the structure, as shown in Figure 5 for application in a regular buck and Figure 7 in the quadratic version. Although the component part count increases for ID and ICD implementation, it also increases the flexibility to achieve multiple intermediate voltage levels to supply different loads.

On comparison of Table 2, related to region 2, and Table 3, related to region 3, it is noticed that the gain of the structures are the same. As mentioned in the previous subsection, this behavior occurs because the antiparallel diode of switch S_2 continue to conduct even if switch S_2 is in 'off'-state until switch S_1 commutates from 'close' to 'open'. This analysis concludes that operational regions 2 and 3 generate similar PWM modes and, consequently, the same intermediate and output voltage levels while region 1 is distinguished by the independence between the stages. Section 4 will further discuss the operational principle of the double switch quadratic buck with ID and ICD structures, for operation in Region 3.

4. Operational Principle of the Double Switch Quadratic Buck Converter in Region 3

Similar to Section 2, where the one-stage I, ID, and ICD topologies have been introduced, this section introduces the PWM modes of the two-stage I, ID, and ICD topologies, where the PWM signal remains the same. The phase shift between δ_2 and δ_1 is considered zero (in phase with each other) for the specific analysis but if differs from zero, the gain behavior shown in Tables 1–3 will differ, since the phase-shift will influence the PWM stages and, consequently, the gain of the structures. The equivalent circuit for the first and second stages, considering all the possibilities between the exchangeable circuits, are present in Figure 8a,b, respectively. In order to evaluate the main characteristics of the topologies, the equivalent circuits presented in Figure 8 are used to mathematically evaluate the voltage and current stresses in all power components. This information is crucial to further dimension the components for practical implementation.

The voltage stress across each power semiconductor are extracted from the PWM mode stages in Figure 8 and summarized in Table 4. The voltage stress across switch S_1 and diode D_1 is affected by the structure applied in PoC2, for instance, if the I structure is applied to PoC2, the voltage stress across switch S_1 is defined by $(1+\delta)V_{IN}$, while if the ID structure is applied in PoC2, the voltage stress across switch S_1 is defined by equation $\frac{(1+3\delta)}{(1+\delta)}V_{IN}$. Similarly, the voltage stress across diodes $D_3/D_5/D_7$, connected to PoC1, are related to the structure connected in PoC1. For instance, maintaining PoC2 fixed with the ID structure, if the ID structure is connected to PoC1, the voltage across the diodes D_3/D_5 is expressed by $\frac{\delta^2}{(1+\delta)^2}V_{IN}$ and, if PoC1 is changed for an ICD structure, the voltage across diodes D_3/D_5 is expressed by $\frac{\delta}{(1+\delta)}V_{IN}$. Switch S_2 and diode D_2 stresses are independent on the structures and are equal to the input voltage V_{IN}. The voltage stress across diodes D_3-D_4-D_5-D_6-D_7-D_8, which are the diodes implemented in the ID and ICD structures, are related to the operational duty cycle, input voltage, and in which PoC the structures are implemented.

Similarly, the PWM modes also provide the RMS current stress in each power semiconductor. To evaluate the RMS current stress in all components, equation (3) is introduced, where i_{x1} is the current that the semiconductor conducts during the first PWM mode and i_{x2} is the current conducted during the second PWM mode. Applying (3), the current stress in each semiconductor is characterized and presented in Table 5. Different than the voltage stress, the current stress of switches S_2 and D_2 depend on the structure connected to PoC2 but are independent of the structure connected to PoC1. For instance, the RMS current stress across switch S_2 is expressed by $\frac{I_{IN}(1-\delta)}{\delta}\sqrt{\frac{1}{\delta}}$ if the I structure is connected to PoC2, and expressed by $\frac{I_{IN}(1-\delta)}{2\delta}\sqrt{\frac{1}{\delta}}$ if the ID structure is connected. It is also noticed that the current is independent on PoC1, in this case.

The analysis abovementioned is important in order to select the components for experimental validation, since the RMS current and voltage stress are the most important characteristics to select MOSFETs and diodes. The next section presents a numerical simulation for the proposed converter performed in PLECS (4.0.2, Plexim GmbH, Zurich, Switzerland).

$$i_{xRMS} = \sqrt{\frac{1}{T_S}\left(\int_0^{DT_S} i_{x1}^2 dt + \int_{DT_S}^{T_S} i_{x2}^2 dt\right)} \tag{3}$$

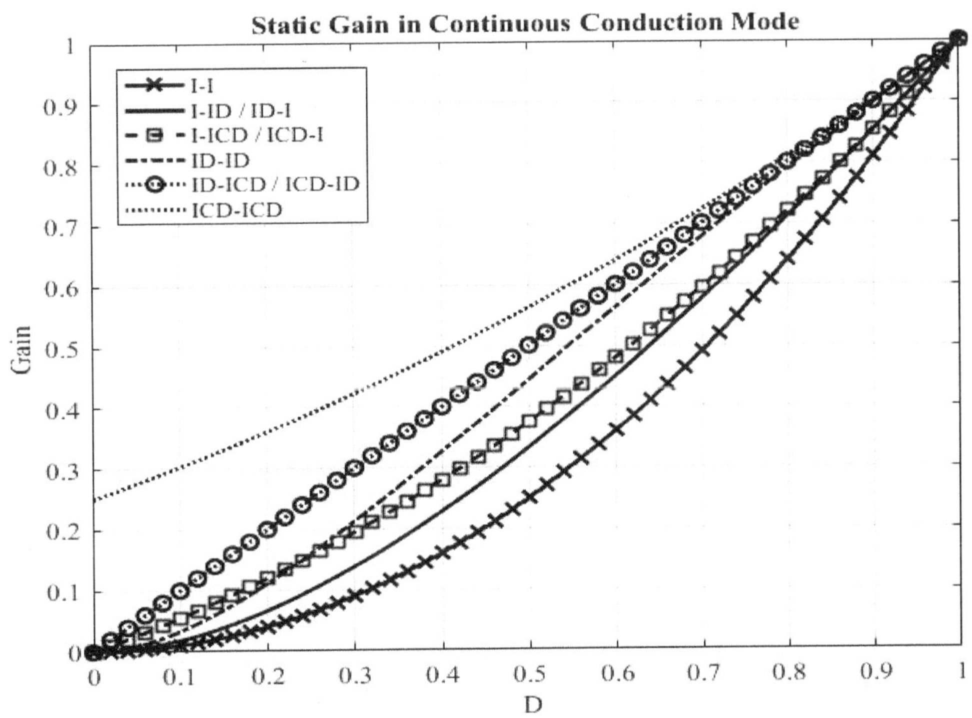

Figure 7. Continuous conduction mode graphical gain representation.

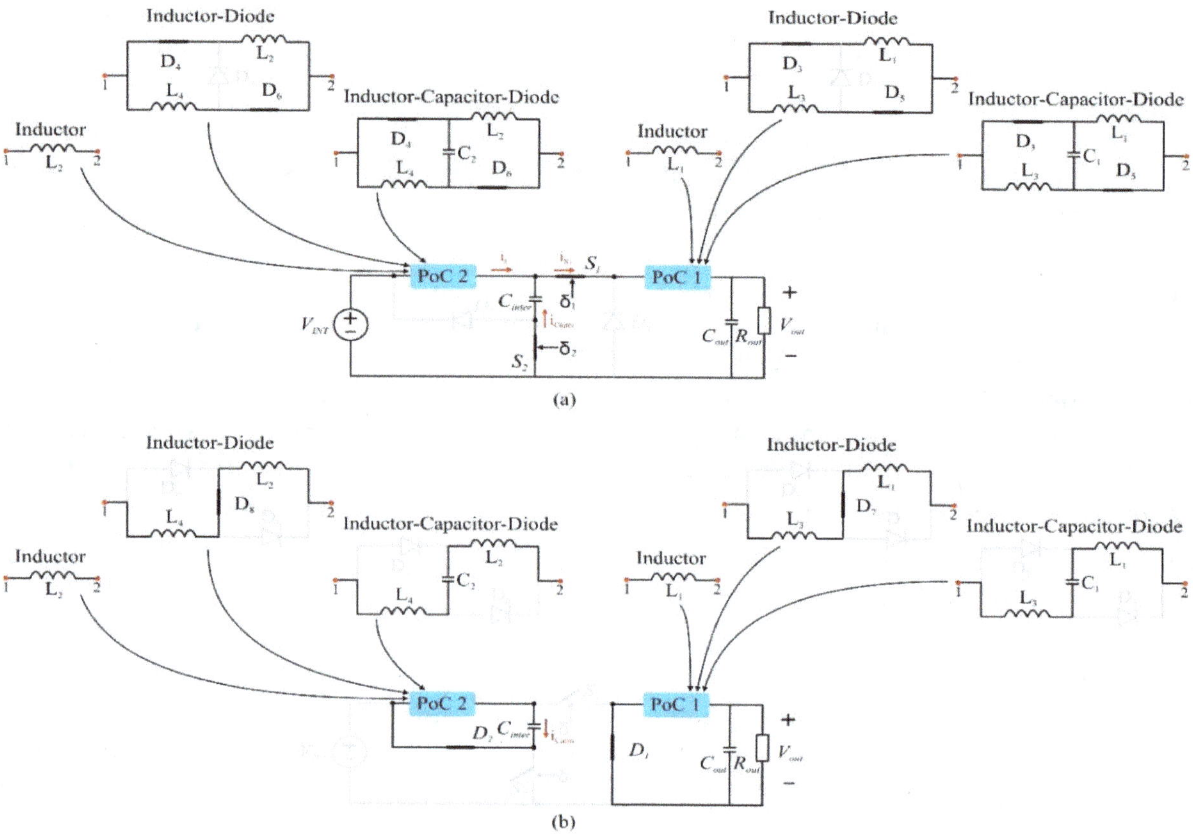

Figure 8. (**a**) Equivalent circuit for PoC 2 and PoC 1 applying I, ID, and ICD structures during the first PWM mode and (**b**) equivalent circuit for PoC 2 and PoC 1 applying I, ID, and ICD structures during the second PWM mode.

Table 4. Voltage stress sum up for all power semiconductors in the power stage. Equations are in function of the input voltage.

PoC2-PoC1	S_1	S_2	D_1	D_2	D_3/D_5	D_4/D_6	D_7	D_8
I-I					-		-	
I-ID	$(1+\delta)V_{IN}$		δV_{IN}		$\frac{\delta^2}{(1+\delta)}V_{IN}$	-	$\frac{(1-\delta)\delta}{(1+\delta)}V_{IN}$	-
I-ICD					$\delta\frac{V_{IN}}{2}$		-	
ID-I					-			
ID-ID	$\frac{(1+3\delta)}{(1+\delta)}V_{IN}$	V_{IN}	$\frac{2\delta}{(1+\delta)}V_{IN}$	V_{IN}	$\frac{\delta^2}{(1+\delta)^2}V_{IN}$	$\frac{\delta}{(1+\delta)}V_{IN}$	$\frac{2(1-\delta)\delta}{(1+\delta)^2}V_{IN}$	$\frac{(1-\delta)}{(1+\delta)}V_{IN}$
ID-ICD					$\frac{\delta}{(1+\delta)}V_{IN}$		-	
ICD-I					-			
ICD-ID	$\frac{(3+\delta)}{2}V_{IN}$		$\frac{(1+\delta)}{2}V_{IN}$		$\delta\frac{V_{IN}}{2}$	$\frac{V_{IN}}{2}$	$\frac{(1-\delta)}{2}V_{IN}$	-
ICD-ICD					$\frac{(1+\delta)}{4}V_{IN}$		-	

Table 5. RMS Current stress sum up for all power semiconductors in the power stage. Equations are in function of the input current.

PoC2-PoC1	S_1	S_2	D_1	D_2	D_3/D_5	D_4/D_6	D_7	D_8
I-I			$\frac{I_{IN}}{\delta^2}\sqrt{(1-\delta)}$		-		-	
I-ID	$\frac{I_{IN}}{\delta}\sqrt{\frac{1}{\delta}}$	$\frac{I_{IN}(1-\delta)}{\delta}\sqrt{\frac{1}{\delta}}$	$\frac{I_{IN}}{2\delta^2}\sqrt{(1-\delta)}$	$\frac{I_{IN}}{\delta}\sqrt{(1-\delta)}$	$\frac{I_{IN}}{2\delta}\sqrt{\frac{1}{\delta}}$	-	$\frac{I_{IN}}{2\delta^2}\sqrt{(1-\delta)}$	-
I-ICD			$\frac{I_{IN}}{\delta(1+\delta)}\sqrt{(1-\delta)}$		$\frac{I_{IN}}{\delta(1+\delta)}\sqrt{\frac{1}{\delta}}$			
ID-I			$\frac{I_{IN}(1+\delta)}{2\delta^2}\sqrt{(1-\delta)}$		-			
ID-ID	$\frac{I_{IN}(1+\delta)}{2\delta}\sqrt{\frac{1}{\delta}}$	$\frac{I_{IN}(1-\delta)}{2\delta}\sqrt{\frac{1}{\delta}}$	$\frac{I_{IN}(1+\delta)}{4\delta^2}\sqrt{(1-\delta)}$	$\frac{I_{IN}}{2\delta}\sqrt{(1-\delta)}$	$\frac{I_{IN}(1+\delta)}{4\delta}\sqrt{\frac{1}{\delta}}$	$\frac{I_{IN}}{2}\sqrt{\frac{1}{\delta}}$	$\frac{I_{IN}(1+\delta)}{4\delta^2}\sqrt{(1-\delta)}$	$\frac{I_{IN}}{2\delta}\sqrt{(1-\delta)}$
ID-ICD			$\frac{I_{IN}}{2\delta}\sqrt{(1-\delta)}$		$\frac{I_{IN}}{2\delta}\sqrt{\frac{1}{\delta}}$		-	
ICD-I			$\frac{2I_{IN}}{\delta(1+\delta)}\sqrt{(1-\delta)}$		-			
ICD-ID	$\frac{2I_{IN}}{(1+\delta)}\sqrt{\frac{1}{\delta}}$	$\frac{I_{IN}(1-\delta)}{(1+\delta)}\sqrt{\frac{1}{\delta}}$	$\frac{I_{IN}}{\delta(1+\delta)}\sqrt{(1-\delta)}$	$\frac{I_{IN}}{(1+\delta)}\sqrt{(1-\delta)}$	$\frac{I_{IN}}{(1+\delta)}\sqrt{\frac{1}{\delta}}$	$\frac{I_{IN}}{(1+\delta)}\sqrt{\frac{1}{\delta}}$	$\frac{I_{IN}}{\delta(1+\delta)}\sqrt{(1-\delta)}$	-
ICD-ICD			$\frac{2I_{IN}}{(1+\delta)^2}\sqrt{(1-\delta)}$		$\frac{2I_{IN}}{(1+\delta)^2}\sqrt{\frac{1}{\delta}}$		-	

5. Numerical Simulations

A numerical simulation was carried out to show the available intermediate voltage levels when an output voltage of 12 V is set as standard output voltage. The 12 V voltage level can supply low voltage/low power loads to, for instance, LEDs or to charge small batteries (unidirectional power flow to charge, since the topologies are unidirectional); this is one of the main reasons to fix the output voltage on this level. The simulation and experimental parameters are presented in Table 6.

Different intermediate voltage levels are generated, depending on the structure connected in PoC2 and the operational duty cycle of the converter, as shown in Table 3. The simulation model was develop in PLECS. The only configuration that is not able to be simulated and experimentally validated in this paper is the ICD-ICD structure, since the input/output voltage level is 48 V/12 V and the theoretical gain of this structure is 0.25 < ICD-ICD < 1, making it impossible to provide the required output voltage.

A simulation is performed for each configuration and the results are shown in Figure 9. The conventional quadratic buck with I-I is show in Figure 9a, generating an intermediate voltage level of 24 V. According to Table 3, the I-ID and ID-I configurations generates the same output voltage but different intermediate voltage levels for operation with the same duty cycle. This behavior is proved via simulation and presented in Figure 9b,d, where the voltage level of 20.25 V (I-ID) and 28.5 V (ID-I) are obtained for a duty cycle of 42.2%.

Similarly, the topologies I-ICD and ICD-I operate with the same duty cycle (36.6%) and input/output voltages, however the intermediate voltage level changes since is applied different structures in PoC2. Simulation results shown in Figure 9c,g confirm the analysis, since an intermediate voltage of 17.6 V (I-ICD) and 32.8 V (ICD-I) are obtained and available to supply loads in different voltage ranges in a flexible way. The structures ID-ICD and ICD-ID are also evaluate via simulation for similar input/output voltages. Intermediate voltages of 19.2 V (ID-ICD) and 30 V (ICD-ID) are obtained and shown in Figure 9f,h, respectively, validating the theoretical analysis presented in Table 3. Additionally, the last configurations present similar overall gain (δ) with the regular buck converter, however they present an additional voltage level that can be used to supply additional loads in different voltage levels. The drawback of the structure, compared with the conventional buck, is the component part count, since the number of inductors and diodes in the power stage increase.

Finally, the ID–ID structure is briefly discussed herein. Figure 9e shows the voltage levels obtained from simulation. Similarly with the conventional I-I quadratic buck, an intermediate voltage level of 24 V is obtained, however an operational duty cycle of 33.3% is used to match the desired output voltage of 12 V.

The simulations were performed with a switching frequency of 300 kHz (switching period of 3.33 μs), since this frequency was used to calculate the I, ID, and ICD inductances for further implementation of the structures. All the inductances were select to present a 20% ripple current limit across the inductors in nominal power. A summary of the theoretical inductor current expressions in each scenario is present in Table 7, in function of the input current I_{IN}. From the PWM mode stages discussed in the previous section and shown in Figure 8a,b, the inductors of the ID and ICD structures charge its terminals in parallel and discharge in series, which guarantees natural balance between the internal currents of the structure. This characteristic enable the non-use of current sensors to control the inductor current, since this control is guarantee directly by the ID and ICD structures, also known as switched-inductor structures. The drawback, as previously mentioned, is the increase in the number of components in the power stage.

With the objective of validating the equations shown in Table 7, a comparison between the theoretical analysis and numerical simulation results was performed and shown in Table 8. The input/output voltage of the circuit is 48 V/12 V and a rated input power of 100 W. The different configurations operate with their respective duty cycles to generate an output voltage of 12 V (check Figure 7 for duty cycle). The comparison shows that the theoretical equations match the simulation results, which proves the validity of the equations to select the correct inductances for experimental verification.

Table 6. Simulated and experimental parameters for validation.

Variable	Symbol	Value/Model
Input Voltage	V_{IN}	48 V
Output Voltage	V_{OUT}	12 V
Input Power	P_{IN}	100 W
Output Capacitance	C_{OUT}	20 uF
Switching Frequency	f_s	300 kHz
Switch Technology	S_1/S_2	STMicroeletronics 100 V 0.0145 Ohm typ 30 A

Table 7. Theoretical expressions for current across the inductors of the structures.

<table>
<tr><th>PoC2-PoC1</th><th>I_{L2}</th><th>I_{L4}</th><th>I_{L1}</th><th>I_{L3}</th></tr>
<tr><td>I-I</td><td rowspan="3">$\frac{I_{IN}}{\delta}$</td><td rowspan="3">-</td><td>$\frac{I_{IN}}{\delta^2}$</td><td>-</td></tr>
<tr><td>I-ID</td><td colspan="2">$\frac{I_{IN}}{2\delta^2}$</td></tr>
<tr><td>I-ICD</td><td colspan="2">$\frac{I_{IN}}{\delta(1+\delta)}$</td></tr>
<tr><td>ID-I</td><td colspan="2" rowspan="3">$\frac{I_{IN}}{2\delta}$</td><td>$\frac{I_{IN}(1+\delta)}{2\delta^2}$</td><td>-</td></tr>
<tr><td>ID-ID</td><td colspan="2">$\frac{I_{IN}(1+\delta)}{4\delta^2}$</td></tr>
<tr><td>ID-ICD</td><td colspan="2">$\frac{I_{IN}}{2\delta}$</td></tr>
<tr><td>ICD-I</td><td colspan="2" rowspan="2">$\frac{I_{IN}}{(1+\delta)}$</td><td>$\frac{2I_{IN}}{\delta(1+\delta)}$</td><td>-</td></tr>
<tr><td>ICD-ID</td><td colspan="2">$\frac{I_{IN}}{\delta(1+\delta)}$</td></tr>
</table>

Table 8. Inductor current comparison: theoretical/simulation.

PoC2-PoC1	I_{L2}	I_{L4}	I_{L1}	I_{L3}
	theo./sim.	theo./sim.	theo./sim.	theo./sim.
I-I	4.16 A/4.164 A	-	8.32 A/8.324 A	-
I-ID	4.95 A/4.955 A		5.868 A/5.871 A	
I ICD	5.628 A/5.662 A		4.12 A/4.12 A	
ID-I	2.464 A/2.465 A		8.304 A/8.306 A	-
ID-ID	3.093 A/3.098 A		6.191 A/6.201 A	
ID-ICD	4.1 A/4.096 A		4.1 A/4.096 A	
ICD-I	1.515 A/1.517 A		8.281 A/8.272 A	-
ICD-ID	1.65 A/1.643 A		6.601 A/6.567 A	

6. Experimental Verification and Test Bench

6.1. *Region 2 ($\delta_2 < \delta_1$)/Region 3 ($\delta_2 = \delta_1 = \delta$)*

This subsection will discuss the results obtained for operation in Region 3 ($\delta_2 = \delta_1 = \delta$). The topologies, discussed in Sections 3 and 4, were realized for further validation. A test bench was built for experimental verification of the configurations discussed herein. Figure 10 shows the test bench setup, together with a simplified bench schematic. It is shown in Figure 10 that a function generator (equipment (4)) is used to generate the PWM signals to drive S_1 and S_2. The experimental results shown herein are in open-loop operation, without a control strategy to regulate the intermediate and output voltage levels. Figure 11 shows the converter realization (middle) surrounded by the PoCs (PoC2 and PoC1) developed for each configuration (the inductors were selected based on the 20% maximal current ripple for each structure).

The PoCs, developed to test the different configurations, were designed separately following the plug-n-play methodology. As shown in the converter realization in Figure 11, the PoC2 and PoC1 are actual points of connection, where different external boards are connected to test different inductor configurations (I, ID, and/or ICD). Points 1 and 2, for PoC2, and 3 and 4, for PoC1, are available to connect the external PoCs. Additionally, Table 6 present the parameters defined for the experimental setup, in which the switching frequency of 300 kHz was select to simulate and experimental validation.

When comparing the waveforms presented in Figure 9a–h (simulation) and in Figure 12a–h (experimental), it can be seen that the voltage levels in all topologies are similar in shape (DC shape) and in amplitude. Figures 9 and 12 present the input, intermediate, and output voltages of the converter for the following configurations (PoC2-PoC1); (a) I-I, (b) I-ID, (c) I-ICD, (d) ID-I, (e) ID–ID, (f) ID-ICD, (g) ICD-I, and (h) ICD-ID. Small deviations are observed between the amplitude values in simulation and in the experimental waveforms due to the nonidealities inherent in the components selected to build the prototype. We also noticed that spikes of voltage appears in all voltage levels. This behavior is common in circuits that operate in high frequency, since internal parasitics of the components and also the parasitic inductances related to tracks and soldering points start to become prominent. Additionally, the commutation between 'on'- and 'off'-stages of switches S_1 and S_2 also generate internal oscillations in the circuit.

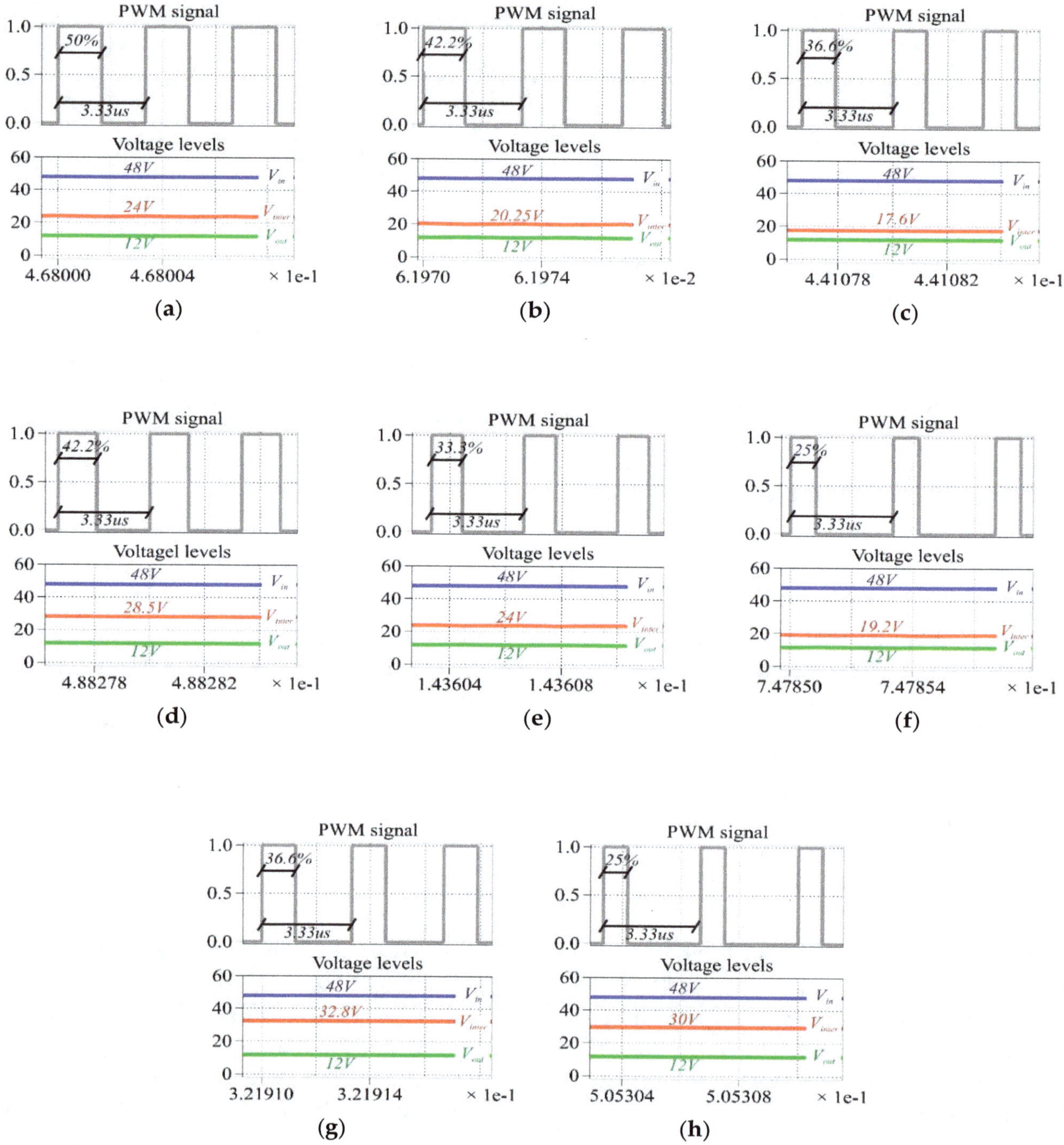

Figure 9. Simulation results showing the different intermediate voltage levels obtained by exchanging the structures for an input/output voltage of 48 V/12 V. PoC2-PoC1: (**a**) I-I; (**b**) I-ID; (**c**) I-ICD; (**d**) ID-I; **e**) ID–ID; (**f**) ID-ICD; (**g**) ICD-I; and (**h**) ICD-ID, respectively.

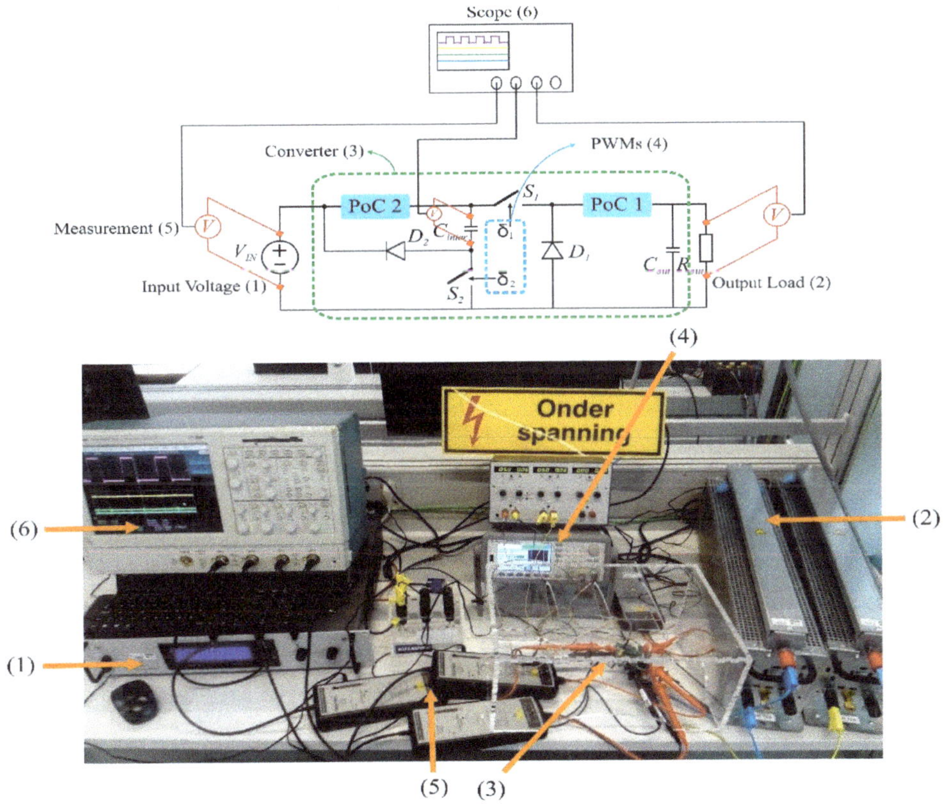

Figure 10. Test bench for experimental realization with (1) input source, (2) output load, (3) quadratic PoC converter, (4) PWM generator, (5) voltage measurements, and (6) scope.

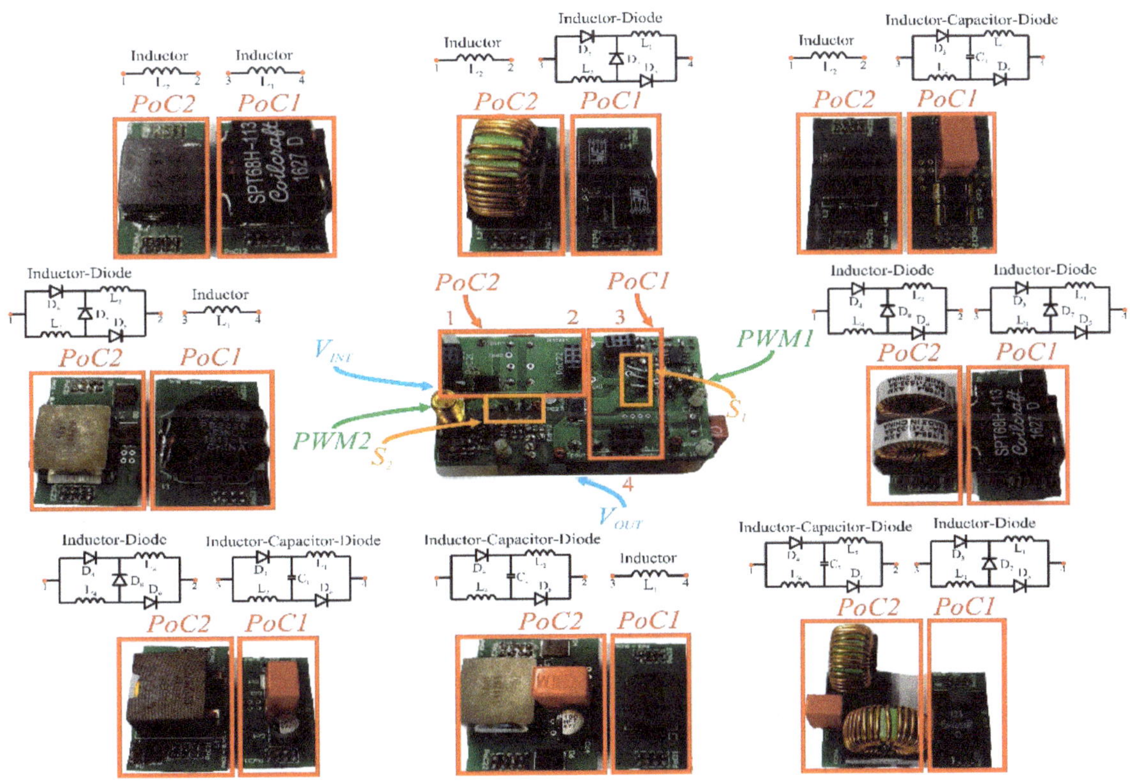

Figure 11. Converter realization (middle) surrounding by PoCs (PoC2 and PoC1) developed to experimental validation of the proposed topologies.

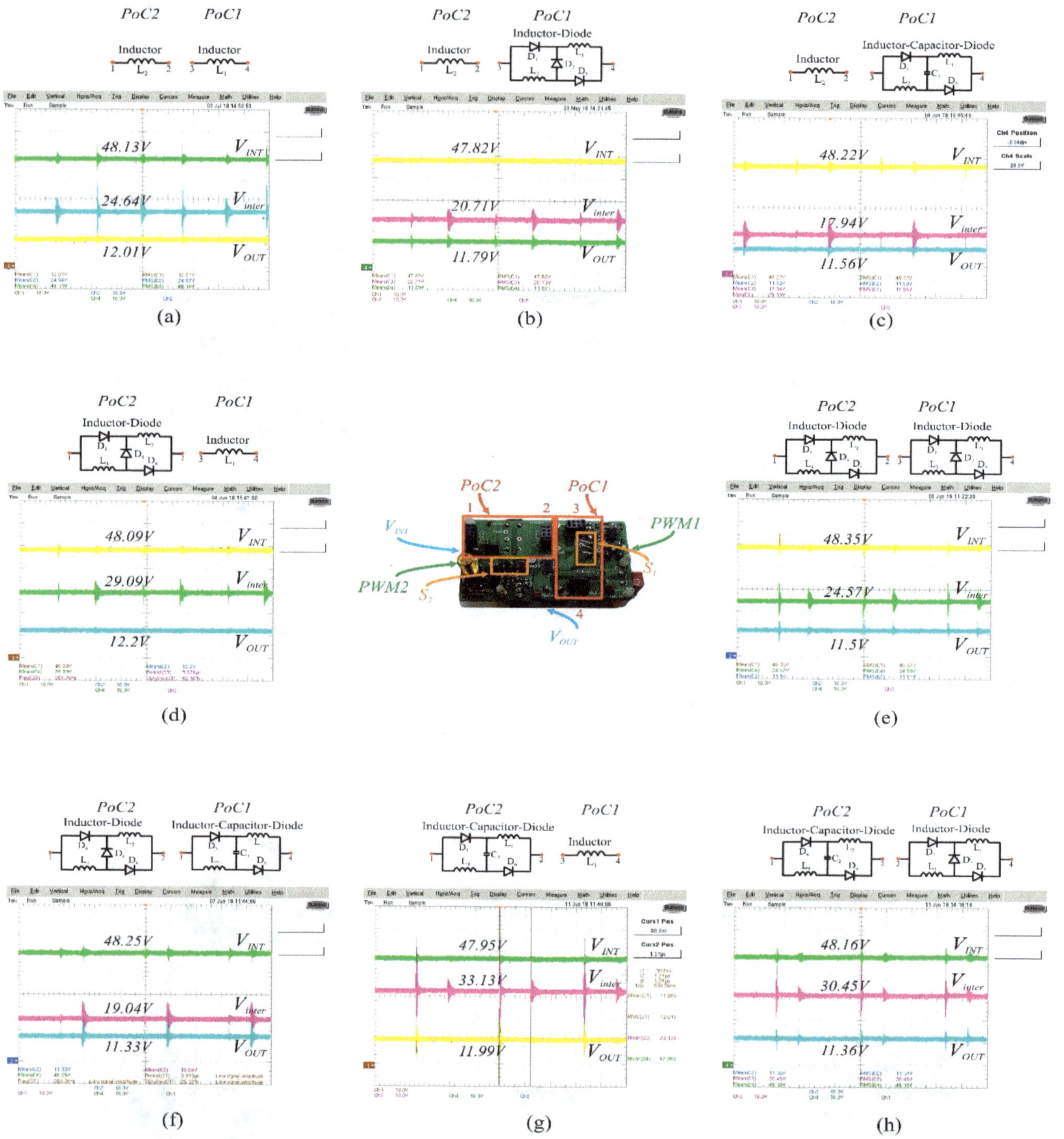

Figure 12. Voltage levels for different structures configurations. From top to bottom and left to right: (**a**) I-I with $\delta = 0.5$; (**b**) I-ID with $\delta = 0.422$; (**c**) I-ICD with $\delta = 0.366$; (**d**) ID-I with $\delta = 0.422$; (**e**) ID-ID with $\delta = 0.333$; (**f**) ID-ICD with $\delta = 0.25$; (**g**) ICD-I with $\delta = 0.366$ and; and (**h**) ICD-ID with $\delta = 0.25$.

As mentioned in Section 3, the gain analysis performed for the operation in region 2 and 3 results in the same gain expressions. The experimental results shown herein are for operation in region 3, but can be extended for the case when region 2 is in operation.

As discussed previously, is was shown that for different operating points and configurations connected in PoC2 and PoC1, different intermediate voltage levels are obtained for an input/output voltage of 48 V/12 V. This voltage, which varies from 33 V to 18 V depending the configuration, is available in the intermediate stage and can be used to supply loads in a broad range. Depending the required intermediate voltage, the best suited configuration can be selected to connect in PoC2 and PoC1. The ICD structure, as shown in Figure 5, limits the gain in the region between [0.5 ; 1], and can be used for a small voltage regulations with a certain accuracy. The ID structure, on the other hand, presents a nonlinear gain behavior in the range between [0 ; 1]. For the same duty cycle, this structure can be used to achieve gains in between the regular buck with I structure and the buck with ICD structure, with the disadvantage of a nonlinear gain.

6.2. Region 1 ($\delta_2 > \delta_1$)

For operation in region 1, an extra PWM mode is performed between the normal PWM modes 1 and 2, during the switching period. This extra PWM mode guarantee the independence between stages 1 and 2, leading to the gain presented in Table 1 for all possible PoC configurations. We noticed that the intermediate gain depends only on the duty cycle δ_2, while the output gain is relate to duty cycles of both stages, since the converters are cascade connected.

Figure 13 shows the PWM modes for operation in region 1 for the I-ICD configuration. Nevertheless, this analysis can be extended for all configurations. As previously mentioned, an extra PWM mode arises during the switching period (PWM mode 2 in this analysis). Switch S_1—the component which is subject to the greatest voltage stress among the power components—as highlighted in Table 4, blocks only the intermediate voltage level during the switch transient between 'on' and 'off' while when operates in regions 2 and 3, the switch blocks the sum between the intermediate voltage and the input voltage. This phenomenon enables the reduction of switching losses in this component once the voltage that the component blocks during the switching transient has a lower value compared with the normal operation, although the maximum voltage that the switch needs to block remains in the same level as shown in Table 4.

Figure 14 presents an experimental verification of the stages discussed previously and shown in Figure 13. Figure 14a presents a waveform with the significant voltage levels of the circuit. As expected, an intermediate voltage of 19.5 V is achieved for a duty cycle of 40% applying the I structure in PoC2. Additionally, Figure 14a also presents the second voltage regulation stage made by the ICD structure connected in PoC1, operating with a duty cycle of 25%, in which a voltage of 11.2 V is achieved at the output of the converter. Figure 14b presents the PWMs applied to the circuit and its switching period, together with the input/output voltage of the circuit. As expected, input and output voltages are in accordance with the theoretical analysis above presented.

Finally, Figure 14c shows the experimental results of the voltage stress across the active semiconductors S_1 and S_2. For the same duty cycle abovementioned, the voltage stress across switch S_2 is independent of the operational duty cycle and equal to the input voltage, as shown in Figure 14c. Additionally, the stress across switch S_1 was also evaluated and presented in Figure 14c. It is verified that, as theoretically explained, the voltage that the switch blocks during transition between 'on' and 'off' stages is the intermediate voltage across C_{inter}, although the maximum voltage stress is related to the sum of input and intermediate voltages (verify PWM mode 3 between points 2–3 in Figure 13). Moreover, the switching losses generated by the overlap between current and voltage during the transition time decrease due to the lower blocking voltage during this transition. For operation in region 1, the addition of switch S_2 with independent duty cycle improves the flexibility of the system and decreases the switching losses across switch S_1.

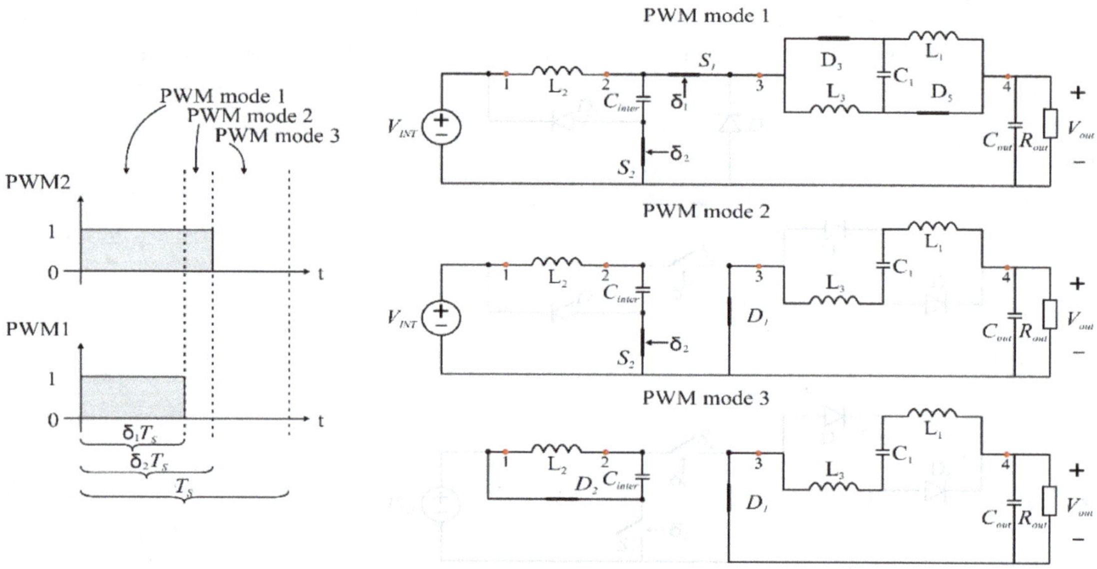

Figure 13. PWM modes with the equivalent circuits during one switching period when $\delta_2 > \delta_1$.

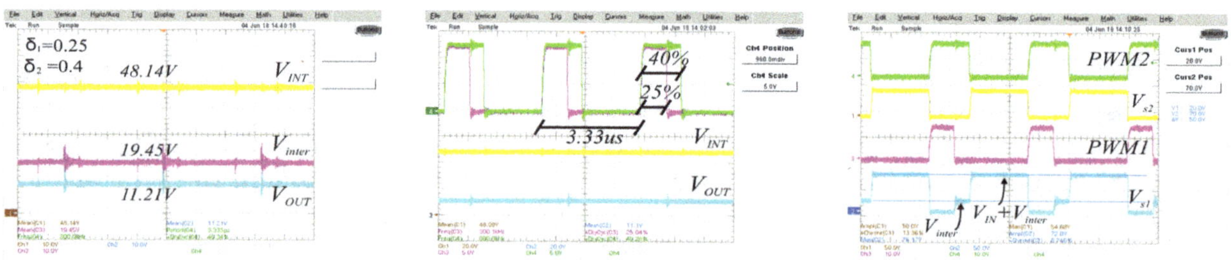

(a) (b) (c)

Figure 14. Experimental results for operation in region 1 for operation with $\delta_2 = 0.4$ and $\delta_1 = 0.25$ and switching frequency of 300 kHz: (**a**) input, intermediate and output voltages; (**b**) input and output voltages together with the respective duty cycles; and (**c**) PWMs and voltage stress across switches S_1 and S_2.

As discussed previously, the introduction of the ID and ICD structures in a step-down configuration (one-stage buck or quadratic buck) enable the possibility to change the voltage level of these structures. When applied in a one-stage buck converter, the ICD structure present similar voltage gain $\left(\frac{1+D}{2}\right)$ as the converter proposed in [50]. The structures, however, present different functionalities, since the abovementioned structure is a switched capacitor based converter with one switch, three diodes, one inductor, and three capacitors in the power stage while the structure, herein presented, is a switched inductor based topology with one switch, three diodes, two inductors, and one capacitor in the power stage. In terms of component part count number, the one-stage ICD structure present lower amount of components compared with the switched capacitor version that present similar gain [50]. The ID structure, on the other hand, presents a nonlinear step-down behavior with one extra diode and one less capacitor in the power stage (compared with the above-mentioned converters).

The ID structure is also explored in step-up configurations [39] and, when applied in a step-down structure, it changes the structure to a one composed by two inductors and two diodes, which leads to a gain of $\frac{D}{1+D}$, performing a higher step-down ratio them the ID structure. A topology with different switched capacitor/inductor structure is proposed for similar voltage gains [44] Comparing these topologies with the quadratic-buck configuration applying the ID and I structures (see Table 3), the gain of these structures is $\frac{2D*D}{1+D}$. The proposed structures, although implement more components in the power stage, present a higher step-down ratio for duty cycles lower than 0.5. For duty cycles above 0.5, the proposed converter present lower step-down capability with higher component part count.

The one-stage buck converter present similar gain (D) compared with the quadratic buck applying ID and ICD structures in PoC1 or PoC2 (see Table 3). For a component part count point of view, the conventional buck converter present more benefits compared to the structures above-mentioned, since the conventional buck converter present the lowest part count structure for a DC–DC step-down configuration. On the other hand, the converters proposed herein present multiple output ports, e.g., intermediate and output voltages, and, according to the analysis presented herein, if a second switch S_2 is implemented, these voltage levels can be independently regulated (operation in Region 1), which is beneficial for an application point of view.

The major advantage that the topologies proposed herein have is the flexibility to adjust the intermediate voltage level according with the project specifications. The experimental results presented in Figures 12 and 14 show that, for the same input/output voltage, different intermediate voltage levels can be obtained depending on the operational regions (Region 1, 2, or 3) and the structure applied to PoC1 and PoC2. The main disadvantage, however, is the number of components in the power stage. When applied in a quadratic buck converters, the ID and ICD structures increase the number of semiconductor and magnetics in the power stage. This feature is undesired in some applications that power density is required, since extra space is required to place all the components. As previously mentioned, the ID and ICD structures are suited for lower voltage regulations and are not applicable if high step-down conversion ratios are required.

7. Conclusions

The analysis in this paper shows the influence of the ID and ICD structures, previously applied in step-up topologies, in a step-down configuration from 48 V to 12 V input-to-output voltage. To achieve the required voltage levels, a quadratic buck structure was implemented since the first analysis showed that the ID and ICD structures have a lower voltage regulation capability compared with the conventional I structure and, if a single stage buck converter is considered for application, the ICD structure cannot achieve the required output voltage level.

A complete theoretical analysis with PWM stages for all possible regions of operation, voltage gain, voltage/current profiles, and simulation is shown for all possible configurations generated by the substitution of the structures in PoC2 and PoC1. Voltage and current stress were evaluated for dimensioning and component selection to subsequently build the prototype for experimental validation. The analysis showed that the ID and ICD structures change the voltage and current profile depending on which PoC is connected. It is proven that the addition of a second switch S_2, together with the operation in Region 1, decreases the voltage stress during turn-off of switch S_1. The main advantage of the proposed approach is that the intermediate voltage level can be selected according to the application, since switch S_2 is implemented and the ID and ICD structures provide different voltage profiles that can be desired for specific operational duty cycles.

The main advantages of the proposed topologies compared to the conventional buck and quadratic buck converters are as follows.

(1) The addition of the second switch S_2 and the possibility to choose between I, ID, and ICD structures to connect in PoC2 and PoC1, increase the flexibility of the circuit, resulting in a broad range of intermediate and output voltage levels.
(2) The circuits with ID and/or ICD feature diode and inductor redundancy, which leads to lower current stress and a fail-safe structure, after all, if one diode/inductor fails in open-circuit, the system can still maintain its functionality but with limited power.
(3) The implementation of S_2 also increase the number of regions of operation, since region 1 is now a possible region of operation. In this region, the blocking voltage of switch S_1 decreases from $V_{IN} + V_{inter}$ to only V_{inter}, as shown in Section 6.

A test bench was assembled and experimental results showing that the ID and ICD structures can be used for small voltage adjustments in a step-down application were presented. Simulation and experimental waveforms were presented to validate the theoretical analysis in Section 4. The ID and ICD proved to be effective structures for applications where small voltage adjustments are required, but did not prove being useful solutions when a high step-down voltage level is required due to its gain characteristics.

8. Patent

A patent related to this work was filled in on the 15th of May 2018, with application number GB1807795.8 at the UKIPO (reference [47]).

Author Contributions: M.D.V.: Conceptualization, Investigation, Methodology, Validation, and Writing—Original Draft. G.V.d.B and S.R.: Validation and Revision. J.D.: Supervisor, Resources, and Revision.

References

1. Vovos, P.N.; Kiprakis, A.E.; Wallace, A.R.; Harrison, G.P. Centralized and Distributed Voltage Control: Impact on Distributed Generation Penetration. *IEEE Trans. Power Syst.* **2007**, *22*, 476–483. [CrossRef]
2. Ault, G.W.; McDonald, J.R.; Burt, G.M. Strategic analysis framework for evaluating distributed generation and utility strategies. *IEE Proc.-Gener.Transm. Distrib.* **2003**, *150*, 475–481. [CrossRef]
3. Kumar, D.; Zare, F.; Ghosh, A. DC Microgrid Technology: System Architectures, AC Grid Interfaces, Grounding Schemes, Power Quality, Communication Networks, Applications, and Standardizations Aspects. *IEEE Access* **2017**, *5*, 12230–12256. [CrossRef]
4. Sannino, A.; Postiglione, G.; Bollen, M.H.J. Feasibility of a DC network for commercial facilities. *IEEE Trans. Ind. Appl.* **2003**, *39*, 1499–1507. [CrossRef]
5. Starke, M.; Li, F.; Tolbert, L.M.; Ozpineci, B. AC vs. DC distribution: Maximum transfer capability. In Proceedings of the 2008 IEEE Power and Energy Society General Meeting—Conversion and Delivery of Electrical Energy in the 21st Century, Pittsburgh, PA, USA, 20–24 July 2008; pp. 1–6.
6. Dragičević, T.; Lu, X.; Vasquez, J.C.; Guerrero, J.M. DC Microgrids—Part II: A Review of Power Architectures, Applications, and Standardization Issues. *IEEE Trans. Power Electron.* **2016**, *31*, 3528–3549. [CrossRef]
7. Han, J.; Oh, Y.-S.; Gwon, G.-H.; Kim, D.-U.; Noh, C.-H.; Jung, T.-H.; Lee, S.-J.; Kim, C.-H. Modeling and Analysis of a Low-Voltage DC Distribution System. *Resources* **2015**, *4*, 713–735. [CrossRef]
8. Salato, M.; Ghisla, U. Optimal power electronic architectures for DC distribution in datacenters. In Proceedings of the 2015 IEEE First International Conference on DC Microgrids (ICDCM), Atlanta, GA, USA, 25–27 May 2015; pp. 245–250.
9. Wang, C.; Jain, P. A quantitative comparison and evaluation of 48V DC and 380V DC distribution systems for datacenters. In Proceedings of the 2014 IEEE 36th International Telecommunications Energy Conference (INTELEC), Vancouver, BC, Canada, 28 September–20 October 2014; pp. 1–7.
10. Madduri, P.A.; Poon, J.; Rosa, J.; Podolsky, M.; Brewer, E.A.; Sanders, S.R. Scalable DC Microgrids for Rural Electrification in Emerging Regions. *IEEE J. Emerg. Sel. Top. Power Electron.* **2016**, *4*, 1195–1205. [CrossRef]
11. Tan, K.; Song, X.; Peng, C.; Liu, P.; Huang, A.Q. Hierarchical protection architecture for 380V DC data center application. In Proceedings of the 2016 IEEE Energy Conversion Congress and Exposition (ECCE), Milwaukee, WI, USA, 18–22 September 2016; pp. 1–8.
12. Kakigano, H.; Miura, Y.; Ise, T. Low-Voltage Bipolar-Type DC Microgrid for Super High Quality Distribution. *IEEE Trans. Power Electron.* **2010**, *25*, 3066–3075. [CrossRef]
13. den Broeck, G.V.; Breucker, S.D.; Beerten, J.; Zwysen, J.; Vecchia, M.D.; Driesen, J. Analysis of three-level converters with voltage balancing capability in bipolar DC distribution networks. In Proceedings of the 2017 IEEE Second International Conference on DC Microgrids (ICDCM), Nuremberg, Germany, 27–29 June 2017; pp. 248–255.
14. Anand, S.; Fernandes, B.G. Optimal voltage level for DC microgrids. In Proceedings of the IECON 2010—36th Annual Conference on IEEE Industrial Electronics Society, Glendale, CA, USA, 7–10 November 2010; pp. 3034–3039.

15. Moussa, S.; Ghorbal, M.J.B.; Slama-Belkhodja, I. DC voltage level choice in residential remote area. In Proceedings of the 2018 9th International Renewable Energy Congress (IREC), Hammamet, Tunisia, 20–22 March 2018; pp. 1–6.
16. Francés, A.; Asensi, R.; García, Ó.; Prieto, R.; Uceda, J. Modeling Electronic Power Converters in Smart DC Microgrids—An Overview. *IEEE Trans. Smart Grid* **2018**, *9*, 6274–6287. [CrossRef]
17. Keshani, M.; Adib, E.; Farzanehfard, H. Micro-inverter based on single-ended primary-inductance converter topology with an active clamp power decoupling. *IET Power Electron.* **2018**, *11*, 73–81. [CrossRef]
18. Liao, C.Y.; Lin, W.S.; Chen, Y.M.; Chou, C.Y. A PV Micro-inverter With PV Current Decoupling Strategy. *IEEE Trans. Power Electron.* **2017**, *32*, 6544–6557. [CrossRef]
19. Ahmed, M.E.S.; Orabi, M.; Abdelrahim, O.M. Two-stage micro-grid inverter with high-voltage gain for photovoltaic applications. *IET Power Electron.* **2013**, *6*, 1812–1821. [CrossRef]
20. Sajadian, S.; Ahmadi, R. Distributed Maximum Power Point Tracking Using Model Predictive Control for Photovoltaic Energy Harvesting Architectures Based on Cascaded Power Optimizers. *IEEE J. Photovolt.* **2017**, *7*, 849–857. [CrossRef]
21. Vinnikov, D.; Chub, A.; Liivik, E.; Roasto, I. High-Performance Quasi-Z-Source Series Resonant DC/DC Converter for Photovoltaic Module-Level Power Electronics Applications. *IEEE Trans. Power Electron.* **2017**, *32*, 3634–3650. [CrossRef]
22. Bergveld, H.J.; Büthker, D.; Castello, C.; Doorn, T.; de Jong, A.; van Otten, R.; de Waal, K. Module-Level DC/DC Conversion for Photovoltaic Systems: The Delta-Conversion Concept. *IEEE Trans. Power Electron.* **2013**, *28*, 2005–2013. [CrossRef]
23. Ravyts, S.; Vecchia, M.D.; Zwysen, J.; van den Broeck, G.; Driesen, J. Comparison Between an Interleaved Boost Converter Using Si MOSFETs Versus GaN HEMTs. In Proceedings of the PCIM Europe 2018, International Exhibition and Conference for Power Electronics, Intelligent Motion, Renewable Energy and Energy Management, Nuremberg, Germany, 5–7 June 2018; pp. 1–8.
24. Ravyts, S.; Vecchia, M.D.; Zwysen, J.; den Broeck, G.V.; Driesen, J. Study on a cascaded DC-DC converter for use in building-integrated photovoltaics. In Proceedings of the 2018 IEEE Texas Power and Energy Conference (TPEC), College Station, TX, USA, 8–9 February 2018; pp. 1–6.
25. Mei, Q.; Shan, M.; Liu, L.; Guerrero, J.M. A Novel Improved Variable Step-Size Incremental-Resistance MPPT Method for PV Systems. *IEEE Trans. Ind. Electron.* **2011**, *58*, 2427–2434. [CrossRef]
26. Thangavelu, A.; Vairakannu, S.; Parvathyshankar, D. Linear open circuit voltage-variable step-size-incremental conductance strategy-based hybrid MPPT controller for remote power applications. *IET Power Electron.* **2017**, *10*, 1363–1376. [CrossRef]
27. Tiwari, R.; Krishnamurthy, K.; Neelakandan, R.B.; Padmanaban, S.; Wheeler, P.W. Neural Network Based Maximum Power Point Tracking Control with Quadratic Boost Converter for PMSG—Wind Energy Conversion System. *Electronics* **2018**, *7*, 20. [CrossRef]
28. Zhang, Y.; Li, X.; Dong, Z.; Liu, Y.f.; Liu, J. High step-up isolated DC-DC converter with multi-cell diode-capacitor network. In Proceedings of the 2017 IEEE Applied Power Electronics Conference and Exposition (APEC), Tampa, FL, USA, 26–30 March 2017; pp. 114–120.
29. Spiazzi, G.; Biadene, D.; Marconi, S.; Bevilacqua, A. Non-isolated High Step-up DC-DC Converter with Minimum Switch Voltage Stress. *IEEE Trans. Power Electron.* **2019**, *34*, 1470–1480. [CrossRef]
30. Schmitz, L.; Martins, D.C.; Coelho, R.F. High Step-Up Non-Isolated ZVS/ZCS DC-DC Converter for Photovoltaic Thin-Film Module Applications. *IEEE J. Emerg. Sel. Top. Power Electron.* **2019**, *7*, 565–575. [CrossRef]
31. Schmitz, L.; Martins, D.C.; Coelho, R.F. Generalized High Step-Up DC-DC Boost-Based Converter with Gain Cell. *IEEE Trans. Circuits Syst. I Regul. Pap.* **2017**, *64*, 480–493. [CrossRef]
32. Wong, Y.-S.; Chen, J.-F.; Liu, K.-B.; Hsieh, Y.-P. A Novel High Step-Up DC-DC Converter with Coupled Inductor and Switched Clamp Capacitor Techniques for Photovoltaic Systems. *Energies* **2017**, *10*, 378. [CrossRef]
33. Gao, Y.; Liu, H.; Ai, J. Novel High Step-Up DC–DC Converter with Three-Winding-Coupled-Inductors and Its Derivatives for a Distributed Generation System. *Energies* **2018**, *11*, 3428. [CrossRef]
34. Pop-Calimanu, I.-M.; Lica, S.; Popescu, S.; Lascu, D.; Lie, I.; Mirsu, R. A New Hybrid Inductor-Based Boost DC-DC Converter Suitable for Applications in Photovoltaic Systems. *Energies* **2019**, *12*, 252. [CrossRef]
35. Chen, S.-J.; Yang, S.-P.; Huang, C.-M.; Chou, H.-M.; Shen, M.-J. Interleaved High Step-Up DC-DC Converter Based on Voltage Multiplier Cell and Voltage-Stacking Techniques for Renewable Energy Applications. *Energies* **2018** *11*, 1632. [CrossRef]

36. Jalilzadeh, T.; Rostami, N.; Babaei, E.; Maalandish, M. Non-Isolated Topology for High Step-Up DC-DC Converters. Available online: https://ieeexplore.ieee.org/document/8390878 (accessed on 14 January 2019).
37. Rodič, M.; Milanovič, M.; Truntič, M.; Ošlaj, B. Switched-Capacitor Boost Converter for Low Power Energy Harvesting Applications. *Energies* **2018**, *11*, 3156. [CrossRef]
38. Tran, V.-T.; Nguyen, M.-K.; Choi, Y.-O.; Cho, G.-B. Switched-Capacitor-Based High Boost DC-DC Converter. *Energies* **2018**, *11*, 987. [CrossRef]
39. Hu, Y.; Li, K.; Yin, Z.; Ioinovici, A. Switched-inductor-based non-isolated large conversion ratio, low components count DC-DC regulators. In Proceedings of the 2015 IEEE Energy Conversion Congress and Exposition (ECCE), Montreal, QC, Canada, 20–24 September 2015; pp. 1398–1405.
40. Ranjana, M.S.B.; Kulkarni, R.M.; Anita, K.; Pooja, C. Non isolated switched inductor SEPIC converter topologies for photovoltaic boost applications. In Proceedings of the 2016 International Conference on Circuit, Power and Computing Technologies (ICCPCT), Nagercoil Andaman and Nicobar Islands, India, 18–19 March 2016; pp. 1–6.
41. Li, D.; Loh, P.C.; Zhu, M.; Gao, F.; Blaabjerg, F. Generalized Multicell Switched-Inductor and Switched-Capacitor Z-Source Inverters. *IEEE Trans. Power Electron.* **2013**, *28*, 837–848. [CrossRef]
42. Salvador, M.A.; Lazzarin, T.B.; Coelho, R.F. High Step-Up DC/DC Converter with Active Switched-Inductor and Passive Switched-Capacitor Networks. *IEEE Trans. Ind. Electron.* **2018**, *65*, 5644–5654. [CrossRef]
43. Qu, K.; Feng, M.; Zhao, J. Inductor-capacitor structure for getting high step-up converter. *Electron. Lett.* **2016**, *52*, 1403–1404. [CrossRef]
44. Xiong, S.; Wong, S.C.; Tan, S.C.; Tse, C.K. A Family of Exponential Step-Down Switched-Capacitor Converters and Their Applications in Two-Stage Converters. *IEEE Trans. Power Electron.* **2014**, *29*, 1870–1880. [CrossRef]
45. Han, J.; von Jouanne, A.; Temes, G.C. A New Approach to Reducing Output Ripple in Switched-Capacitor-Based Step-Down DC-DC Converters. *IEEE Trans. Power Electron.* **2006**, *21*, 1548–1555. [CrossRef]
46. Uno, M.; Kukita, A. PWM Switched Capacitor Converter with Switched-Capacitor-Inductor Cell for Adjustable High Step-Down Voltage Conversion. *IEEE Trans. Power Electron.* **2019**, *34*, 425–437. [CrossRef]
47. Dalla Vecchia, M.; Driesen, J. Quadratic Buck Converter. UKIPO Patent GB1807795, 14 May 2018.
48. Angulo-Garcia, D.; Angulo, F.; Osorio, G.; Olivar, G. Control of a DC-DC Buck Converter through Contraction Techniques. *Energies* **2018**, *11*, 3086. [CrossRef]
49. Demian, A.E.; Treviso, C.H.G.; Gallo, C.A.; Tofoli, F.L. Non-isolated DC-DC converters with wide conversion range used to drive high-brightness LEDs. In Proceedings of the 2009 Brazilian Power Electronics Conference, Bonito, MS, Brazil, 27 September–1 October 2009; pp. 598–605.
50. Vecchia, M.D.; Lazzarin, T.B. A hybrid switched capacitor dc-dc buck converter. In Proceedings of the 2015 IEEE 13th Brazilian Power Electronics Conference and 1st Southern Power Electronics Conference (COBEP/SPEC), Fortaleza, CE, Brazil, 29 November–2 December 2015; pp. 1–6.

4

SiC-Based High Efficiency High Isolation Dual Active Bridge Converter for a Power Electronic Transformer

Mariam Saeed *, María R. Rogina, Alberto Rodríguez, Manuel Arias and Fernando Briz

Department of Electrical Engineering, University of Oviedo, 33204 Asturias, Spain; rodriguezrmaria@uniovi.es (M.R.R.); rodriguezalberto@uniovi.es (A.R.); ariasmanuel@uniovi.es (M.A.); fernando@isa.uniovi.es (F.B.)

* Correspondence: saeedmariam@uniovi.es

Abstract: This paper discusses the benefits of using silicon carbide (SiC) devices in a three-stage modular power electronic transformer. According to the requirements to be fulfilled by each stage, the second one (the DC/DC isolation converter) presents the most estimable improvements to be gained from the use of SiC devices. Therefore, this paper is focused on this second stage, implemented with a SiC-based dual active bridge. Selection of the SiC devices is detailed tackling the efficiency improvement which can be obtained when they are co-packed with SiC antiparallel Schottky diodes in addition to their intrinsic body diode. This efficiency improvement is dependent on the dual active bridge operation point. Hence, a simple device loss model is presented to assess the efficiency improvement and understand the reasons for this dependence. Experimental results from a 5-kW Dual Active Bridge prototype have been obtained to validate the model. The dual active bridge converter is also tested as part of the full PET module operating at rated power.

Keywords: SiC devices; antiparallel diode; dual active bridge; power electronic transformer; high-frequency transformer

---

1. Introduction

A line-frequency transformer (LFT) is a key element in transmission and distribution for traditional centralized generation-based systems. Their main functionality is to interface different voltage levels in the grid [1]. LFTs are a well-established, relatively cheap, and reliable technology. However, they fail to cope with modern grid demands, such as the integration of distributed resources and energy storage systems, as well as power flow control.

The power electronic transformer (PET), also called a solid-state transformer (SST), was introduced in 1970 [2]. It is considered an alternative to LFT, as it connects two AC voltage ports while providing galvanic isolation [1]. PET is a semiconductor-based energy conversion system based on fast-switching devices, which potentially enables a significant reduction in volume and weight [3]. Moreover, thanks to the controllability of the power devices, the PET provides additional functionalities, such as reactive power, harmonics and imbalances compensation, ride-through capabilities, and smart protections.

The power semiconductor technology used in PETs has been traditionally based on Silicon (Si). However, the fast advances in wide-band-gap (WBG), specially the Silicon Carbide (SiC), power semiconductors has attracted the attention to their use in the medium voltage (MV) modular three-stage PETs [4], mainly due to their high blocking voltage along with their superior switching behavior [5,6].

Several examples of using SiC devices in different PET topologies exist in literature [7–13]. In the majority of applications, 1.2/1.7 kV commercial SiC MOSFETs are used for LV side devices, while for HV side, 10 kV non-commercial MOSFETs are used [6,8]. In some works, SiC was used in all the PET stages, such as in the TIPS (transformer-less intelligent power substation), which is an all-SiC three-stage PET topology [12]. In [14], a detailed comparison was carried out between two cases:

(1) using 10 kV and 3.3 kV SiC MOSFETs and, (2) using 10 kV and 3.3 kV Si IGBTs, for different PET topologies. Authors concluded the importance of further research into the combined use of Si IGBT with SiC MOSFETs for three-stage PETs. Moreover, authors hinted at the importance of analyzing the relevance of including a SiC antiparallel diode to justify its additional cost implication.

Accordingly, the contribution of this paper is thought in two aspects:

1. Studying the requirements of the devices in each of the PET stages to assess the benefit of integrating SiC and, consequently, identifying the stages making best use of it.
2. Selection of the SiC devices for the second PET stage, based on a dual active bridge converter (DAB). As will be explained, this stage presents the most benefits from using SiC devices. Special attention will be paid to the device operation intrinsic to the converter topology. The use of SiC MOSFETs co-packaged with an additional SiC antiparallel diode is also investigated. A loss model is developed to estimate the potential efficiency improvement in this case. Both devices (i.e., with only intrinsic body diode and with additional SiC diode) are experimentally compared at rated power to validate the proposed model.

The paper is organized as follows: Section 2 describes the selected PET topology. Section 3 studies the integration of SiC devices in the different stages of the PET and identifies the practical limitations to the enhancements introduced by the use of these devices. Section 4 discusses the DAB converter development where the selection of the devices is detailed and the improvements introduced by using a SiC antiparallel diode is studied. Experimental results are provided in Section 5.

2. PET Topology

The selected PET topology is shown in Figure 1, and was previously discussed in [15–20]. It is a three-stage modular PET, with a Cascaded H-Bridge (CHB) converter acting as the front end AC/DC converter providing a HVAC link (V_{acHV}). Each CHB cell integrates a DAB converter to provide the isolation between the HV and LV sides. Moreover, the integration of the DAB to the CHB cell provides the capability of bidirectional power transfer at the cell level [18,19]. The LV outputs of all DABs are connected in parallel to provide a high-current LVDC link (V_{dcLV}), which is connected to a three-phase four-leg (3P4L) converter providing the LVAC link (V_{acLV}).

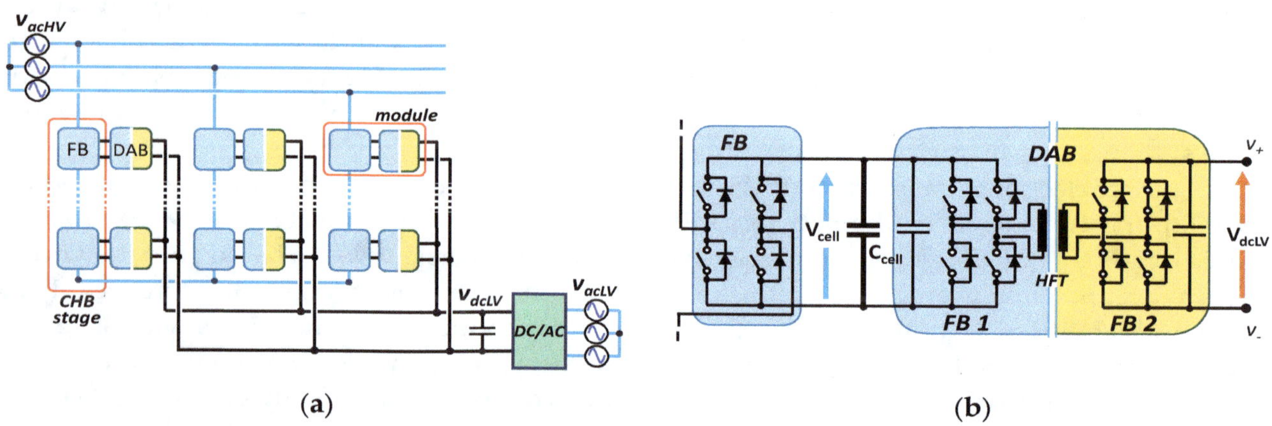

Figure 1. Selected Power Electronic Transformer (PET) structure: (**a**) Three-stage modular PET topology based on a Cascaded H-Bridge (CHB) converter [17]. (**b**) One PET module composed of a Dual Active Bridge (DAB) and the full bridge of the CHB.

As is clear in Figure 1, the structure is fully modular as it is formed by several identically-stacked cells. Thanks to this modular structure, the number of cells of the CHB is chosen in such a way that the cell voltage (V_{cell}) is equal to the required V_{dcLV}. In this way the design of this DAB is, to some extent, simplified using a unity transformation ratio (i.e., no step-up/down).

This PET topology requires high galvanic isolation between HV and LV sides. The isolation is provided by the DAB high frequency transformer (HFT). Table 1 shows the characteristics of the developed PET.

Table 1. Main Power Electronic Transformer (PET) characteristics.

Element	Parameter	Value
PET	Rated power	105 kW
	Number of cells	21 (7 per phase)
	HV/LV grid voltage (L-L)	6 kV/400 V
DAB	Rated power	5 kW
	Switching frequency (f_{sw})	30 kHz
	HFT isolation	24 kV
	Input voltage ($V_i = V_{cell}$)	800 V
	Transformer turns ratio (n)	1:1
	Leakage inductance (L_{lk})	423 μH
CHB	Rated power	5 kW
	C_{cell}	600 μF(film)

3. Use of WBG for PET

The superior material properties of WBG semiconductors allow power devices to operate at higher temperatures, voltages and switching frequency in comparison to Si counterparts [4,21]. Among WBG materials, SiC presents the most mature technology for high voltage devices [4]. Both 1.2 kV and 1.7 kV SiC MOSFETs are already available on the market with a wide range of current ratings [5,22,23]. The use of SiC MOSFETs not only introduces a relevant improvement to the efficiency of fast switching power converters, but also enables going to higher switching frequencies at high blocking voltage which cannot be achieved using available Si IGBTs. However, at limited switching frequency requirements (<10 kHz), especially for high power (>100 A), Si IGBTs are still the preferred choice due to cost-effectivity and reliability in addition to SiC MOSFET higher dv/dt, di/dt, and EMI issues [24].

Accordingly, to analyze the merits of integrating SiC devices in the PET, the device requirements for each of the three stages of the PET are identified below.

3.1. Device Requirements per PET Stage

In the addressed PET topology, since no step-up/down is needed, the power devices employed on both transformer sides (i.e., HV and LV sides) have the same voltage rating. These include devices of the CHB full bridge (FB), the two DAB full bridges (FB1, FB2), and the DC/AC converter devices (see Figure 1). While this is true, the specifications in terms of current ratings and commutation requirements differ significantly.

CHB: its power devices do not need to commutate fast due to the multilevel nature of the topology [25,26]. Moreover, C_{cell} size is not determined by the cell switching frequency. Therefore, SiC switching devices are not of special merit here, even when going to higher V_{cell} (i.e., to reduce the number of stacked modules), still Si IGBT would be the selected option [14]. On the other hand, SiC free-wheeling antiparallel diodes are quite interesting in this case since the CHB full bridge is required to handle positive and negative currents (i.e., not only during the dead time). Therefore, SiC diodes can effectively improve the CHB efficiency due to their reduced conduction forward voltage drop as well as lower reverse recovery time compared to Si diodes [27].

DAB: high switching frequencies can provide significant reduction of the size and weight of the converter magnetics and the input and output capacitors. Additionally, going to higher PET module voltage is not achievable using Si devices, unless the switching frequency is reduced to few kHz. Regarding the antiparallel diode, the advantages of using SiC are still a controversial issue [14,28] and, therefore, will be analyzed in this work.

3P4L DC/AC: commutation requirements are not high (in the range of few kHz [29]), and since it interfaces the LVAC grid, high blocking voltages are not required. Therefore, Si devices are a good candidate for this stage.

Considering the above discussion, for this PET, 1.7 kV Si IGBT devices with SiC freewheeling diodes are used for the CHB FB. As for the DAB, 1.2 kV SiC MOSFETs are used and possible enhancements introduced by a SiC antiparallel diode will be discussed in detail in Section 4. Finally, 1.7 kV Si IGBTs are used for the 3P4L converter.

In the next section, the possible achievable enhancements gained by employing SiC in the DAB are discussed highlighting the practical limitations intrinsic to this PET topology.

3.2. Benefits and Practical Limitations of Using SiC MOSFETs for the DAB Converter

The use of SiC MOSFETs for the DAB converter in this PET structure has two main benefits:

1. It enables increasing the switching frequency of the DAB and, therefore, decreasing the size of the HFT [30,31].
2. The high blocking voltages enable increasing V_{cell} and, therefore, for a given V_{acHV}, the number of stacked PET modules can be reduced (see Figure 1) [7,17]. This is also advantageous for the HFT. Since each module handles more power, this leads to a higher transformer power density. In order to see clearly the effect of this, Figure 2 shows the HFT power density as a function of the cell voltage where the PET total power and V_{acHV} are fixed (i.e., the number of PET modules vary). A comparative analysis regarding this relation is previously presented in [17] showing an improvement in the HFT power density as V_{cell} increases.

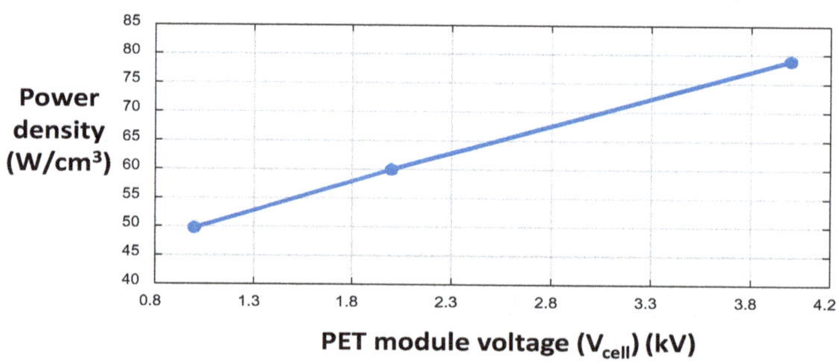

Figure 2. Theoretically calculated High Frequency Transformer (HFT) power density for three designs varying the Power Electronic Transformer (PET) module voltage (V_{cell}) and the handled power [17].

However, these SiC potential benefits may be compromised by certain practical implementation constraints.

On one hand, increasing V_{cell} has several adverse effects. High DC link voltages create practical problems for feeding the control circuitry in each cell. As commercial auxiliary power supplies (APS) do not provide the required isolation [29], each module circuitry has to be supplied from its DC link, where the HV and LV sides have separate APSs [32]. Commercial APSs can be used for voltages under 1 kV, otherwise, custom solutions must be implemented, such as the modular ISOP topology proposed in [29]. Consequently, various aspects must be considered for the selection of V_{cell}.

On the other hand, the size reduction of the HFT, in this particular case, is constrained by the high isolation required by the PET. This isolation imposes minimum clearance distances between windings, which compromises the window utilization factor resulting in a physical limit on further size reduction.

4. SiC-Based DAB Converter

The DAB (see Figure 3) is selected for the intermediate stage of the PET, as it provides galvanic isolation as well as bidirectional power flow [33,34]. The DAB is based on two active bridges interfaced through an HFT, which provides the required galvanic isolation. This converter provides bidirectional operation by controlling the phase shift between the AC voltages generated by both bridges (V_1 and V_2). Also, this converter can have relatively high efficiency due to the soft-switching operation of all the devices at nominal conditions (zero-voltage switching, ZVS) [35,36].

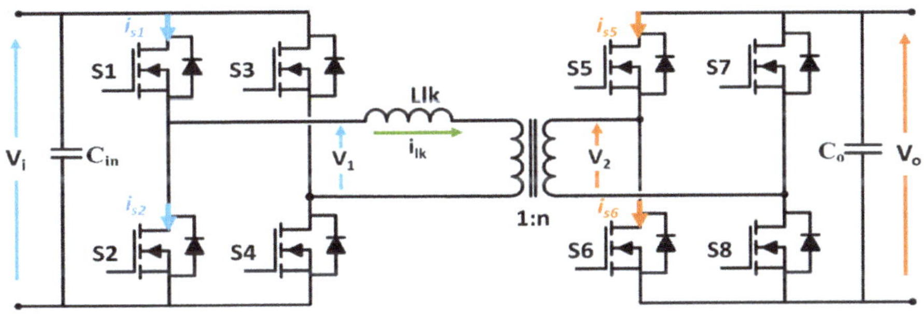

Figure 3. Dual active bridge (DAB) converter schematic.

4.1. SiC Device Selection

Regarding the selection of the SiC devices for the DAB, 6.5, 10, and 15 kV SiC MOSFETs have been reported for laboratory prototypes [37–42], but are far from being a viable commercial alternative yet. Current ratings offered in commercially available 1.7 kV SiC MOSFETs are still limited [23], and their commutation characteristic must be improved. The 1.2 kV SiC devices remain as the most mature technology available in the SiC device market. Consequently, these devices have been selected for the DAB. The DC link voltage is, therefore, set to 800 V.

Seven 1.2 kV SiC commercial MOSFETs were selected; two power modules and five discrete N-channel SiC MOSFETs. A comparative analysis of these devices was carried out in a boost converter operating in continuous conduction mode (CCM) (see Figure 4).

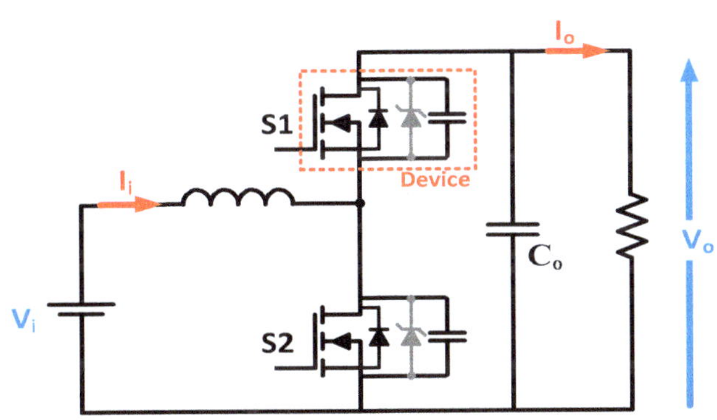

Figure 4. Boost converter test bench schematic.

The boost converter was chosen as a preliminary test bench due to its simplicity and rapid prototyping, but foremost, due to its similar operation to a DAB converter as it has two switching devices in a leg with an inductance connected to the middle point. However, the differences in operation between a CCM boost and a DAB are well understood and, therefore, perspective devices resulting from the first selection stage are then tested in a DAB converter prototype.

The test bench used commercial driver boards from CREE and a commercial FPGA-based controller platform (BASYS2). Tests in the boost converter were done at 2 kW 400/800 V for switching frequencies of 30, 50, and 100 kHz and a dead time of 500 ns. Table 2 summarizes the results. The efficiency is calculated using the input and output DC voltages and currents of the converter measured using digital multimeters. It is observed that all the devices show a high efficiency barely affected by the increase in switching frequency which was increased by a factor of more than three.

Table 2. SiC MOSFET comparative analysis in a boost converter.

Manufacturer	Reference	Package	Rated Current @100 °C (A)	R_{DS} (mΩ)	C_{out} (pF)	Efficiency (%)			Price per Bridge (€)
						30 kHz	50 kHz	100 kHz	
ST	SCT30N120	TO-247-3	34	80	130	97.77	98.16	97.81	100
ROHM	SCH2080KE + SBD	TO-247-3	28	80	175	97.71	98.02	97.69	132
ROHM	SCT2080KE	TO-247-3	28	80	77	97.82	98.17	97.93	80
CREE	C2M0040120D	TO-247-3	40	40	150	97.76	98.10	97.84	128
CREE	C2M0025120D	TO-247-3	60	25	220	97.56	97.88	97.42	252
CREE	CAS120M12BM2 + SBD	6-pack 45 mm	59	25	400	97.49	97.54	97.50	412
CREE	CCS050M12CM2 + SBD	Half-bridge 62 mm	138	13	900	96.44	95.8	-	660

Since the performance of all seven MOSFETs is comparable, the selection of the adequate option was mainly based on the size and the price. The specifications for the DAB converter are shown in Table 1. Accordingly, the peak current handled by the devices can be calculated from Equation (1) [35], where T is half the switching period, d is the phase shift, L_{lk} is the leakage inductance, v_o is the output voltage, v_i is the input voltage and n is the HFT turns ratio:

$$I_{p\_lk} = \frac{T}{2L_{lk}}\left(2\frac{v_o}{n}d + v_i - \frac{v_o}{n}\right) \quad (1)$$

The current peak is calculated for the maximum phase-shift, this is selected to be 0.35 according to [35]. Based on Equation (1), this current is approximately equal to 11 A. The device current rating is selected to be twice the magnitude of the peak current handled by the devices to keep a safety margin. Therefore, the minimum required current rating is 22 A. This eliminates the two modules, and the CREE 60 A discrete, as the size and price are not justified in this case. The remaining four discrete devices are almost equally favored except for the ROHM SCH2080KE, as it includes a SiC Schottky barrier diode (SBD) co-packaged with the MOSFET.

4.2. Antiparallel SBD for a DAB Converter

Observing the efficiency of the boost converter using ROHM SCH2080KE versus for example ROHM SCT2080KE (without an additional SBD), it is consistently higher without SiC SBD. For the boost converter, this is logical as an additional antiparallel diode increases the output capacitance (see Table 2) and since hard-switching occurs, this increases the switching losses. Although, this makes sense and is simple to understand for the boost, for the DAB, it is more complicated as ZVS is implemented. That being the case, it is not valid to make a selection between both devices unless the additional diode behavior is studied for the DAB operation to identify if it improves or worsens its efficiency. This issue is addressed as follows, including: (1) understanding the potential effects introduced by a SiC antiparallel diode in a DAB converter, (2) developing a simple analytical loss model to estimate the possible efficiency improvement introduced by the SiC diode at a certain DAB operating point, and (3) validation of the proposed model using experimental results in a DAB prototype.

The two devices used in the analysis are the ROHM devices (i.e., SCH2080KE and SCT2080KE) as it is the same die but one packed with a SiC antiparallel SBD [22]. Characteristics of both devices are provided in Table 3. The diode forward voltage, $V_{F\text{-}diode}$, is obtained from the datasheet at the value of I_{p_lk} (see Equation (1)) where the employed phase shift is that corresponding to 5 kW.

Table 3. Specifications of the compared devices.

Characteristics		SCH2080KE +SBD	SCT2080KE
$R_{ON\text{-}MOSFET}$		125 mΩ	
E_{off} (R_g = 10 Ω)		110 µJ	120 µJ
E_{on} (R_g = 10 Ω)		330 µJ	275 µJ
Anti-parallel diode	Type	SiC SBD	Body diode
	R_D (Estimated)	45 mΩ	320 mΩ
	V_{knee}	0.85 V	1.4 V
	$V_{F\text{-}Diode}$ (@ I_{p_lk})	1.2 V	4.5 V

4.2.1. Advantages and Disadvantages of an Additional SBD

In order to provide a qualitative preliminary understanding of the possible effects of the SiC SBD on the DAB efficiency, the waveforms of the DAB are shown in Figure 5.

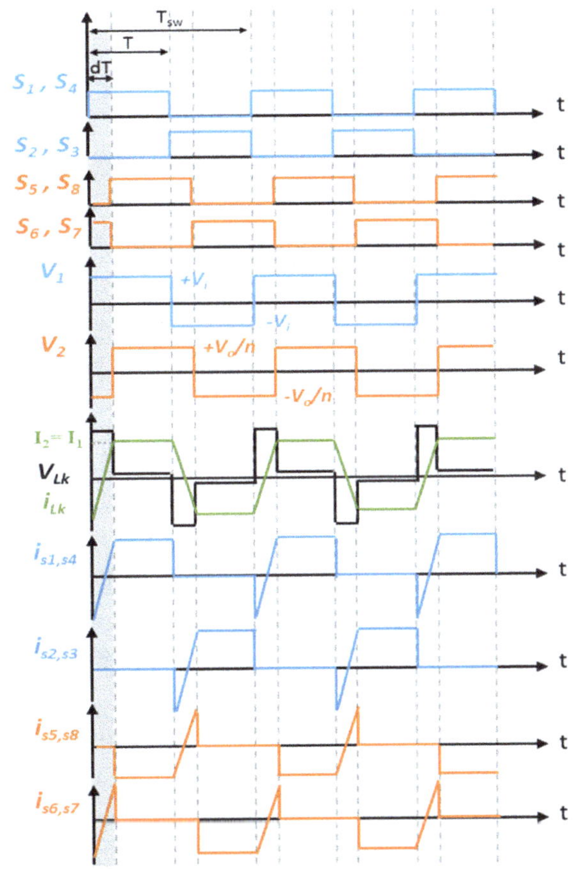

Figure 5. Dual Active Bridge (DAB) converter waveforms.

The potential advantages of using SCH2080KE +SBD in comparison to SCT2080KE are analyzed as follows:

- Since $V_{F\text{-}Diode}$ of the additional SiC SBD is more than three times lower than that of the body diode (see Table 3), the power losses due to diode conduction during the dead times are lower using SCH2080KE +SBD.
- The diode can enter into conduction while the MOSFET is ON. This can be seen from Figure 5, when the current through the MOSFET is negative (i.e., source to drain), a forward voltage drop,

$V_{F\text{-}MOSFET}$, is applied to the diode. If $V_{F\text{-}MOSFET}$ is higher than the diode knee voltage (V_{knee}), then the diode conducts. If this case is true, current is shared between the MOSFET and the diode and, therefore, conduction losses are reduced.

- Unlike boost converter operation, soft switching is adopted in DAB devices at turn ON and, therefore, the additional output capacitance introduced by the additional SiC diode, illustrated in Figure 4, does not significantly penalize the switching losses (see Table 2).

However, these advantages can be compromised by several situations resulting in almost no advantages of having a SiC SBD, these cases are summarized as follows:

- From Figure 6, diodes do not operate during the whole period of the dead time. During (A), the output parasitic capacitances of the MOSFETs (see Figure 4) in one leg are exchanging the voltage (i.e., one is discharging while the other is charging) and during (B), diodes conduct due to circulating currents. In some cases, (B) can tend to zero depending on the value of the current charging/discharging the parasitic capacitors and on the dead time.
- If the current through the MOSFETs is not high enough to produce a $V_{F\text{-}MOSFET} > V_{knee}$, then the diode will never conduct during MOSFET ON time.
- The turn OFF switching losses should be analyzed as the enhancement introduced by the SiC SBD in the conduction losses can, in some cases, be compromised by the increase in turn OFF switching losses (due to the extra capacitance).

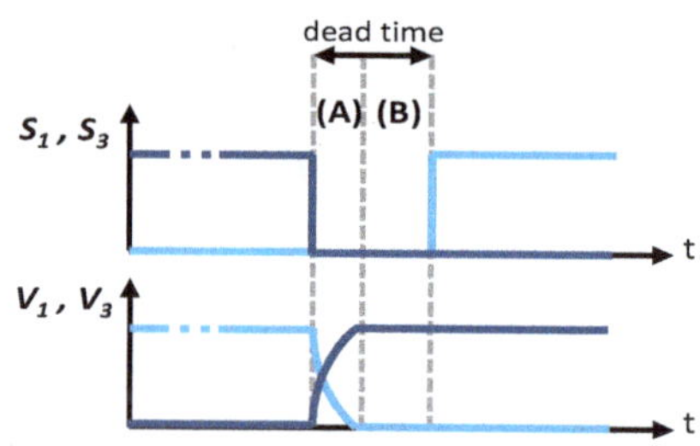

Figure 6. Waveforms of the gates and drain to source voltages of two MOSFETs in one leg during the dead time period.

As a conclusion to the previous discussion, it is important to develop a simple model to determine, for a certain DAB operation point, if an extra antiparallel SiC diode co-packed with the SiC MOSFET is worthy or avoiding it is better. This is introduced in the next section.

4.2.2. Model to Assess the Efficiency Improvement Using a SiC Diode

Diode Conduction Intervals

Regarding diode conduction interval during the dead time, in order to estimate the time duration (B) (see Figure 6), the capacitor charging time during (A) is estimated using Equation (2), where $C_{o(er)}$ is the MOSFET effective output capacitance given by the data sheet and $V_1 = V_i = V_o$. Accordingly, i_{lk} is considered flat ($I_1 = I_2$ in Figure 5) and, therefore, as an approximation, the peak current (I_{p_lk}) is considered to be equal during all the switching transitions.

$$t_{(A)} = 2\frac{V_1\, C_{o(er)}}{I_{p\_lk}} \tag{2}$$

Since $t_{(B)}$ = dead time − $t_{(A)}$, therefore diode operates for 570 ns for the case of the SiC SBD and 588 ns for the body diode. This time represents around 95% of the total dead time (600 ns) and, therefore, the reduction in conduction losses during diode operation is relevant to consider.

Regarding the diode conduction interval during MOSFET ON time, first, the diode characteristics are identified from the MOSFET datasheet (see Table 3). Figure 7 shows the forward voltage drop for both diodes ($V_{F\text{-}Diode}$) as a function of the current through the diode. Additionally, the voltage drop on the MOSFET due to its ON resistance ($V_{F\text{-}MOSFET}$) as a function of the current it is conducting is illustrated.

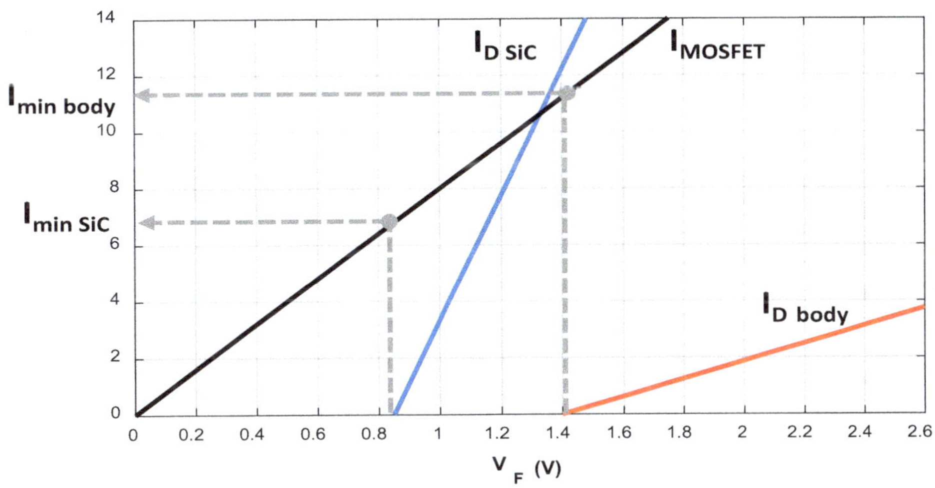

Figure 7. MOSFET voltage drop and diode characteristic curve for the SiC Schottky Barrier Diode (SBD) and the body diode.

It is possible to see that each diode conducts when the current through the MOSFET is above a certain minimum value. It is clearly lower in the case of the SiC SBD. This can be estimated from Equation (3), where $R_{ON\text{-}MOSFET}$ stands for ON-state drain-to-source MOSFET resistance:

$$I_{min} = \frac{V_{knee}}{R_{ON-MOSFET}} \tag{3}$$

Based on the data in Table 3, $I_{min\ SiC}$ is 6.8 A while I_{min} body is 11.2 A. Since the theoretical calculated $I_{p\_lk} = 9$ A, therefore it is not possible that the body diode enters into conduction while the MOSFET is ON. On the other hand, the SiC diode would conduct when the current exceeds 6.8 A.

Estimation of the Power Losses

Power losses are estimated for both cases of the DAB using MOSFETs with a SiC antiparallel diode (SCH2080KE) and with only the body diode (SCT2080KE). Theoretical power loss estimation will be compared to the experimental efficiency results presented in the next section.

First, the periods where the diode can conduct (when it is forward biased) are identified as shown in Figure 8. It can be seen that, the diodes of all the ON MOSFETs can conduct during (1) and (4) as the current through all the devices is negative (see Figure 5), while during (3) and (6) only the diodes of the secondary bridge can conduct.

Accordingly, to simplify the power loss estimation the following assumptions are made:

- As $I_{min\ SiC}$ is around 75% of $I_{pk\text{-}lk}$, therefore, it is assumed that the diode operates only during period (3) and (6) when the peak current is passing through the devices (periods (1) and (4) are neglected).
- The power loss estimation is performed only during the periods with different operation for the case of the SiC SBD and the body diode. In other words, only the difference in losses between both cases is considered. These intervals are: (3), (6) and the dead times.

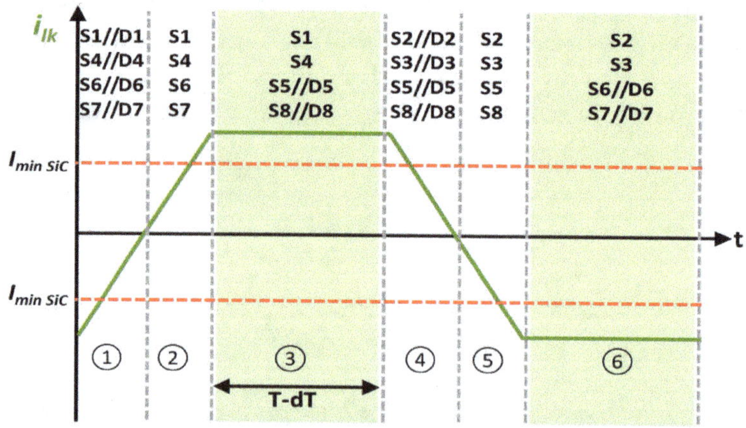

Figure 8. Current through the power transfer inductance of the Dual Active Bridge (DAB) showing periods where the antiparallel diode can enter into conduction mode during MOSFETs conduction.

During intervals (3) and (6), the conduction and switching losses of all the devices are estimated, while during the dead times, only the diodes conduction losses are estimated.

(a) *Losses during dead time:*

During one switching period (T_{sw}), four intervals of dead time take place. During each interval two diodes conduct. Accordingly, the diodes total conduction losses are estimated using Equation (4), where $V_{F\text{-}Diode}$ is the diode forward voltage at I_{p_lk} obtained from the diode characteristic curve given by the datasheet (linearized in Figure 7).

$$P_{deadtime} = 8 \cdot V_{F-Diode} \cdot I_{p\_lk} \cdot \frac{t_{(B)}}{T_{sw}} \tag{4}$$

(b) *Losses during interval (3) and (6) (Figure 8):*

Losses during these two intervals are divided into: (1) turn OFF switching losses (P_{sw}) and (2) MOSFET and diode conduction losses (P_{MOSFET} and P_{Diode}). The switching losses are straightforward, estimated from MOSFET E_{OFF} using Equation (5).

$$P_{sw} = 8 \cdot E_{OFF} \cdot f_{sw} \tag{5}$$

On the other hand, conduction losses are not straight forward. As noticed from Figure 8, in both (3) and (6) intervals, the primary bridge devices are different from the secondary ones. For example, during (3), for the primary, only the MOSFETs conduct (i.e., all the current flows through the MOSFETs, S1 & S4). In this case, four MOSFETs of the primary bridge conduct during (3) and (6) (two in each interval). Therefore, the total primary conduction losses can be easily estimated using Equation (6).

$$P_{MOSFET-prim} = 4 \cdot {I_{p\_lk}}^2 \cdot R_{ON-MOSFET} \cdot \frac{T - dT}{T_{sw}} \tag{6}$$

However, for the secondary, the current is shared between the diode and the MOSFET (S5, D5 and S8, D8). Accordingly, the current conducted by each element has to be estimated. Since the voltage drop on the MOSFET is equal to the diode $V_{F\text{-}Diode}$, then:

$$I_M \cdot R_{ON-MOSFET} = V_{knee} + I_D \cdot R_D \tag{7}$$

where I_M and I_D are the currents through the MOSFET and the diode respectively and R_D is the diode dynamic resistance.

Since the current is shared by both elements, then, I_M and I_D can be obtained from Equations (7) and (8):

$$I_M + I_D = I_{p\_lk} \quad (8)$$

Finally, the total conduction losses of the secondary bridge during T_{sw} can be calculated from Equations (9) and (10):

$$P_{MOSFET-sec} = 4 \cdot I_M{}^2 \cdot R_{ON-MOSFET} \cdot \frac{T - dT}{T_{sw}} \quad (9)$$

$$P_{Diode-sec} = 4 \cdot I_D \cdot V_{F-Diode}|_{I_D} \cdot \frac{T - dT}{T_{sw}} \quad (10)$$

The values of the power loss components estimated in this section are summarized in Table 4 where the enhancement to the DAB efficiency is estimated to be approximately 0.22% given the rated power is 5 kW.

Table 4. Theoretically-estimated power losses.

Interval	Losses		SCH2080KE +SBD	SCT2080KE
Dead time (600 ns)	Diode conduction ($P_{deadtime}$)		1.6 W	6.2 W
(3) & (6)	Switching (P_{sw})		26.4 W	28.8 W
	Conduction	$P_{MOSFET\text{-}prim}$	17.4 W	17.4 W
		$P_{MOSFET\text{-}sec}$	10.4 W	17.4 W
		$P_{Diode\text{-}sec}$	2.8 W	-
Total losses			58.6 W	69.8 W

4.2.3. Experimental Validation of the Proposed Loss Model

A DAB test bench was constructed and tests were performed at the nominal operation defined in Table 1 and the results are summarized in Table 5.

Table 5. Experimental results of the Dual Active Bridge (DAB) with and without SiC Schottky Barrier Diode (SBD).

Characteristics	SCH2080KE +SBD	SCT2080KE
Antiparallel diode	SiC SBD	Body diode
Efficiency (η)	98.1%	97.8%
Phase-shift (d)	0.29	
Dead time	600 ns	
I_{p_lk}	Calculated from Equation (1): 9 A Measured (experimental): 9.8 A	

It was observed that the measured efficiency of the DAB converter composed of MOSFETs with an additional SiC antiparallel SBD is higher than that with only the body diode. The enhancement observed from the experimental results is around 0.3%, which validates the calculation introduced previously resulting in 0.22%. These results validate the proposed hypothesis and approach.

5. Experimental Results of the PET Module

The developed full-scale PET module is shown in Figure 9. Its structure is that schematically shown in Figure 1b. Table 6 summarize the main components of the PET module.

Figure 9. The developed Power Electronic Transformer (PET) module.

Table 6. Main Power Electronic Transformer (PET) module components.

Element	Characteristics	Reference
DAB	ROHM 1.2 kV, 28 A SiC MOSFET CREE 2-channel drivers	SCH2080KE CGD15HB62P1
CHB	Infineon 1.7 kV Si IGBT + SiC diode Infineon two-channel drivers	Not commercial 2ED300C17-S
Controller	Xilinx FPGA (Spartan 3E) custom board	XC3S250E-4TQG144I

5.1. Developed HFT

One of the key aspects in the design of this HFT is the high galvanic isolation required between its primary and secondary sides, being 24 kV in this PET. This presents significant challenges compared to the ones considered in literature [43,44]. Moreover, the DAB power transfer inductance (L_{lk}) is magnetically integrated in the HFT by making use of its series leakage inductance [43]. This reduces size and cost but, on the other hand, imposes additional constrains on the HFT design due to the required accuracy of L_{lk}.

The HFT design is, therefore, a tradeoff between four variables: transferred power capability, temperature rise (i.e., losses), size and cost. Several design iterations are performed to achieve the required isolation with an acceptable tradeoff between these variables. The HFT design was previously presented in [45]. The main experimental validation tests are provided in this work for completeness.

Figure 10a shows laboratory developed HFT for test purposes and Figure 10b shows the final factory encapsulated HFT. A UU core structure is used typically selected in literature for separate winding [46,47]. The core ferrite material is selected to be Ferroxcube® 3C90 based on the DAB switching frequency [43]. An epoxy resin providing 15 kV/mm and exhibiting acceptable thermal conductance of 3 W/mK was used for encapsulation.

Figure 11 shows a schematic representation of the HFT design from ANSYS PEmag® software (Canonsburg, PA, USA).

Figure 10. High Frequency Transformer (HFT) prototype. (a) Non-encapsulated laboratory developed using 3D printed bobbins. (b) Encapsulated final HFT.

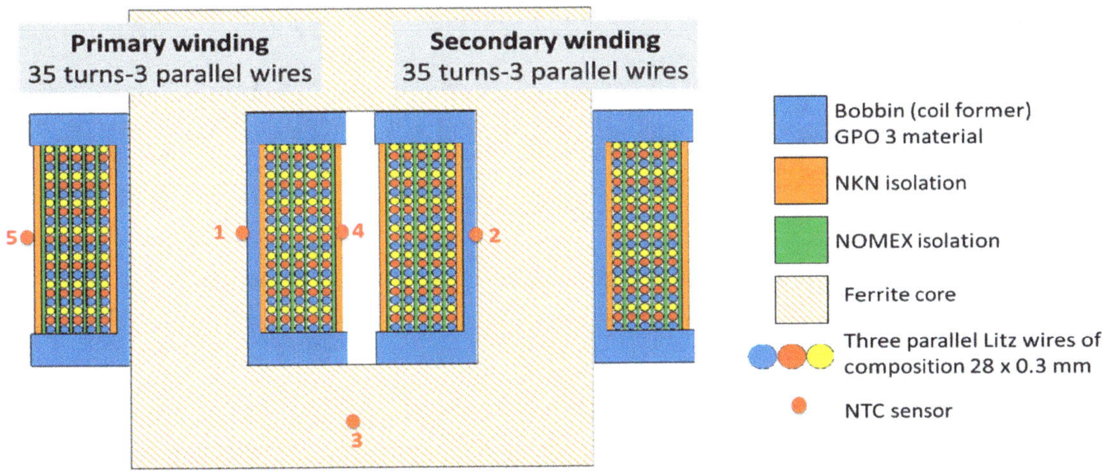

Figure 11. High Frequency Transformer (HFT) design taken from ANSYS PEmag®.

To verify the achieved isolation, a high potential test is done for both porotypes and results are compared. A Hipot tester is used to apply a voltage potential of up to 24 kV between both windings. The leakage current flowing from the winding with the higher potential, through air/resin, is recorded and shown in Figure 12. A significant diversion between both prototypes is clear at higher voltage potentials. Having lower leakage currents means that successive partial discharge due to high dv/dt is avoided which would lead to eventual insulation breakdown [48].

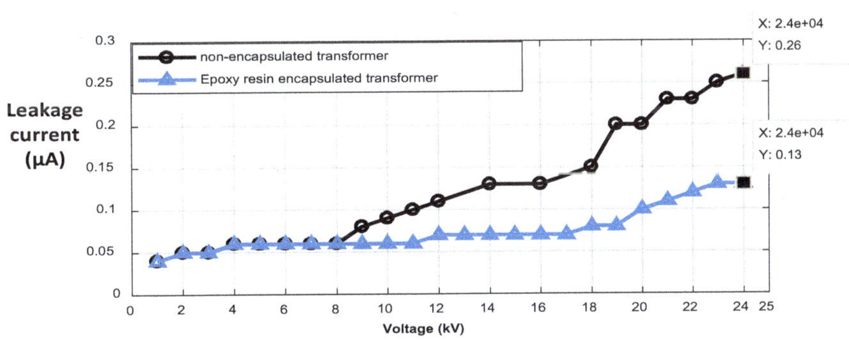

Figure 12. Experimental results. Leakage current measured between High Frequency Transformer (HFT) windings, as a function of the voltage difference between them.

To locate the hottest spot and verify its temperature rise, five NTC sensors were mounted inside the HFT (see Figure 11). The location of this spot differs in encapsulated (NTC 1) and non-encapsulated prototypes (NTC 4). The temperature profile of the encapsulated HFT hottest spot during a four-hour

DAB test at rated operation is shown in Figure 13. The steady state temperature, under natural convection, was recorded to be 60 °C.

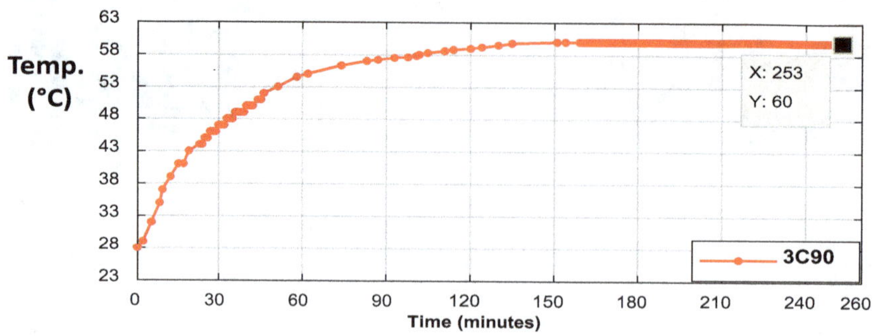

Figure 13. Experimental results. Hottest spot temperature profile for the encapsulated High Frequency Transformer (HFT) using 3C90 ferrite core measured using a Negative Temperature Coefficient (NTC) sensor.

A summary of the final HFT design is shown in Table 7.

Table 7. High Frequency Transformer (HFT) design summary.

Quantity	Value
Isolation	24 kV
Core structure	UU100/57/25
Ferrite material	Ferroxcube® 3C90
Encapsulation	Epoxy resin (15 kV/mm) (ROYAPOX 912 THC/2)
Bobbins	GPO-3 (20 kV/mm)
Number of turns (N)	35
Temperature rise at NTC 4	40 °C

5.2. Experimental Validation

Figure 14 shows the performed test connection diagram. A DC power supply is connected to the LVDC side to provide V_{dcLV} and the power is transferred from FB2 to FB1. The DAB current control regulates the transferred power using single phase shift (SPS) modulation while the CHB full bridge regulates the cell capacitor voltage (V_{cell}).

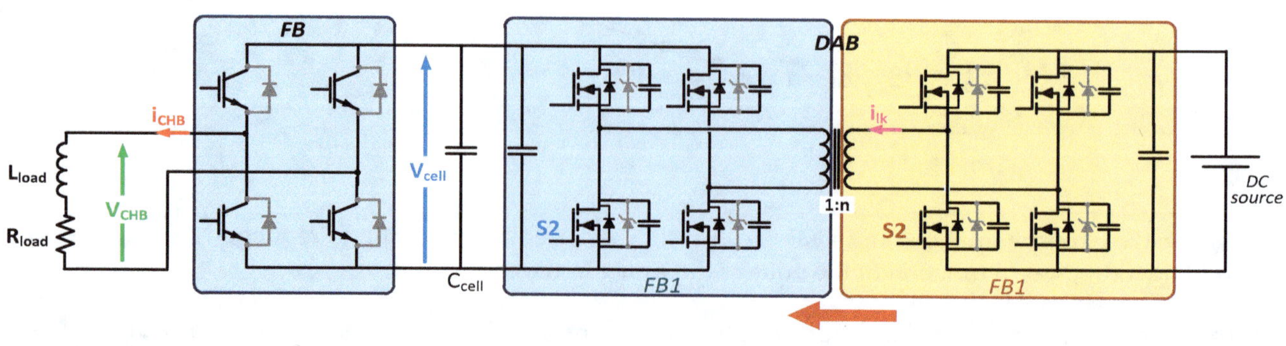

Figure 14. Schematic of the Power Electronic Transformer (PET) module during the performed experimental test.

Figure 15 shows the experimental results for the PET module nominal operation (described in Table 1). Figure 15a shows the DAB waveforms, where FB2 gates are leading FB1 gates and the phase shift between both controls the magnitude of the transferred power [35].

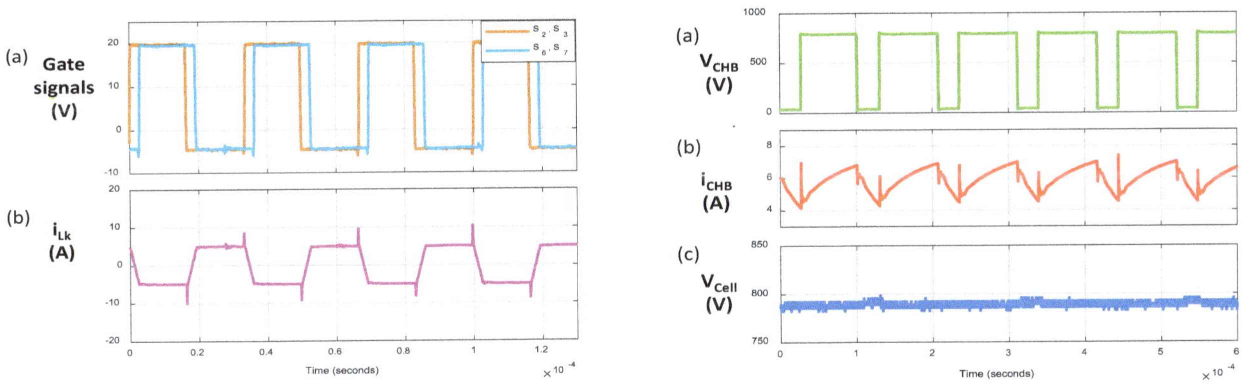

(a) **(b)**

Figure 15. Experimental results. (**a**) Dual Active Bridge (DAB) waveforms. Two gate signals of DAB FB1 and FB2 and leakage inductance current (i_{lK}). (**b**) Cascaded H-Bridge (CHB) full bridge waveforms. CHB full bridge output voltage (V_{CHB}), CHB full bridge output current (i_{CHB}), and the cell capacitor voltage (V_{cell}).

Figure 15b shows the CHB waveforms, where V_{CHB} and i_{CHB} are the CHB full-bridge output voltage and current respectively. V_{cell} is regulated at 800 V by the CHB voltage control.

6. Conclusions

This paper analyzes the use of SiC devices in three-stage modular PETs. It has been shown that SiC MOSFETs can be especially advantageous in the isolation stage, as they combine high blocking voltage and high switching frequencies, leading to higher efficiency, size reduction of the isolation transformer, and higher power density. On the other hand, limited merit is achieved using SiC in the front end AC/DC and the LV side DC/AC converters. However, this partially depends on the application of the PET and the selected topology.

The paper details the selection of SiC MOSFETs for the DAB DC/DC isolation stage experimentally comparing different commercial devices. The benefits of a SiC antiparallel SBD is investigated. It is concluded that the SiC diode improves the efficiency of the DAB as it reduces the total conduction losses of the device. However, this improvement strongly depends on two critical design aspects: the employed dead time interval and the time interval during which the current is driven by both the MOSFET and the diode at the same time (due to the voltage drop in the MOSFET on resistance directly biasing the diode). Generally, the efficiency improvement can be more relevant for a flatter inductor current shape and when the DAB is working around its nominal operation point. A loss model was presented to assess the introduced efficiency improvement and was experimentally validated.

Experimental results showing the operation of the full-scale DAB converter at rated conditions as part of the PET module are provided.

Author Contributions: Conceptualization, A.R., M.A. and F.B.; methodology, A.R., M.A. and F.B.; software, M.S.; validation, M.S. and M.R.R.; formal analysis, M.S. and M.R.R.; investigation, M.S. and M.R.R.; writing—review and editing, M.S., M.R.R., A.R., M.A. and F.B.; supervision, A.R., M.A. and F.B.; funding acquisition, A.R., M.A. and F.B. All authors have read and agreed to the published version of the manuscript.

Nomenclature

$\mathbf{V_{acHV}}$	HVAC grid voltage
$\mathbf{V_{acLV}}$	LVAC grid voltage
$\mathbf{V_{dcLV}}$	LVDC link voltage
$\mathbf{C_{dcLV}}$	LVDC link capacitor
$\mathbf{V_{cell}}$	Cell capacitor voltage
$\mathbf{C_{cell}}$	Cell capacitor
$\mathbf{f_{sw}}$	Switching frequency
T	Half the switching period of the DAB
d	Phase shift of the DAB represented as a ratio of T
$\mathbf{L_{lk}}$	DAB leakage inductance
$\mathbf{V_{knee}}$	Diode knee voltage
$\mathbf{V_{F\text{-}diode}}$	Diode forward voltage
$\mathbf{I_{pk\text{-}lk}}$	Peak current through the leakage inductance
$\mathbf{R_{ON\text{-}MOSFET}}$	MOSFET ON resistance
$\mathbf{C_{o(er)}}$	MOSFET effective output capacitance

References

1. van der Merwe, J.W.; Mouton, H.d.T. The solid-state transformer concept: A new era in power distribution. In Proceedings of the AFRICON 2009, Nairobi, Kenya, 23–25 September 2009; pp. 1–6.
2. McMurray, W. Power Converter Circuits Having a High-Frequency Link. U.S. Patent 3517300, 23 June 1970.
3. Ronan, E.R.; Sudhoff, S.D.; Glover, S.F.; Galloway, D.L. A power electronic-based distribution transformer. *IEEE Trans. Power Del.* **2002**, *17*, 537–543. [CrossRef]
4. Rothmund, D.; Ortiz, G.; Guillod, T.; Kolar, J.W. 10kV SiC-based isolated DC-DC converter for medium voltage-connected Solid-State Transformers. In Proceedings of the IEEE Applied Power Electronics Conference and Exposition (APEC), Charlotte, NC, USA, 15–19 March 2015; pp. 1096–1103.
5. Millán, J.; Godignon, P.; Perpiñà, X.; Pérez-Tomás, A.; Rebollo, J. A Survey of Wide Bandgap Power Semiconductor Devices. *IEEE Trans. Power Electron.* **2014**, *29*, 2155–2163. [CrossRef]
6. Millán, J.; Friedrichs, P.; Mihaila, A.; Soler, V.; Rebollo, J.; Banu, V.; Godignon, P. High-voltage SiC devices: Diodes and MOSFETs. In Proceedings of the International Semiconductor Conference (CAS), Sinaia, Romania, 12–14 October 2015; pp. 11–18.
7. Dong, D.; Agamy, M.; Bebic, J.Z.; Chen, Q.; Mandrusiak, G. A Modular SiC High-Frequency Solid-State Transformer for Medium-Voltage Applications: Design, Implementation, and Testing. *IEEE J. Emerg. Sel. Top. Power Electron.* **2019**, *7*, 768–778. [CrossRef]
8. Raju, R.; Dame, M.; Steigerwald, R. Solid-state transformer using silicon carbide-based modular building blocks. In Proceedings of the IEEE International Conference of Power Electronics and Drive Systems (PEDS), Honolulu, HI, USA, 12–15 December 2017; pp. 1–7.
9. Yao, R.; Zheng, Z. Design and control of a SiC-based three-stage AC to DC power electronic transformer for electric traction applications. *J. Eng.* **2019**, *2019*, 2947–2952. [CrossRef]
10. Wang, G.; Huang, X.; Wang, J.; Zhao, T.; Bhattacharya, S.; Huang, A.Q. Comparisons of 6.5kV 25A Si IGBT and 10-kV SiC MOSFET in Solid-State Transformer application. In Proceedings of the IEEE Energy Conversion Congress and Exposition (ECCE), Atlanta, GA, USA, 12–16 September 2010; pp. 100–104.
11. Madhusoodhanan, S.; Tripathi, A.; Patel, D.; Mainali, K.; Kadavelugu, A.; Hazra, S.; Bhattacharya, S.; Hatua, K. Solid-State Transformer and MV Grid Tie Applications Enabled by 15 kV SiC IGBTs and 10 kV SiC MOSFETs Based Multilevel Converters. *IEEE Trans. Ind. Appl.* **2015**, *51*, 3343–3360. [CrossRef]
12. Hatua, K.; Dutta, S.; Tripathi, A.; Baek, S.; Karimi, G.; Bhattacharya, S. Transformerless intelligent power substation design with 15 kV SiC IGBT for grid interconnection. In Proceedings of the IEEE Energy Conversion Congress and Exposition (ECCE), Phoenix, AZ, USA, 17–22 September 2011; pp. 4225–4232.
13. Abu-Siada, A.; Budiri, J.; Abdou, A.F. Solid State Transformers Topologies, Controllers, and Applications: State-of-the-Art Literature Review. *Electronics* **2018**, *7*, 298. [CrossRef]

14. Evans, N.M.; Lagier, T.; Pereira, A. A preliminary loss comparison of solid-state transformers in a rail application employing silicon carbide (SiC) MOSFET switches. In Proceedings of the 8th IET International Conference on Power Electronics, Machines and Drives (PEMD), Glasgow, UK, 19 April 2016; pp. 1–6.
15. Lopez, M.; Briz, F.; Zapico, A.; Rodriguez, A.; Diaz-Reigosa, D. Control strategies for MMC using cells with power transfer capability. In Proceedings of the IEEE Energy Conversion Congress and Exposition (ECCE), Montreal, QC, Canada, 20–24 September 2015; pp. 3570–3577.
16. Cuartas, J.M.; de la Cruz, A.; Briz, F.; Lopez, M. Start-up, functionalities and protection issues for CHB-based solid state transformers. In Proceedings of the IEEE International Conference on Environment and Electrical Engineering and IEEE Industrial and Commercial Power Systems Europe (EEEIC/I&CPS Europe), Milan, Italy, 6–9 June 2017; pp. 1–5.
17. López, M.; Briz, F.; Saeed, M.; Arias, M.; Rodríguez, A. Comparative analysis of modular multiport power electronic transformer topologies. In Proceedings of the IEEE Energy Conversion Congress and Exposition (ECCE), Milwaukee, WI, USA, 18–22 September 2016; pp. 1–8.
18. Briz, F.; Lopez, M.; Rodriguez, A.; Arias, M. Modular Power Electronic Transformers: Modular Multilevel Converter versus Cascaded H-Bridge Solutions. *IEEE Ind. Electron. Mag.* **2016**, *10*, 6–19. [CrossRef]
19. Briz, F.; López, M.; Zapico, A.; Rodríguez, A.; Díaz-Reigosa, D. Operation and control of MMCs using cells with power transfer capability. In Proceedings of the IEEE Applied Power Electronics Conference and Exposition (APEC), Charlotte, NC, USA, 15–19 March 2015; pp. 980–987.
20. European Union. "Silicon Carbide Power Technology for Energy Efficient Devices (SPEED)", Ref. FP7-NMP3-LA-2013-604057, EU–FP7, Large Scale Integrating Collaborative Research Project. Available online: https://cordis.europa.eu/project/id/604057 (accessed on 4 March 2020).
21. Wensong, Y. *Wide Band-Gap Devices for Solid State Transformer Applications*; Tutorial; FREEDM Systems Center: Raleigh, NC, USA, 2017.
22. ROHM Semiconductor Electronics Industry Company. Available online: www.rohm.com (accessed on 2 March 2020).
23. Wolfspeed, Inc. Available online: www.wolfspeed.com (accessed on 2 March 2020).
24. Zhang, L.; Yuan, X.; Wu, X.; Shi, C.; Zhang, J.; Zhang, Y. Performance Evaluation of High-Power SiC MOSFET Modules in Comparison to Si IGBT Modules. *IEEE Trans. Power Electron.* **2019**, *34*, 1181–1196. [CrossRef]
25. Malinowski, M.; Gopakumar, K.; Rodriguez, J.; Perez, M.A. A Survey on Cascaded Multilevel Inverters. *IEEE Trans. Ind. Electron.* **2010**, *57*, 2197–2206. [CrossRef]
26. Rodriguez, J.; Lai, J.-S.; Peng, F.Z. Multilevel inverters: A survey of topologies, controls, and applications. *IEEE Trans. Ind. Electron.* **2002**, *49*, 724–738. [CrossRef]
27. Sahib, A.; Abdulbaqi, I.M.; Ahmed, A.H. Analysis of a Four-Leg Inverter Feeding a Nonlinear Load using SVPWM Technique. In Proceedings of the Second Al-Sadiq International Conference on Multidisciplinary in IT and Communication Science and Applications (AIC-MITCSA), Baghdad, Iraq, 30–31 December 2017; pp. 19–24.
28. Martin, D.; Killeen, P.; Curbow, W.A.; Sparkman, B.; Kegley, L.E.; McNutt, T. Comparing the switching performance of SiC MOSFET intrinsic body diode to additional SiC schottky diodes in SiC power modules. In Proceedings of the IEEE 4th Workshop on Wide Bandgap Power Devices and Applications (WiPDA), Fayetteville, AR, USA, 7–9 November 2016; pp. 242–246.
29. Rodriguez, A.; Rogina, M.R.; Saeed, M.; Lamar, D.G.; Arias, M.; Lopez, M.; Briz, F. Auxiliary power supply based on a modular ISOP flyback configuration with very high input voltage. In Proceedings of the IEEE Energy Conversion Congress and Exposition (ECCE), Milwaukee, WI, USA, 18–22 September 2016; pp. 1–7.
30. Huber, J.E.; Kolar, J.W. Solid-State Transformers: On the Origins and Evolution of Key Concepts. *IEEE Ind. Electron. Mag.* **2016**, *10*, 19–28. [CrossRef]
31. McLyman, C.W.T. *High Reliability Magnetic Devices: Design and Fabrication*; Marcel Dekker: New York, NY, USA, 2002.
32. Saeed, M.; Cuartas, J.M.; Rodriguez, A.; Arias, M.; Briz, F. Energization and Start-Up of CHB-Based Modular Three-Stage Solid-State Transformers. *IEEE Trans. Ind. Appl.* **2018**, *54*, 5483–5492. [CrossRef]
33. de Doncker, R.W.; Divan, R.W.; Kheraluwala, M.H. A three-phase soft-switched high power-density dc/dc converter for high-power applications. *IEEE Trans. Ind. Appl.* **1991**, *27*, 63–73. [CrossRef]
34. Kheraluwala, M.H.; Gascoigne, R.W.; Divan, D.M.; Baumann, E.D. Performance characterization of a high-power dual active bridge dc-to-dc converter. *IEEE Trans. Ind. Appl.* **1992**, *28*, 1294–1301. [CrossRef]

35. Rodríguez, A.; Sebastian, J.; Lamar, D.G.; Hernando, M.M.; Vazquez, A. An overall study of a dual active bridge for bidirectional DC/DC conversion. In Proceedings of the IEEE Energy Conversion Congress & Exposition (ECCE), Atlanta, GA, USA, 12–16 September 2010; pp. 1129–1135.
36. Calderon, C.; Barrado, A.; Rodriguez, A.; Alou, P.; Lazaro, A.; Fernandez, C.; Zumel, P. General Analysis of Switching Modes in a Dual Active Bridge with Triple Phase Shift Modulation. *Energies* **2018**, *11*, 2419. [CrossRef]
37. Wang, G.; Huang, A.Q.; Wang, F.; Song, X.; Ni, X.; Ryu, S.H.; Grider, D.; Schupbach, M.; Palmour, J. Static and dynamic performance characterization and comparison of 15 kV SiC MOSFET and 15 kV SiC n-IGBTs. In Proceedings of the IEEE 27th International Symposium on Power Semiconductor Devices & IC's (ISPSD), Hong Kong, China, 10–14 May 2015; pp. 229–232.
38. Kumar, A.; Parashar, S.; Sabri, S.; van Brunt, E.; Bhattacharya, S.; Veliadis, V. Ruggedness of 6.5 kV, 30 a SiC MOSFETs in extreme transient conditions. In Proceedings of the IEEE 30th International Symposium on Power Semiconductor Devices and ICs (ISPSD), Chicago, IL, USA, 13–17 May 2018; pp. 423–426.
39. Wang, L.; Zhu, Q.; Yu, W.; Huang, A.Q. A medium voltage bidirectional DC-DC converter combining resonant and dual active bridge converters. In Proceedings of the IEEE Applied Power Electronics Conference and Exposition (APEC), Charlotte, NC, USA, 15–19 March 2015; pp. 1104–1111.
40. Wang, F.; Wang, G.; Huang, A.; Yu, W.; Ni, X. Design and operation of A 3.6kV high performance solid state transformer based on 13kV SiC MOSFET and JBS diode. In Proceedings of the IEEE Energy Conversion Congress and Exposition (ECCE), Pittsburgh, PA, USA, 14–18 September 2014; pp. 4553–4560.
41. Pala, V.; Brunt, E.V.; Cheng, L.; O'Loughlin, M.; Richmond, J.; Burk, A.; Allen, S.T.; Grider, D.; Palmour, J.W.; Scozzie, C.J. 10 kV and 15 kV silicon carbide power MOSFETs for next-generation energy conversion and transmission systems. In Proceedings of the IEEE Energy Conversion Congress and Exposition (ECCE), Pittsburgh, PA, USA, 14–18 September 2014; pp. 449–454.
42. Yun, C.-G.; Cho, Y. Active Hybrid Solid State Transformer Based on Multi-Level Converter Using SiC MOSFET. *Energies* **2019**, *12*, 66. [CrossRef]
43. Mu, M.; Xue, L.; Boroyevich, D.; Hughes, B.; Mattavelli, P. Design of integrated transformer and inductor for high frequency dual active bridge GaN Charger for PHEV. In Proceedings of the IEEE Applied Power Electronics Conference and Exposition (APEC), Charlotte, NC, USA, 15–19 March 2015; pp. 579–585.
44. Hoang, K.D.; Wang, J. Design optimization of high frequency transformer for dual active bridge DC-DC converter. In Proceedings of the XXth International Conference on Electrical Machines (ICEM), Marseille, France, 2–5 September 2012; pp. 2311–2317.
45. Saeed, M.; Rogina, M.R.; Lopez, M.; Rodriguez, A.; Arias, M.; Briz, F. Design and construction of a DAB using SiC MOSFETs with an isolation of 24 kV for PET applications. In Proceedings of the IEEE 19th European Conference on Power Electronics and Applications (EPE'17 ECCE Europe), Warsaw, Poland, 11–14 September 2017; pp. P.1–P.10.
46. Ortiz, G.; Biela, J.; Kolar, J.W. Optimized design of medium frequency transformers with high isolation requirements. In Proceedings of the IEEE IECON—36th Annual Conference on Industrial Electronics Society, Glendale, AZ, USA, 7–10 November 2010; pp. 631–638.
47. Zhao, S.; Li, Q.; Lee, F.C. High frequency transformer design for modular power conversion from medium voltage AC to 400V DC. In Proceedings of the IEEE Applied Power Electronics Conference and Exposition (APEC), Tampa, FL, USA, 26–30 March 2017; pp. 2894–2901.
48. Chen, Q.; Raju, R.; Dong, D.; Agamy, M. High Frequency Transformer Insulation in Medium Voltage SiC enabled Air-cooled Solid-State Transformers. In Proceedings of the IEEE Energy Conversion Congress and Exposition (ECCE), Portland, OR, USA, 23–27 September 2018; pp. 2436–2443.

5

Computer-Aided Design of Digital Compensators for DC/DC Power Converters

Pablo Zumel *, Cristina Fernández, Marlon A. Granda, Antonio Lázaro and Andrés Barrado

Grupo de Sistemas Electrónicos de Potencia, Departamento de Tecnología Electrónica, Escuela Politécnica Superior, Universidad Carlos III de Madrid, 28911 Leganes, Madrid, Spain; cfernand@ing.uc3m.es (C.F.); mgranda@ing.uc3m.es (M.A.G.); alazaro@ing.uc3m.es (A.L.); andres.barrado@uc3m.es (A.B.)

* Correspondence: pzumel@ing.uc3m.es

Abstract: Digital control of high-frequency power converters has been used extensively in recent years, providing flexibility, enhancing integration, and allowing for smart control strategies. The core of standard digital control is the discrete linear compensator, which can be calculated in the frequency domain using well-known methods based on the frequency response requirements (crossover frequency, f_c, and phase margin, *PM*). However, for a given compensator topology, it is not possible to fulfill all combinations of crossover frequency and phase margin, due to the frequency response of the controlled plant and the limitations of the compensator. This paper studies the performance space (f_c, *PM*) that includes the set of achievable crossover frequencies and phase margin requirements for a combination of converter topology, compensator topology, and sensors, taking into account the effects of digital implementation, such as delays and limit cycling. Regarding limit cycling, two different conditions have been considered, which are related to the design of the digital compensator: a limited compensator integral gain, and a minimum gain margin. This approach can be easily implemented by a computer to speed up the calculations. The performance space provides significant insight into the control design, and can be used to compare compensator designs, select the simplest compensator topology to achieve a given requirement, determine the dynamic limitations of a given configuration, and analyze the effects of delays in the performance of the control loop. Moreover, a figure of merit is proposed to compare the dynamic performance of the different designs. The main goal is to provide a tool that identifies the most suitable compensator design in terms of the dynamic performance, the complexity of the implementation, and the computational resources. The proposed procedure to design the compensator has been validated in the laboratory using an actual DC/DC converter and a digital hardware controller. The tests also validate the theoretical performance space and the most suitable compensator design for a given dynamic specification.

Keywords: power converters; digital control; design space; frequency domain

---

1. Introduction

Digital controllers have conquered new areas where analog control was predominant, such as high-frequency DC/DC converters. Features such as special operation modes [1,2], complex control and modulation strategies [3], self-identification [4], autotuning [5–7], communications [8], compensator flexibility, and technology integration [9–11] are the added values provided by digital control.

Digital control of high-frequency converters has been approached using different control techniques. State feedback techniques have been used in [12–14], allowing for the arbitrary placement of the closed loop poles. Full state feedback (inductor current and output voltage) is used in [12] to implement the control law, providing improved performance in a boost converter acting as a power factor corrector. In [13,14], state feedback is applied to multiphase power converters to ensure current-sharing. These techniques are suitable for keeping different quantities in control, typically current and voltage. Therefore, they can be considered as an alternative in multiple loop control structures.

Reference [15] proposes a time-domain design method. The digital compensator is derived from specified rise time and overshoot. In [16], the compensator is designed based on the Internal Model Control design method used in motor control. In [17], complex zero compensation is used to improve transient response. In [18], the root locus method is applied to the calculation of a digital controller for a multiphase buck converter.

Frequency domain techniques are very popular when designing analog linear compensators for DC/DC converters [19]. The compensator is calculated by determining the frequency response of the plant and the dynamic requirements regarding their crossover frequency (f_c) and phase margin (PM). A classical design approach, such as the K-factor method [20] ensures that the magnitude of the open-loop gain is 0 dB at the crossover frequency, and the phase of the open-loop gain provides the required phase margin at that frequency. These design techniques can also be applied to digital linear compensators, considering the discrete nature of the compensator and its implementation, which implies additional factors concerning the analog approach, such as sampling frequency, time delays, and quantization issues. However, the designed controller may not fulfill the expected specifications due to the limitations of the design method itself and the digital implementation—that is, there are (f_c, PM) combinations which are impossible to reach.

In [21], an autotuning system, illustrated with a buck converter, automatically calculated digital PI (proportional-integral), PD (proportional-derivative), and PID (proportional integral derivative) compensators. The (f_c, PM) requirements and the knowledge of the converter dynamics issued from the autotuning system were the basis for the compensator calculation, but this design approach does not consider in detail the attainability of f_c and PM requirements. In [22], the digital compensator was calculated directly in the z domain, taking a given value of f_c and PM as target performance, without a previous analysis of which of their values could be reached with the proposed compensator type. In [23], the compensator calculation was integrated into the design and optimization process. The design space of the power converter was analyzed considering that the controller was a digital PID, including saturation effects. Different design spaces were proposed, with each one for a particular (f_c, PM) specification.

In all the cited cases, the compensator was calculated from the desired values of (f_c, PM) and the rest of the bandwidth was ignored. Thus, due to the frequency response of the system, the open-loop gain may cross 0 dB at frequencies different to the crossover frequency, yielding a design that does not meet the dynamic specifications and which could even be unstable. Because of this limitation, each design must be checked after its calculations. If the design does not meet the requirements, it is difficult to decide whether to modify the dynamic specifications, change the topology of the compensator, or modify the sensor.

Thus, it is very interesting and useful for the designer to know the limited set of possible achievable dynamic specifications for a combination of a power converter, compensator, and sensor. This set determines the performance space of the controller regarding the achievable crossover frequency, f_c, and phase margin, PM.

References [24,25] present studies of the design space of analog compensators. However, there are issues related to digital implementation that have an impact on the calculation of the design space and require further investigation, such as sampling frequency, time delays, and limit cycling [26,27]. The consideration of these effects involves additional boundaries and substantially modifies the aspect of the performance space.

This paper proposes a design tool integrating frequency-based design techniques to: (a) determine all the possible solutions without establishing, a priori, the design target (f_c, PM); (b) assess the influence of different parameters of the compensator on the dynamic performance of the system;

(c) determine the simplest compensator design regarding the dynamic performance for a given dynamic specification, the complexity of the implementation and the computational resources.

This work presents the algorithms to calculate this new performance space and analyze their impact on the controller design. The graphical representation of the performance space in the axis (f_c, PM) can provide the designer with a deep insight on the influence of design parameters through sensitivity analysis. The contribution of this paper regarding prior work is the exhaustive analysis of the digital compensator design (especially in the case of PID); consideration of the limit cycling conditions; and experimental validation.

Section 2 deals with the proposed approach to designing the digital compensator, the description of the limit cycling conditions, and the proposal of a criterion to compare the performance of compensators that provide the same crossover frequency and phase margin. The performance space is introduced in Section 3, defined as a set of feasible and stable designs characterized by their dynamic performance regarding (f_c, PM). Section 4 presents the experimental validation of the proposal. Section 5 summarizes the conclusions of the work.

2. Elements for the Calculation of the Digital Compensator

2.1. Model of the System in the Frequency Domain

There are different approaches to calculating a digital linear compensator to control a power converter. One of them is based on the calculation of a continuous compensator that fulfills the dynamic requirements [28,29]. This continuous compensator is discretized by means of conventional methods, such as bilinear or Tustin transformation, to obtain the discrete compensator. This approach provides robustness to the design and confidence for analog designers. However, in some cases, the resulting controller is more complex than required because it includes additional poles that are not necessary for the digital controller (e.g., high-frequency poles are not required since the controlled quantity is sampled at the switching frequency).

A second approach is based on calculating the exact model of the converter in the discrete domain, obtaining its z transfer function [30,31]. This alternative allows for the direct calculation of the discrete compensator using conventional design techniques for discrete control systems. However, this modeling technique is less extended among the conventional design procedures.

The approach used in this work is based on the frequency response of every block, with its continuous or discrete nature being independent and the frequency range limited to half the switching frequency. Although this is an approximation, it works well, as described in later sections, and provides flexibility to the designer. The elements of the block diagram (Figure 1) are as follows:

- The discrete compensator $C(\omega)$ is the element to be calculated. Its frequency response is calculated by the direct substitution of z by $e^{j\omega\tau_{samp}}$, where τ_{samp} is the sampling period. Despite the sampling period τ_{samp} and switching period τ not necessarily being equal, in this work, they are considered the same for the sake of simplicity ($\tau_{samp} = \tau$);
- Switching power converter $G_{vd}(\omega)$. The more accurate approach to obtaining the frequency response of the power converter (plant) is exact discrete modeling [30,31]. However, good results can also be obtained by using averaged modeling techniques [19,32] or experimental measurements. If a continuous averaged model is used, the small alias approach should be ensured. In the case of exact discrete modeling or experimental results, the effect of a large ripple is inherently considered;
- Time delay τ_{delay}. All time delays are lumped in a single block despite their origin: modulator [32], analog to digital converter (ADC) and computation. This approximation is valid in most cases. More detailed descriptions could consider the time delay block split in several parts, but this approach is numerically equivalent to the consideration of a single block;
- Static gains of modulator (G_{PWM0}) and analog to digital converter (G_{ADC0});
- Sensor transfer function $G_S(\omega)$.

The key transfer function to calculate the compensator is the uncompensated loop transfer function T_U given in Equation (1), where $\omega < \pi/\tau$. The open-loop transfer function T is defined in Equation (2).

$$T_U(\omega) = G_{PWM0}{\cdot}e^{-j\omega\cdot\tau_{delay}}G_{ADC0}G_{vd}(\omega)G_S(\omega) \tag{1}$$

$$T(\omega) = T_U(\omega)C(\omega) \tag{2}$$

2.2. Compensator Calculation

The calculation of the compensator is based on the specified phase margin and crossover frequency in Equations (3) and (4). Therefore, there are two constraints: the first concerning the definition of the cross-over frequency, f_c, since the magnitude of the compensated transfer function has to be equal to 1 at f_c Equation (3); and the second concerning the definition of phase margin, PM Equation (4).

$$|T_U(\omega){\cdot}C(\omega)| = 1 = 0\ dB\ if\ \omega = \omega_c = 2\pi{\cdot}f_c \tag{3}$$

$$arg(T_U(\omega_c){\cdot}C(\omega_c)) = -\pi + PM \tag{4}$$

Without loss of generality, two different compensator types are considered: PI (5) and PID (6).

$$C_{PI}(z) = K\frac{\left(z - e^{-2\pi\cdot f_z\cdot\tau}\right)}{(z-1)} = K\frac{(z - r_z)}{(z-1)} \tag{5}$$

$$C_{PID}(z) = K\frac{\left(z - e^{-2\pi\cdot f_{z1}\cdot\tau}\right)\left(z - e^{-2\pi\cdot f_{z2}\cdot\tau}\right)}{(z-1)z} = K\frac{(z - r_{z1})(z - r_{z2})}{(z-1)z} \tag{6}$$

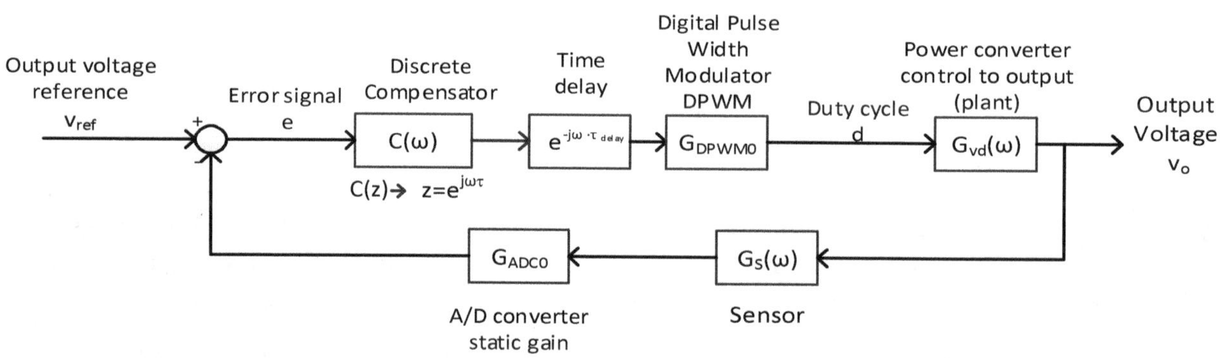

Figure 1. Block diagram of a switched power converter with digital control.

As previously explained, the compensator is calculated from the frequency response of the different blocks of the system. The frequency response of the compensator depends on the topology (PI or PID) and can be calculated by Equations (7) or (8).

$$C_{PI}(\omega) = C_{PI}(z)|_{z=e^{j\omega\tau}} \tag{7}$$

$$C_{PID}(\omega) = C_{PID}(z)|_{z=e^{j\omega\tau}} \tag{8}$$

In the proposed form, the PI regulator has only two parameters (K and f_z) that can be determined using the constraints (3) and (4). In the case of PI and PID, if the value of f_z issued from calculations is negative or complex, the solution must be discarded since it yields zeros outside the unit circle or complex coefficients in the difference equation, respectively.

However, the PID compensator has three parameters (K, f_{z1} and f_{z2}), while there are only two constraints from dynamic requirements (f_c and PM). Therefore, an additional constraint must be established to calculate the compensator. In this work, two possibilities are considered as an additional constraint to determine the coefficients of a PID compensator:

- PID1: the ratio between the frequency of one of the zeros f_{z2} and the crossover frequency f_c is given by the designer, as expressed in Equation (9). This approach is a generalization of the proposed calculation criterion given in [33]. In this case, it is possible to derive an analytical expression of the compensator design parameters K and f_{z1} as a function of the uncompensated transfer function value at f_c, PM, and K_1, as depicted in Equations (14)–(16).

$$\frac{f_{z2}}{f_c} = K_1 \tag{9}$$

- PID2: the ratio between the frequency of both zeros is given by the designer, as expressed in Equation (10). When this ratio is equal to one, both zeros are located at the same frequency, and there are analytical expressions to calculate the compensator parameters Equations (17)–(19). However, for other values of K_2, it is not possible to derive analytical expressions to calculate the parameters directly, because the resulting equation is transcendental. Therefore, numerical methods are required to solve the nonlinear resulting expressions.

$$\frac{f_{z2}}{f_{z1}} = K_2 \tag{10}$$

In the following paragraphs, the equations to calculate each of the proposed compensators are summarized. Thus, the expressions for the calculation of proportional-integral (PI) compensator are Equations (11)–(13). Equation (11) is the transfer function, Equation (12) is the difference equation and Equation (13) is the zero location.

$$C_{PI}(z) = \frac{d(z)}{e(z)} = K\frac{(z - r_z)}{(z-1)} = K\frac{(1 - r_z z^{-1})}{(1 - z^{-1})} \tag{11}$$

$$d[n] = d[n-1] + K{\cdot}e[n] - K{\cdot}r_z{\cdot}e[n-1] \tag{12}$$

$$r_z = \cos(\omega_c\tau) - \frac{\sin(\omega_c\tau)}{\tan\ \ PM - \pi - arg(T_U(j\omega_c)) + \text{atan}\ \frac{\sin(\omega_c\tau)}{\cos(\omega_c\tau)-1}} \tag{13}$$

where:

$$r_z = e^{-2\pi f_z\tau}; K = \frac{1}{\frac{\left|e^{j\omega_c\cdot\tau} - e^{-2\pi f_z\tau}\right|}{\left|e^{j\omega_c\tau}-1\right|}\left|T_U(\omega_c)\right|}$$

Expressions for the calculation of proportional-integral-derivative PID1 compensator are Equations (14)–(16). The transfer function is Equation (14), the difference equation is Equation (15) and Equation (16) provides the location of zero 1 (note that the frequency of the other zero is determined by the value of K_1).

$$C_{PID1}(z) = \frac{d(z)}{e(z)} = K\frac{(z - r_{z1})(z - r_{z2})}{(z-1)z} = K\frac{(1 - r_{z1}z^{-1})(1 - r_{z2}z^{-1})}{(1 - z^{-1})} \tag{14}$$

$$d[n] = d[n-1] + K{\cdot}e[n] - K(r_{z1} + r_{z2}){\cdot}e[n-1] + K(r_{z1}{\cdot}r_{z2}){\cdot}e[n-2] \tag{15}$$

$$r_{z1} = \frac{B{\cdot}D - A + (C - B{\cdot}E){\cdot}K_z}{(B{\cdot}E - C - B{\cdot}K_z)} \tag{16}$$

where:

$$A = \sin(2\omega_c\tau);\ B = \tan\ \ PM - \pi + \arg\ \ e^{j\omega_c\tau} - 1\ \ + \omega_c\tau - \arg(T_U(\omega_c))\ \ ; C = \sin(\omega_c\tau);$$

$$D = \cos(2\omega_c\tau);\ E = \cos(\omega_c\tau);\quad K_z = e^{-\omega_c K_1 \tau}; r_{z1} = e^{-2\pi f_{z1}\tau}; r_{z2} = e^{-2\pi f_{z2}\tau}$$

$$K = \frac{1}{\frac{\left|e^{j\omega_c\cdot\tau} - e^{-2\pi f_{z1}\tau}\right|\cdot\left|e^{j\omega_c\cdot\tau} - e^{-2\pi f_{z2}\tau}\right|}{\left|e^{j\omega_c\tau} - 1\right|\cdot\left|e^{j\omega_c\tau}\right|}\left|T_U(\omega_c)\right|}$$

Finally, expressions for the calculation of PID2 compensator are Equations (17)–(19). The transfer function if Equation (17), the difference equation is Equation (18). Equation (19) is applicable when both zeros are located at the same frequency ($K_2 = 1$).

$$C_{PID2}(z) = \frac{d(z)}{e(z)} = K\frac{(z - r_{z1})(z - r_{z2})}{(z-1)z} = K\frac{1 - r_{z1}z^{-1}\quad 1 - r_{z2}z^{-1}}{(1 - z^{-1})} \tag{17}$$

$$d[n] = d[n-1] + K{\cdot}e[n] - K(r_{z1} + r_{z2}){\cdot}e[n-1] + K(r_{z1}{\cdot}r_{z2}){\cdot}e[n-2] \tag{18}$$

$$r_z = \cos(\omega_c\tau) - \frac{\sin(\omega_c\tau)}{\tan\ \frac{1}{2}\ PM - \pi - arg(T_U(j\omega_c)) + \mathrm{atan}\ \frac{\sin(\omega_c\tau)}{\cos(\omega_c\tau)-1}\ + \omega_c\tau} \tag{19}$$

where:

$$r_z = r_{z1} = r_{z2} = e^{-2\pi f_z\tau};\qquad K = \frac{1}{\frac{\left|e^{j\omega_c\cdot\tau} - e^{-2\pi f_z\tau}\right|^2}{\left|e^{j\omega_c\tau}-1\right|\cdot\left|e^{j\omega_c\tau}\right|}\left|T_U(\omega_c)\right|}$$

In the case of PID2 compensator with two zeros at different frequency ($r_{z1} \neq r_{z2}$), a numerical approach is used to find the frequency of the zeroes.

2.3. Analysis of the Calculated Compensators

The expressions (11)–(19) provide the compensator parameters to meet the requirements of gain and phase only at the crossover frequency. However, as mentioned in the introduction, because of the frequency response of the plant, the behavior of the open-loop gain at other frequencies can render the solution unstable or not feasible, or even result in a different dynamic performance than expected. Thus, it is interesting to analyze these possibilities considering the whole bandwidth of the loop transfer function.

Five combinations of crossover frequency and phase margin have been considered to analyze the different possibilities. The resulting linear compensators yield the open-loop gain transfer functions illustrated in Figure 2. The plant considered in the calculations is a buck converter (input voltage Vin = 12 V, output voltage Vo = 3 V; output inductance $L_O = 1 \times 10^{-6}$ H; output capacitance $C_O = 47 \times 10^{-6}$ F; output power Po = 10 W; switching frequency fsw = 1000×10^3 Hz; total time delay td = (1/fsw) × 0.5). In the plots, the continuous line represents the theoretical frequency response, while the crosses correspond to the frequency response obtained from time-domain simulations using the discrete compensator performing and ACsweep with the commercial simulator PSIM.

Case 1 (target $f_c = 84$ kHz, target $PM = 45^\circ$) is a feasible solution, since the magnitude of the open-loop gain crosses 0 dB only once, the phase crosses $-180°$ only when the magnitude is lower than 0 dB, and the obtained zero frequency is real and positive.

Case 2 (target $f_c = 79$ kHz, target $PM = 15^\circ$) is a conditionally stable solution, because the phase is lower than $-180°$ at frequencies where the magnitude of the open-loop gain is higher than 0 dB. The point -1 is outside the Nyquist contour, but a decrease in the gain of the system could make the system unstable. Thus, this solution is discarded despite the resulting system not being formally unstable, and the crossover frequency and phase margin should not be included inside the performance space.

Case 3 (target $f_c = 20$ kHz, target $PM = 75^\circ$) is unstable because the phase is lower than $-180°$ when the magnitude is higher than 0 dB. Therefore, this solution must be discarded.

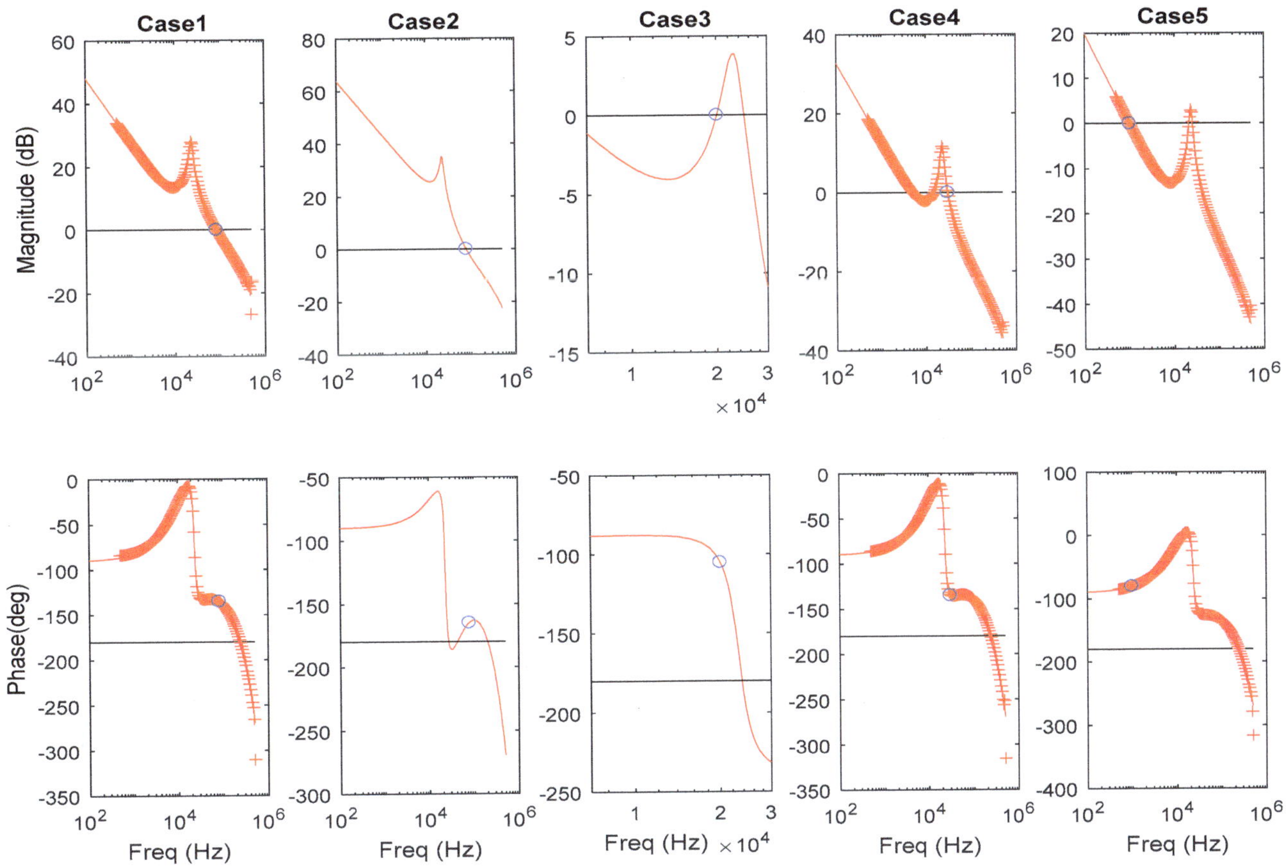

Figure 2. Open-loop transfer function for different combinations of crossover frequency and phase margin. Solid line: theoretical freq. response; crossed line: freq. response from the time-domain simulation with a PSIM simulator.

Case 4 (target $f_c = 20$ kHz, target $PM = 75^{\circ}$) is stable, but the effective crossover frequency is lower than expected due to the two crossings through 0 dB that appear approximately at 5 kHz and 15 kHz. The result is that the settling time of the system does not correspond with the target crossover frequency, and the system is slower than expected. Thus, with this design procedure and digital compensator, it is not possible to achieve the specified combination of crossover frequency and phase margin. This combination should be considered out of the performance space.

Case 5 (target $f_c = 1$ kHz, target $PM = 100^{\circ}$) is stable, but the effective crossover frequency is higher than expected. This means that the dynamic response of the system does not match the requirements. Moreover, the effective phase margin is lower. Therefore, this solution is also discarded.

The main conclusion of this analysis is that there are combinations of dynamic specifications that yield designs which should be considered out of the performance space. Reasons to discard the resulting designs are as follows:

- The magnitude of the resulting open-loop gain crosses 0 dB more than once: the solution is discarded since the effective phase margin is different than expected. Thus, the resulting design does not fulfill the dynamic specifications;
- The phase of the resulting open-loop gain crosses -180° with a magnitude greater than 0 dB: the solution is discarded because it is unstable or conditionally stable.

2.4. Limit Cycling Conditions

One of the characteristics of digitally controlled power converters is the limit cycles that can appear under different conditions. This issue is still a subject of research, especially in the case of transient conditions; but in the literature, different conditions have been established to avoid limit cycling [26]. Some of them are related to the relative resolution between the PWM (pulse width modulator) and the ADC. These are out of the scope of this work, since they do not depend on the compensator calculation, but rather their hardware implementation. However, there are other two conditions related to the design of the linear compensator.

The first condition [26,27] is related to the compensator integral gain K_i, defined in Equation (20). The expressions of the integral gain K_i for PI and PID compensators are given in Equations (21) and (22) respectively.

$$K_i = \lim_{z \to 1}(z-1)C(z) \tag{20}$$

$$K_i = K \cdot (1 - r_z) \tag{21}$$

$$K_i = K(1 - r_{z1})(1 - r_{z2}) \tag{22}$$

This integral gain K_i must be limited in order for a single unit impulse of the error signal to produce a step change in the controlled quantity (e.g., output voltage). This condition is expressed in Equation (23). Theoretically, the parameter a is equal to 1, but in practice a, safety factor is considered, which can typically be $a = 0.5$.

$$0 < T_U(\omega = 0) \cdot K_i < a \tag{23}$$

Once the digital compensator is calculated, it is easy to check if it meets Equation (23) to determine whether the solution should be discarded and considered out of the performance space. However, when analyzing Equations (21) and (22), interesting trends can be identified. In a PI compensator, as K and r_z are directly determined by the constraints imposed by f_c and PM, the only possibility to avoid limit cycling is to change (f_c, PM); that is, to modify the dynamic requirements. On the other hand, in the case of a PID compensator (topologies PID1 and PID2), the additional degree of freedom can be used to avoid limit cycling without changing the f_c and PM requirements, but by changing the parameters of K_1 or K_2 in Equations (9) and (10).

This effect is illustrated in Figure 3. The plots show the value of $T_U(\omega = 0) \cdot K_i$ depending on the parameters K_1 or K_2 and for a given combination of f_c and PM. When K_1 or K_2 are small enough, the plotted product is lower than $a = 1$, so a limit cycling condition is avoided. The conclusion is that by decreasing the frequency of one of the PID zeros, the gain of the system at low frequency is limited enough to avoid the integral gain limit cycling. However, one zero at low frequency can result in a poor low frequency response, as will be shown in the next section. Therefore, a trade-off must be found.

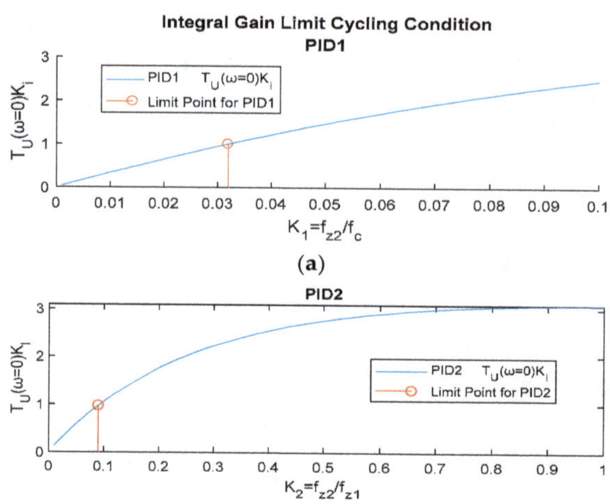

(b)

Figure 3. Analysis of the integral gain limit cycling condition for the PID1 (**a**) and PID2 (**b**) compensators.

The second limit cycling condition refers to the requirement of a minimum gain margin, GM, to avoid limit cycling due to the sampling effect in Equation (24), where the parameter α is the security margin.

$$GM > 4.2\text{ dB} - 20\cdot\log(\alpha) \tag{24}$$

This condition depends essentially on the crossover frequency f_c. When a high f_c is desired, close to the Nyquist frequency, the gain margin is limited, since the slope of T is limited for a PID compensator beyond f_c. Thus, changing the value of the PID parameter has a very slight effect on the GM, since beyond f_c, the value of T is very similar even if there are significant differences for low frequencies, as illustrated in the open-loop gain displayed in Figure 4a.

2.5. Difference among Solutions with the Same Crossover Frequency and Phase Margin

According to the previous section, for a given f_c and PM, there is only one possible design if all the parameters are fixed: the type of compensator, sampling frequency, parameters K_1 or K_2, and so forth. Following the previously described criteria, it is possible to determine which designs are included in the performance space and which are not. However, it would be interesting to provide a method to compare the different solutions and indicate the best one in terms of the complexity of the implementation, the computational resources, and so forth. Thus, a PI compensator is preferred to a PID for the same crossover frequency and phase margin, since the difference equation has fewer coefficients (expressions (11)–(19)). The compensator with a lower sampling period is preferred, since its implementation needs fewer digital resources.

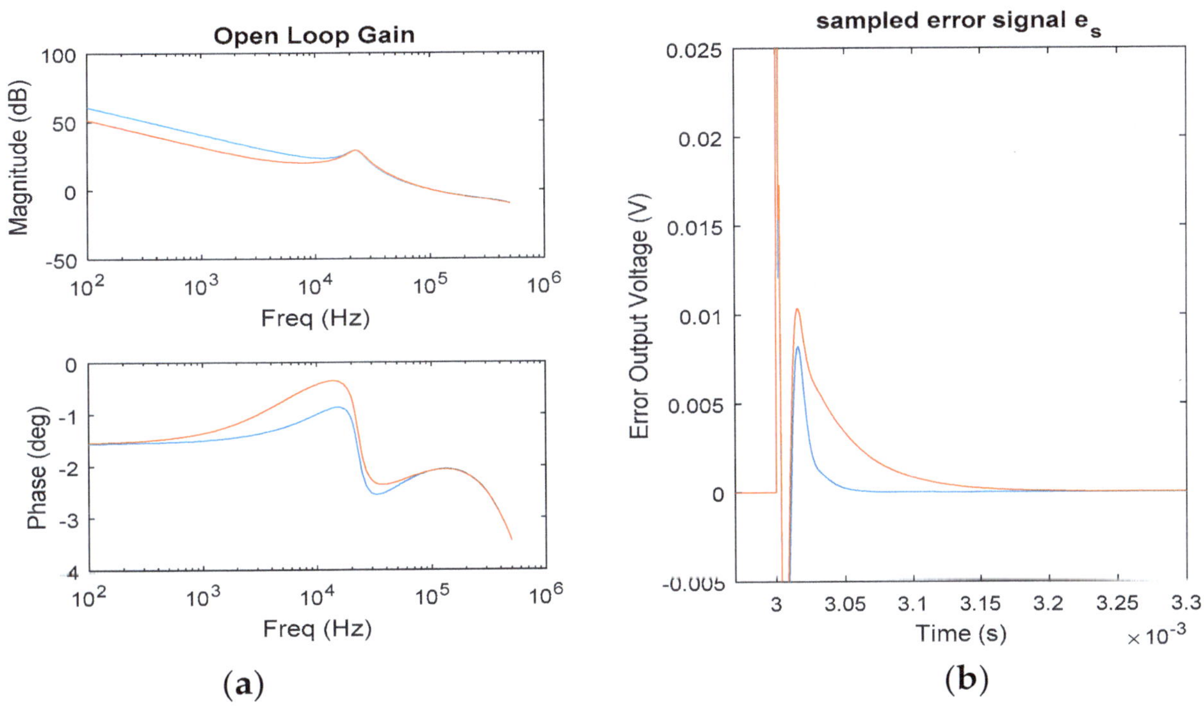

Figure 4. (**a**) Open-loop gain T, corresponding to two different designs of PID2 with the same (f_c, PM) and different parameter values of K_2; (**b**) error signal in the time domain when a step in the reference voltage is applied.

However, for the PID compensator, there are different possible designs, since an additional design constraint is used either in PID1 and PID2. Therefore, a performance index should be established to choose between compensators with the same architecture and different parameters.

Figure 4 illustrates this issue. Two different PID2 compensators with different values of K_2 have been calculated. Both achieve the same f_c and PM, but their frequency response is different in the rest of the bandwidth (Figure 4a). The difference in magnitude is approximately 9 dB in the range of 100–1000 Hz for this illustration example. Thus, the transient response for a reference step is also different. When evaluating the error output voltage signal in the time domain—that is, the difference between the output voltage reference (reference signal) and the actual output voltage (measured controlled signal) in Figure 4b, the design with a lower magnitude in the range of 100–1000 Hz exhibits a longer settling time.

To compare different designs with the same f_c and PM as those appearing in Figure 4, the performance index L is defined in Equation (25). This index is the norm of the difference between the closed loop transfer function of the system $CL(f_k)$ compared with the ideal one, which is equal to 1 at any frequency; f_k is the k-th element of the frequency vector, considering that the proposed design approach is numerical. $CL(f_k)$ is obtained from the open-loop transfer function $T(f_k)$. A weighting factor $\frac{1}{f_k}$ is included to give more importance to the low frequency response. The factor $\frac{f_k - f_{k-1}}{f_{max} - f_{min}}$ is considered to take the non-linear spacing in the frequency vector into account.

$$L = \sqrt{\sum_{k=1}^{N} \left| (CL(f_k) - 1) \cdot \frac{1}{f_k} \right|^2 \cdot \frac{f_k - f_{k-1}}{f_{max} - f_{min}}} = \sqrt{\sum_{k=1}^{N} \left| \left(\frac{T(f_k)}{1 + T(f_k)} - 1 \right) \cdot \frac{1}{f_k} \right|^2 \cdot \frac{f_k - f_{k-1}}{f_{max} - f_{min}}} \quad (25)$$

The ideal value of this index is $L = 0$, which means that the closed-loop transfer function is equal to 1. Comparing two different solutions—that is, two different compensators, the best solution corresponds to the minimum value of index L. It is difficult to find an exact relationship between the proposed index L and the conventional time domain indexes. However, in all tested cases, a higher value of L means a higher error in the time domain, as illustrated in Figure 5, where six different PID designs for the same (f_c, PM) are compared. The parameters changing from one design to a different one are K_1 and K_2, as defined in Equations (9) and (10) respectively—that is, the separation between zeros of the compensator. Designs A have higher values of K_1 and K_2 than designs B, and both have higher values of K_1 and K_2 than designs C. In Figure 5a, the frequency response for the different designs is shown. In this particular case, designs PID2B and PID1B are very similar (for example, PID2B is 0.08 dB greater than PID1B at 400 Hz). In Figure 5b, the RMS (root mean square) value of the error signal in the time domain is plotted for the six designs, considering an output voltage reference step and normalizing the values to the lowest RMS value. In Figure 5c, the value of the proposed L index normalizing the value to the lower L index value is shown. Although there is no linear relationship between the time-domain error RMS value and L index value, the lower the L index value, the lower the error RMS value.

Time-response performance indexes are an extended tool to compare compensator performance [34]. However, since the presented design approach is based on the frequency domain, it is desirable to establish a performance index based on the frequency response to quickly compare different compensator performances. Note that if the design method requires the comparison of many solutions, saving computational resource is important.

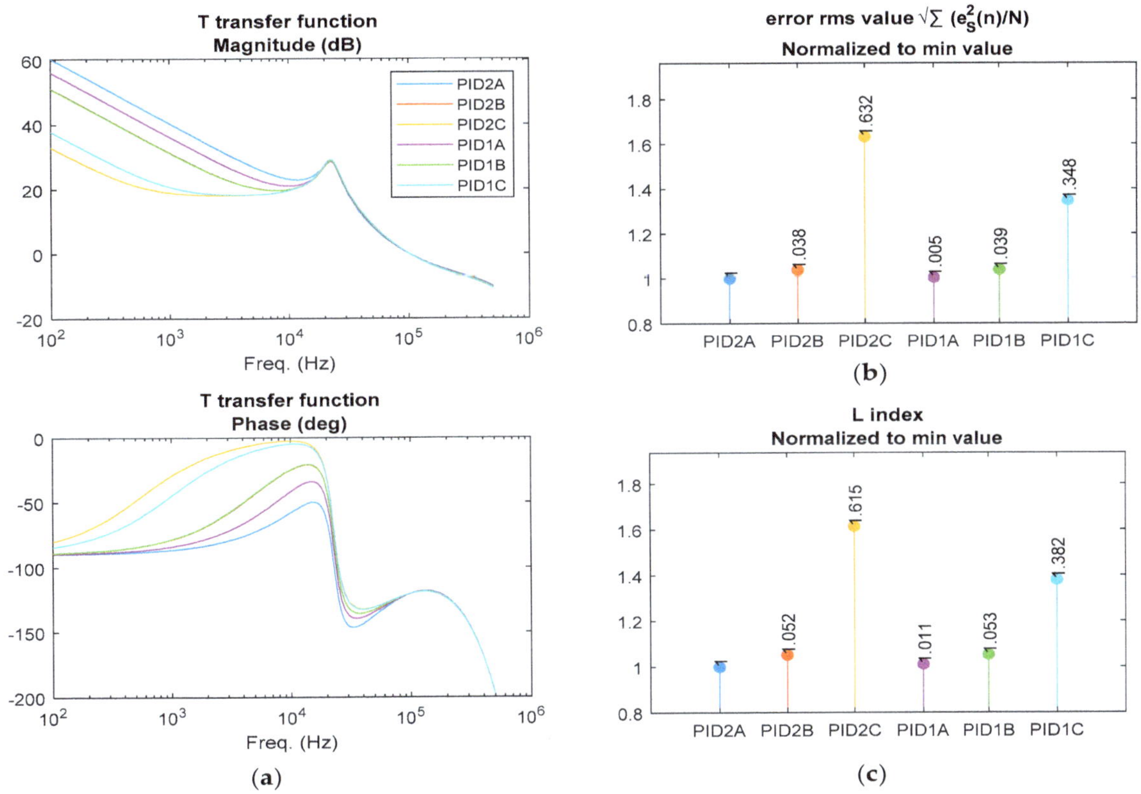

Figure 5. Comparison of solutions with the same (f_c, PM) pair corresponding to different compensator types and parameters. (**a**) Open-loop transfer function; (**b**) RMS value of the error signal obtained from the simulation; (**c**) L index obtained from the open-loop transfer function.

3. Determination of the Performance Space (f_c, PM)

The previous section introduced the frequency-based model of the system, the calculation of different compensator topologies, the criteria to determine whether the designs fulfill the dynamic requirements, and an index to compare compensators that provide the same dynamic specifications (f_c, PM).

However, significant advantages can be obtained if the proposed calculation procedure is automated and applied to find a performance space in such a way that all the possible combinations of (f_c, PM) that can be fulfilled are found. Thus, the performance space is graphically represented in a plot with frequency units in the horizontal axis and phase units in the vertical axis. In [25], this plot is called a solution map. In this paper, the design space is the set of parameters of the compensator that have to be calculated to obtain given results regarding (f_c, PM)—that is, to obtain a given performance. Those parameters of the compensator are, finally, the coefficients of the difference Equations (12), (15) and (18), and the performance is the pair (f_c, PM). In the following paragraphs, an example illustrates the generation of the performance space (f_c, PM) and the influence of different factors, such as the type of compensator and limit cycling conditions. The specifications of the power converter used as our example are as follows: buck topology, input voltage Vin = 12 V, output voltage Vo = 3; output inductance $L_O = 1 \times 10^{-6}$ H; output capacitance $C_O = 47 \times 10^{-6}$ F; capacitor equivalent series resistance $R_{ESRC} = 20 \times 10^{-3}$ Ω; output power Po = 10 W; switching frequency fsw = 1000×10^3 Hz; total time delay td = (1/fsw) × 0.5.

In Figure 6a, the uncompensated open-loop transfer function T_u defined in Equation (1) is shown, including time delays, static gains, and the converter-transfer function (Figure 1). In Figure 6b, the described performance space for the example buck converter and a PID1 compensator is plotted. Limit cycling conditions are not involved in these first calculations. Every point represents a combination of f_c and PM, and the valid combinations are grouped into different areas, considering the different possibilities discussed in Section 2.3:

- The white area corresponds to solutions with non-real or negative frequencies;
- The pink area is the set of solutions which are unstable or conditionally stable. This area is particularly large at high frequencies;
- The blue area is the set of valid solutions that corresponds to feasible stable designs with the same f_c and PM as plotted in the performance space.

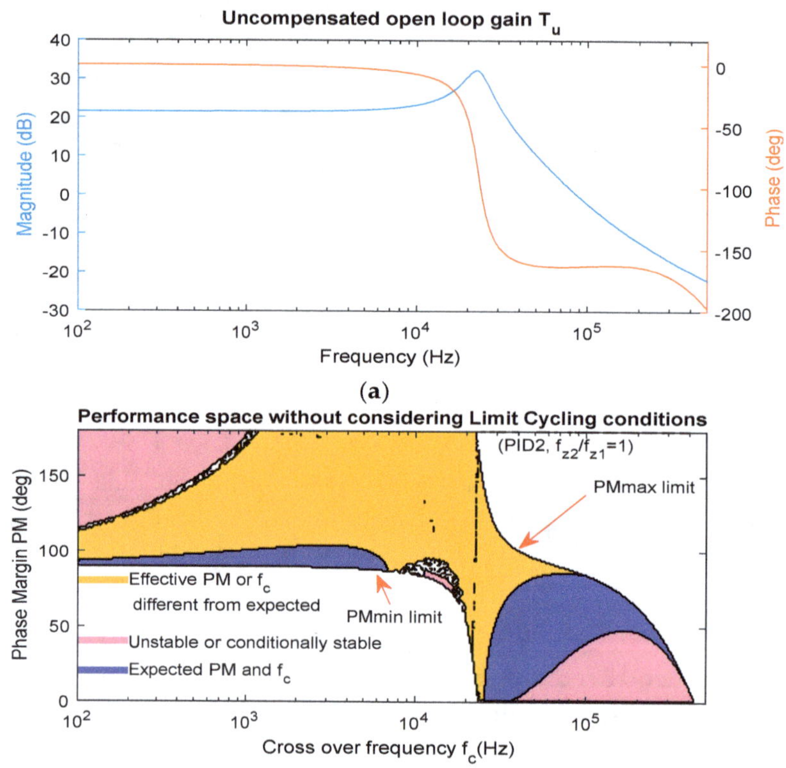

(a)

(b)

Figure 6. (**a**) Bode plot of the uncompensated transfer function, Tu; (**b**) performance space for a PID1 compensator, without considering limit cycling conditions.

The chosen plant, in this case, exhibits a relatively high Q—that is, the resonance peak is around 10 dB above the low frequency magnitude of the G_{vd} transfer function. This fact is relevant to producing pairs (f_c, PM) corresponding to cases 2, 3, 4, and 5 in Figure 2.

The algorithm to calculate the performance space is described in the flowchart of Figure 7. It has been elaborated using the theoretical basis provided in Section 2. The maximum PM that can be achieved (PMmax limit in Figure 6b) can be theoretically calculated as the phase of the uncompensated open-loop transfer function T_U, plus the maximum phase boost that the compensator can provide. On the other hand, there is a limit at low frequencies for the PM (PMmin limit in Figure 6b), determined by the phase of the uncompensated open-loop transfer function T_U and the minimum phase provided by the compensator (-90 degree in any case). These two boundaries are relevant to reduce the calculations and shorten the calculation time of the performance space [35]. Once the initial limits of PM have been established, a double sweep is carried out. For each frequency, the value of PM is initialized and the validity conditions are checked, including those described in Section 2.2, 0 and 0. If they are met, the value of PM is increased, and the validity conditions are checked again.

Once a non-valid solution is achieved for a given frequency value, the PM loop is interrupted, and the algorithm goes to the next frequency value.

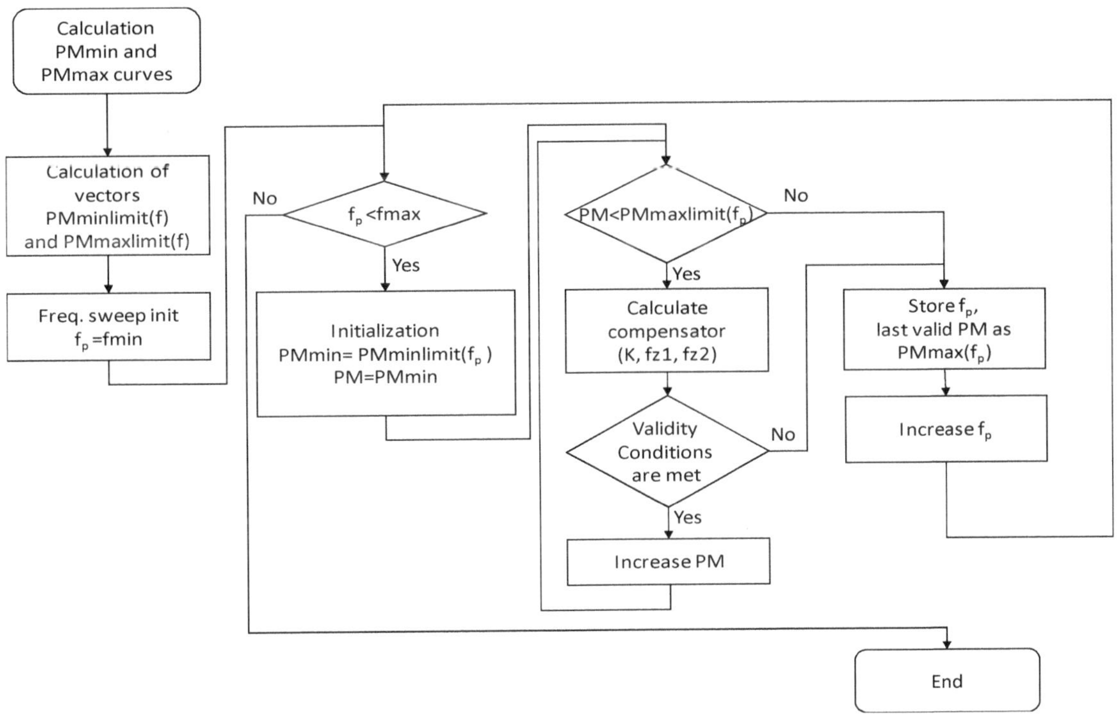

Figure 7. Flowchart of the procedure to calculate the limits of the performance space (f_c, PM).

The flowchart in Figure 7 describes the procedure to calculate the limits of the performance space. Moreover, it can be also used to plot the complete performance space (f_c, PM) if PMminlimit(f) and PMmaxlimit(f) are set to the minimum and maximum PM values, respectively, and the validity conditions do not interrupt the loop but are used to classify the solutions.

The most valuable contribution of this algorithm is the ability to analyze all possible designs very quickly and compare different compensator types or the influence of design parameters. The following discussion illustrates this added value.

Limit cycling conditions are analyzed in Figure 8 in the case of PID compensators. Different types of compensators and design parameters (ratio of zero frequency, K_1, and K_2) are considered by running the algorithm of Figure 7 once per compensator type and design parameter K_1 or K_2. Note that limit cycling conditions, as described in Section 2.4, can be easily included in the calculations.

The area limited by the red line in Figure 8 corresponds to the pairs (f_c, PM) that do not meet the integral gain limit cycling condition. This area is smaller, as the separation between the zero frequencies increases (ratio f_{z2}/f_{z1} for PID2 and ratio f_{z2}/f_c for PID1 decrease). The area limited by the green line in the plots of Figure 8 is the set of solutions that do not meet the limit cycling condition referring to the gain margin. This condition is more related to the frequency itself than to the type of compensator or the value of its design parameter. No significant gains are obtained when the frequencies of the zeros are very different.

Merged results for the PID1 and PID2 compensators are shown in Figure 9. Areas limited by red or green lines in Figure 8 are now discarded designs due to the existence of limit cycles. The area limited by the blue line is the performance space free of limit cycling. In the case of PID2, there is a remarkable difference when the frequency of the zeros changes: the lower the ratio f_{z2}/f_{z1}, the wider the allowable performance space. The integral gain limit cycling condition has, in this case, the strongest influence on the allowable designs. PID1 exhibits similar behavior, although the differences are slightly less significant. Both PID1 and PID2 provide similar results when the ratios f_{z2}/f_{z1} or

f_{z2}/f_c are very small (0.01 in this case). This fact could be used as a criterion to minimize the possibility of a limit cycle, using a high separation between the frequencies of the zeros (low values of f_{z2}/f_{z1}, and f_{z2}/f_c). However, the response of the system at low frequencies can be degraded, as depicted in Section 2.5 and as it is confirmed in the experimental measurements. A trade-off should be achieved between the reduction of the area with limit cycling due to integral gain and the degradation of the low-frequency response.

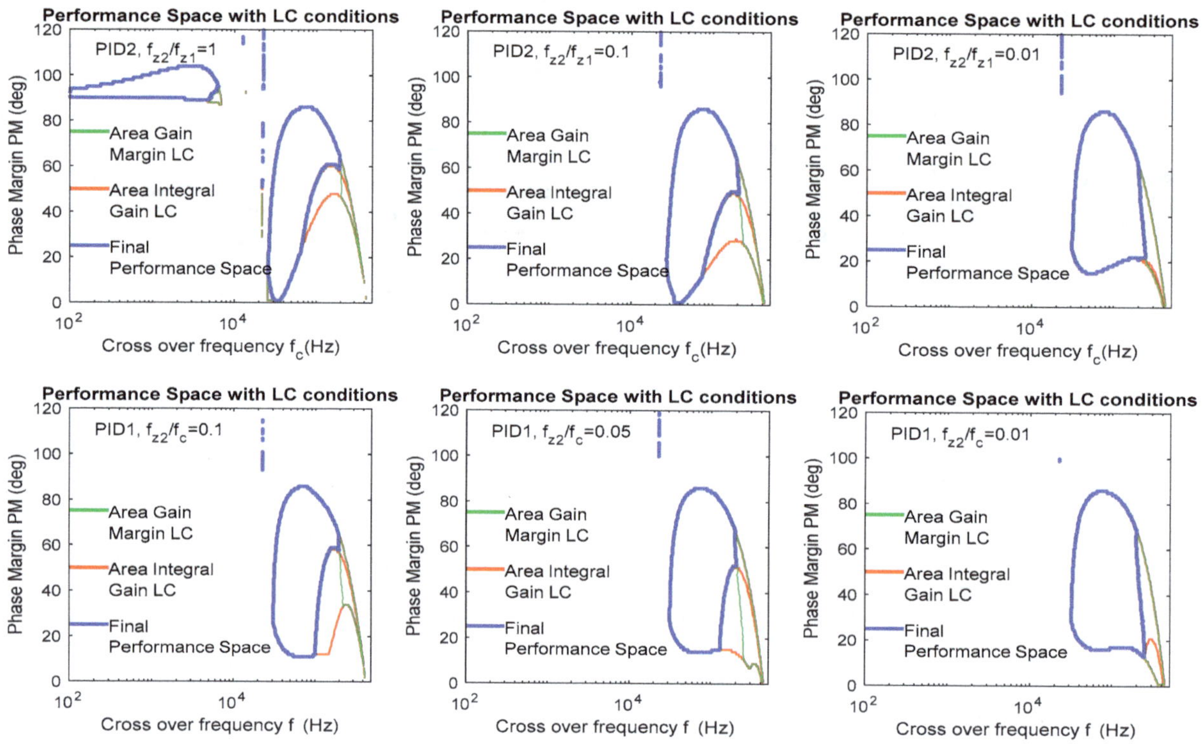

Figure 8. Performance space for PID compensators, remarking areas discarded due to the limit cycling (LC) conditions.

Figure 10 illustrates the limit cycling. Time-domain simulations with a PSIM simulator have been carried out. The ADC and the DPWM (digital PWM) are quantized, but the DPWM has a much higher resolution than the ADC. Therefore, limit cycling concerning the relative resolution of ADC and DPWM [27] is avoided. Small steps in the output voltage reference are introduced to force transient intervals. After those periods, the system enters into a steady state. If there are still oscillations in the output voltage, there is a limit cycle. As predicted by the performance space of Figure 9, two of the six analyzed designs (PID2 $f_{z2}/f_{z1} = 1$ and PID1 $f_{z2}/f_c = 0.1$) are outside the performance space in the area corresponding to limit cycling conditions. The other designs do not suffer from limit cycling, agreeing with Figure 9. Limit cycling failures occur in the case of the lower zero-frequency separation, confirming the trends identified in the previous paragraphs.

A final performance space has been calculated and plotted in Figure 11. This plot summarizes the best possible design for every dynamic requirement $(f_c,\ PM)$. Note that all considered compensator topologies cover some area in the performance space.

The first criterion to select the best solution is simplicity: PI is preferred to PID because of the reduced number of coefficients, Equations (12), (15) and (18). The second criterion is the value of index L, as defined in Section 2.5—that is, in the case of PID compensators achieving a feasible design for a given specified $(f_c,\ PM)$, the preferred solution is the one with the lower index L.

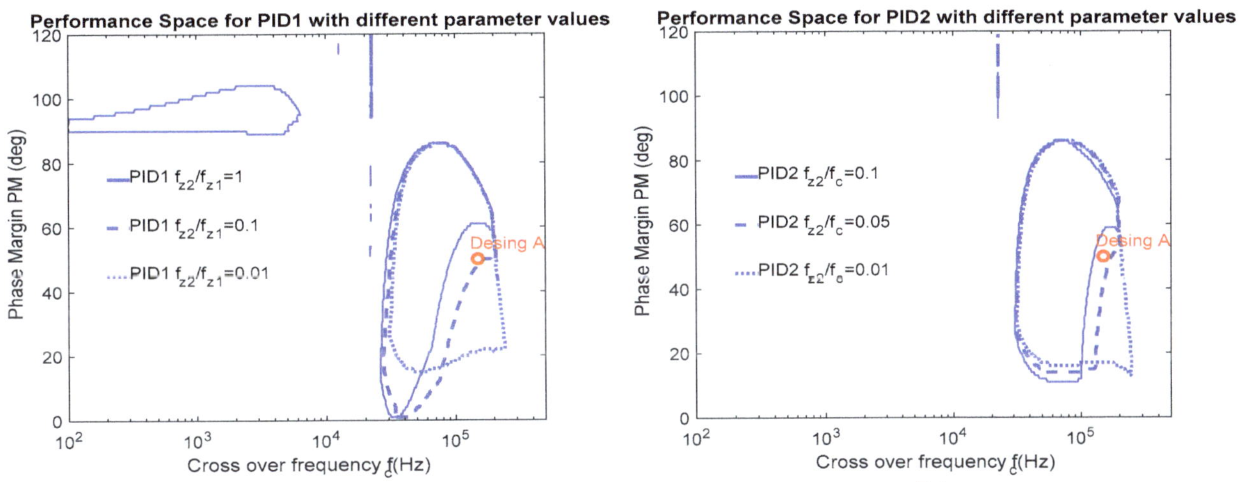

(a) (b)

Figure 9. Performance space for the two different PID compensators, considering different values of the design parameter for each of them: (**a**) PID1; (**b**) PID2.

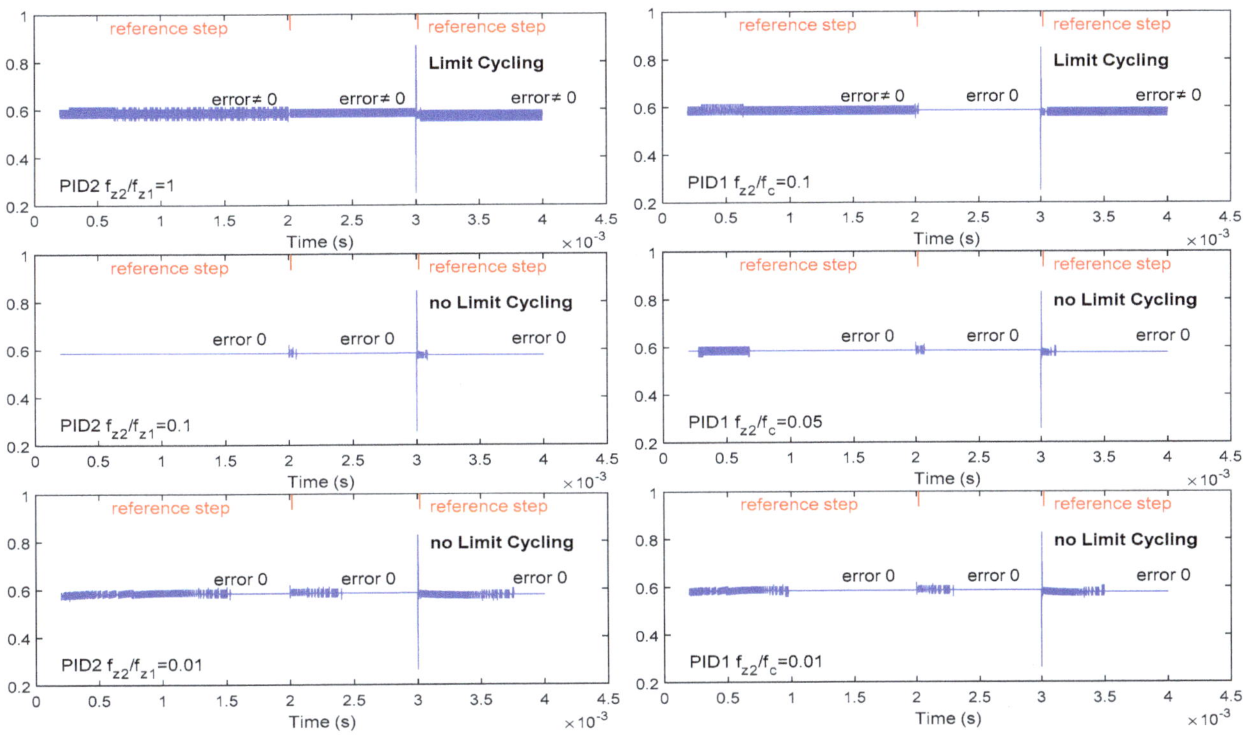

Figure 10. Time response to reference steps for different compensators corresponding to design A (Figure 8). A limit cycle exists for PID1 $f_{z2}/f_{z1} = 1$ and PID2 $f_{z2}/f_c = 0.1$, as predicted by the performance space.

Analyzing this example, the PI is obviously the best solution for a low-frequency requirement. Even if the PID can provide low values of f_c, the PI has a simpler implementation that implies less calculation in the difference equation. Designs using the PI compensator beyond the resonance frequency of the plant are limited to low values of PM, which in practice have no significant interest.

The preferred option for values of f_c beyond the resonance frequency is the PID2 with two zeros at the same frequency. The limitation due to the integral gain limit cycling condition can be overcome using either PID2 or PID1, with higher separation between zero frequencies.

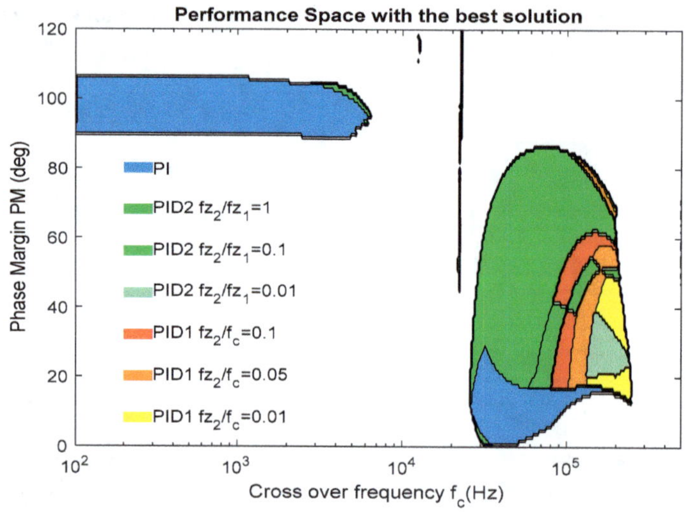

Figure 11. Performance space considering seven different compensators and selecting the best possible solution among them for each point of the performance space (f_c, PM).

All results presented in this section have been obtained by implementing in Matlab the algorithm to generate the performance space (Figure 7). Apart from the comparison of different compensators, the same procedure can be used to perform a sensitivity analysis of other parameters, such as time delays and sampling frequency [35].

4. Experimental Validation

An actual prototype has been built and tested in the laboratory to validate the proposed design procedure, the models, and the assumptions considered in this work. The digital controller has been implemented in a Zybo board, including a Xilinx Zynq-7010 device. An external ADC was used (ADS7476A), limiting the resolution to 10 bits. The specification of the power converter is as follows: input voltage Vin = 8 V, output voltage Vo = 4 V; output inductance $L_O = 76 \times 10^{-6}$ H; output capacitance $C_O = 100 \times 10^{-6}$ F; load resistance Ro = 10 Ω; switching frequency fsw = 100×10^3 Hz. The transfer function of the power converter G_{vd} (Figure 1) has not been calculated, but was measured with a frequency response analyzer. Therefore, all parasitic components (inductor and capacitor series resistances, MOSFET conduction resistance, etc.) are considered in the compensator calculation. The uncompensated loop gain T_u is shown in Figure 12, as well as the performance space, without considering the limit cycling conditions. There are some differences with the performance space presented in Figure 6b due to the differences in the uncompensated open-loop transfer function:

- The total time delay produces a significant phase loss beyond the resonance frequency of the uncompensated open loon gain, limiting the feasible solutions at high frequency (PMmax limit). This phase loss is compensated partially by the effect of the equivalent series resistance of the output capacitor;
- The Q factor is lower, and so in the performance space, there are possible solutions at crossover frequencies close to the resonance frequency.

The combination of the performance spaces for the different type of compensators and different design parameters is shown in Figure 13. In this case, PI ad PID2 with $f_{z2}/f_{z1} = 1$ are the compensator topologies that cover the major part of the area of possible designs. The PID1 topology provides possible designs at high f_c and low PM, where the PID2 with $f_{z2}/f_{z1} = 1$ does not meet the limit cycling condition regarding the integral gain.

The resulting open-loop gain has been measured in the actual prototype for three dynamic specifications, named as design point 1, 2, and 3, detailed in Figure 13. For each design point, both PID1 and PID2 have been designed with $f_{z2}/f_c = 0.1$ and $f_{z2}/f_{z1} = 1$, respectively. Measurements and theoretical predictions are plotted in Figure 14, which match very well, especially near the crossover frequency. Another interesting comparison between PID1 and PID2 is the magnitude of the open-loop gain for the same design point. Bellow the cross-frequency in each case, PID2 reaches a higher magnitude than PID1, which means a lower L index. That is why, in the performance space plotted in Figure 13, PID2 is a better solution than PID1 for the three illustrated design points. This result is also illustrated in the time-domain waveforms appearing in Figure 15. The output voltage has been measured when a voltage reference step is applied. In the left plot, the six different possibilities for design-point 3 are shown. Although the same crossover frequency and phase margin are selected for the compared waveforms, the system exhibits a different time-response due to the zero location, as shown in previous sections. PID2 with $f_{z2}/f_{z1}= 1$ is the fastest response, matching with the choice to plot the performance space in Figure 13. In the right side of Figure 15, the time-response for design-point 1 for a single compensator is shown to illustrate that, despite the high cross-over frequency, no limit cycles are produced after the transitions, as predicted by the theory.

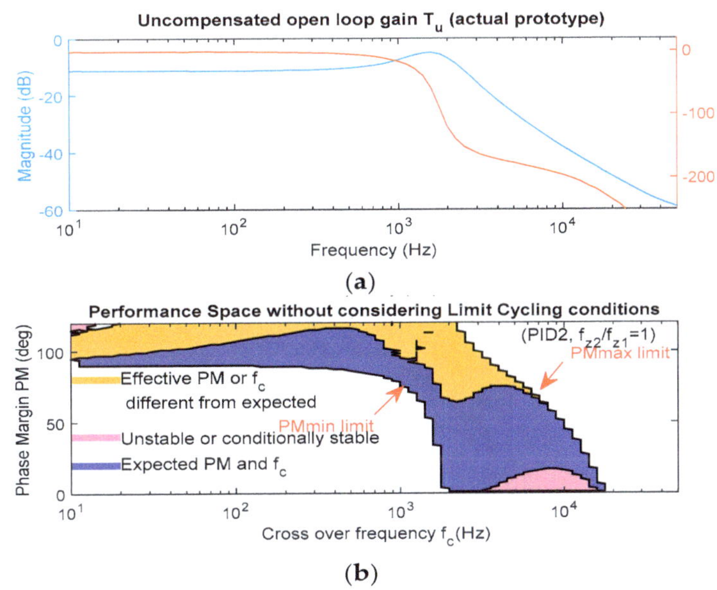

Figure 12. (**a**) The actual prototype's uncompensated open loop gain; (**b**) performance space without considering limit cycling conditions.

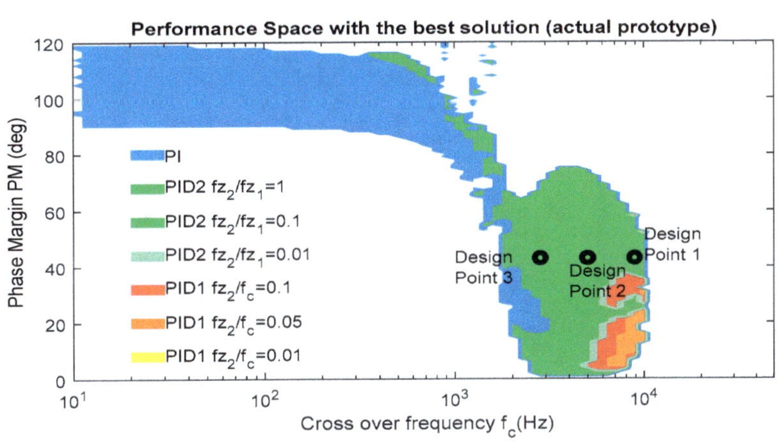

Figure 13. Performance space for the actual prototype with the best solution for each point.

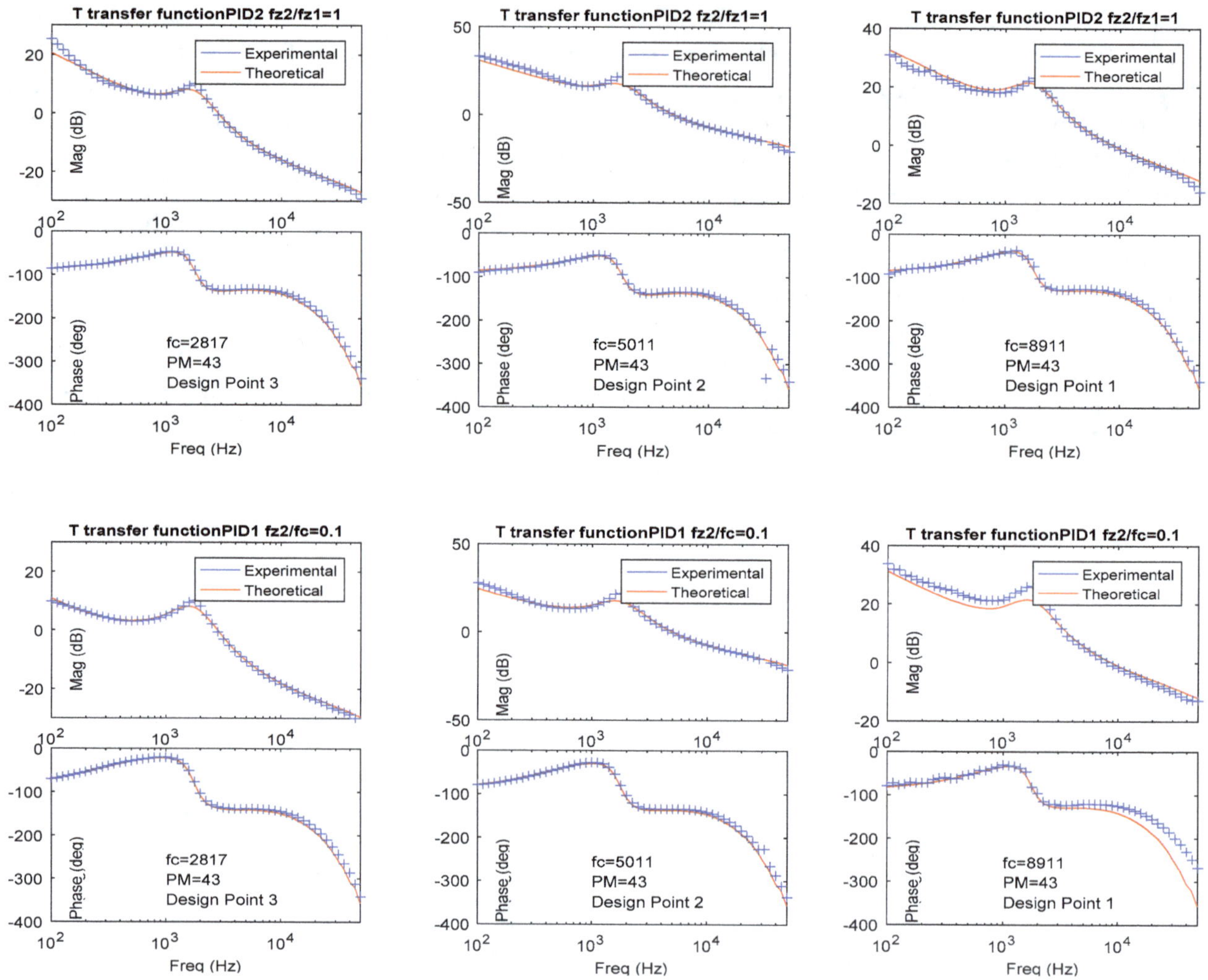

Figure 14. Comparison between the experimental measurements and the theoretical predictions for different PID compensators, considering the three design points in Figure 13.

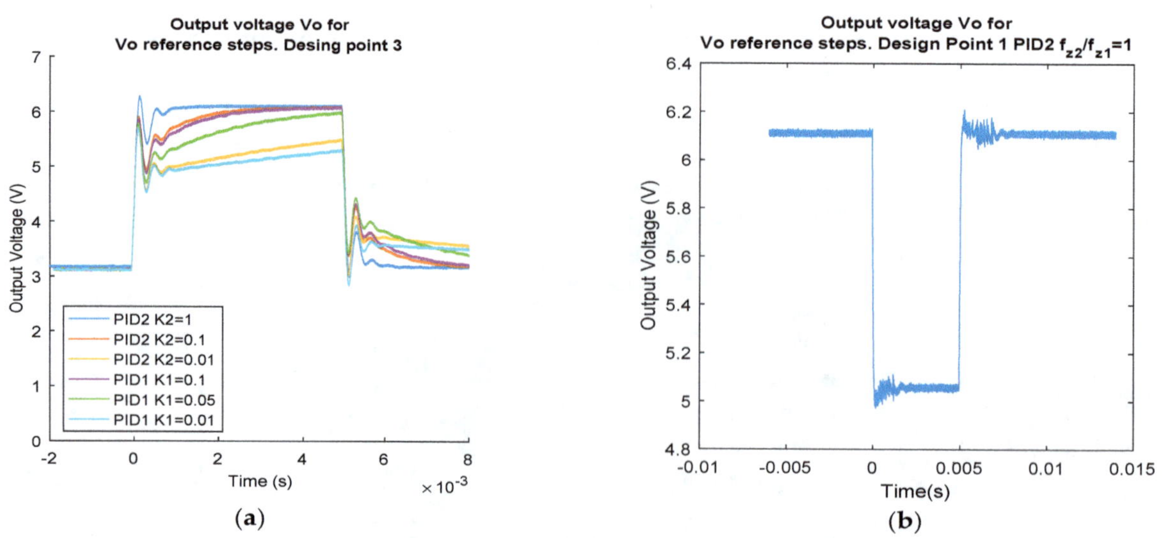

Figure 15. Experimental time-domain waveforms when there is a step in the voltage reference. (**a**) Step response for different designs with f_c = 2817 Hz, PM = 43° (Design point 3); (**b**) step response for design-point 1, using 9 bits in the ADC (note that there is no limit cycling).

5. Conclusions

A frequency-based design approach for the calculation of a digital compensator for DC/DC power-switching converters has been presented. It is based on the analytical calculation of the coefficients of the compensator from the dynamic specifications phase margin, PM, and cross-over frequency, f_c, considering the sampling frequency, delay effects, and limit cycling conditions. The approach is focused on the identification of the dynamic specifications that in fact can be achieved, and its graphical representation on a (f_c, PM) axis. Designs that are not feasible, do not fulfill the requirements, are unstable, or do not meet the limit cycling conditions, are excluded from the performance space.

The design procedure has been automatized. A simple algorithm can generate all feasible designs and plot the performance space. It allows for the identification of the dynamic limitations of a given power converter with a given compensator topology. The most important benefit provided by this approach is the ability to perform a quick and straightforward comparison of compensator topologies or sensitivity analysis of a specific parameter, ensuring the feasibility of the proposed designs.

Three compensator topologies have been considered in the calculations: PI and PID, with two different criteria to establish the frequency of the two zeros (PID1 and PID2). The design procedure has been applied in a particular example, considering seven different compensator types and choosing the more suitable design for each point (f_c, PM) using a proposed figure of merit calculated from the frequency response.

The described analysis based on the performance space also allows for determination of the simplest compensator design in terms of the dynamic performance, the complexity of the implementation, and the computational resources. In the analyzed example, PI is the best option for low-frequency dynamic requirements. In general, the option covering a larger area in the performance space is PID with two zeros at the same frequency. However, this performance space can be limited by the integral limit cycling condition. Other PID designs with separated zeros can overcome this limitation, but the time response should be assessed. The experimental results of a lab prototype agree with the theoretical predictions.

Author Contributions: Conceptualization, P.Z., C.F., A.L. and A.B.; Data curation, M.A.G.; Funding acquisition, C.F. and A.B.; Methodology, P.Z. and C.F.; Software, M.A.G.; Validation, M.A.G.; Writing—original draft, P.Z. and C.F.; Writing—review & editing, M.A.G., A.L. and A.B.

References

1. Sathishkumar, P.; Krishna, T.N.; Khan, M.A.; Zeb, K.; Kim, H.J. Digital Soft Start Implementation for Minimizing Start Up Transients in High Power DAB-IBDC Converter. *Energies* **2018**, *11*, 956. [CrossRef]
2. Costabeber, A.; Mattavelli, P.; Saggini, S. Digital Time-Optimal Phase Shedding in Multiphase Buck Converters. *IEEE Trans. Power Electron.* **2010**, *25*, 2242–2247. [CrossRef]
3. Rodríguez, J.; Lamar, G.D.; Aller, G.D.; Miaja, F.P.; Sebastián, J. Efficient Visible Light Communication Transmitters Based on Switching-Mode dc-dc Converters. *Sensors* **2018**, *18*, 1127. [CrossRef] [PubMed]
4. Shirazi, M.; Morroni, J.; Dolgov, A.; Zane, R.; Maksimovic, D. Integration of Frequency Response Measurement Capabilities in Digital Controllers for DC/DC Converters. *IEEE Trans. Power Electron.* **2008**, *23*, 2524–2535. [CrossRef]
5. Shirazi, M.; Zane, R.; Maksimovic, D. An Autotuning Digital Controller for DC/DC Power Converters Based on Online Frequency-Response Measurement. *IEEE Trans. Power Electron.* **2009**, *24*, 2578–2588. [CrossRef]
6. Stefanutti, W.; Mattavelli, P.; Saggini, S.; Ghioni, M. Autotuning of Digitally Controlled DC–DC Converters Based on Relay Feedback. *IEEE Trans. Power Electron.* **2007**, *22*, 199–207. [CrossRef]
7. Costabeber, A.; Mattavelli, P.; Saggini, S.; Bianco, A. Digital Autotuning of DC—DC Converters Based on a Model Reference Impulse Response. *IEEE Trans. Power Electron.* **2011**, *26*, 2915–2924. [CrossRef]
8. Stefanutti, W.; Saggini, S.; Mattavelli, P.; Ghioni, M. Power Line Communication in Digitally Controlled DC-DC Converters Using Switching Frequency Modulation. *IEEE Trans. Ind. Electron.* **2008**, *55*, 1509–1518. [CrossRef]

9. Park, Y.J.; Khan, Z.H.; Oh, S.J.; Jang, B.G.; Ahmad, N.; Khan, D.; Abbasizadeh, H.; Shah, S.A.; Pu, Y.G.; Hwang, K.C.; et al. Single Inductor-Multiple Output DPWM DC-DC Boost Converter with a High Efficiency and Small Area. *Energies* **2018**, *11*, 725. [CrossRef]
10. Xie, K.; Hu, G.; Yi, H.; Lyu, Z.; Xiang, Y. A Novel Digital Control Method of a Single-Phase Grid-Connected Inverter Based on a Virtual Closed-Loop Circuit and Complex Vector Representation. *Energies* **2017**, *10*, 2068. [CrossRef]
11. Thi Kim Nga, T.; Park, S.M.; Park, Y.J.; Park, S.H.; Kim, S.; Van Cong Thuong, T.; Lee, M.; Hwang, K.; Yang, Y.; Lee, K.Y. A Wide Input Range Buck-Boost DC–DC Converter Using Hysteresis Triple-Mode Control Technique with Peak Efficiency of 94.8% for RF Energy Harvesting Applications. *Energies* **2018**, *11*, 1618. [CrossRef]
12. Jackson, D.K.; Leeb, S.B. A digitally controlled amplifier with ripple cancellation. *IEEE Trans. Power Electron.* **2003**, *18*, 486–494. [CrossRef]
13. Bae, H.-S.; Yang, J.-H.; Lee, J.-H.; Cho, B.-H. Digital state feedback current control using the pole placement technique. *J. POWER Electron.* **2007**, *7*, 213–221.
14. Bae, H.S.; Yang, J.H.; Lee, J.H.; Cho, B.H. Digital state feedback control and feed-forward compensation for a parallel module DC-DC converter using the pole placement technique. In Proceedings of the IEEE Applied Power Electronics Conference and Exposition—APEC, Austin, TX, USA, 24–28 February 2008; pp. 1722–1727.
15. Peretz, M.M.; Ben-Yaakov, S. Time-Domain Design of Digital Compensators for PWM DC-DC Converters. *IEEE Trans. Power Electron.* **2012**, *27*, 284–293. [CrossRef]
16. Miao, B.; Zane, R.; Maksimović, D. Automated digital controller design for switching converters. In Proceedings of the PESC Rec. IEEE Annual Power Electronics Specialist Conference, Recife, Brazil, 12–16 June 2005; pp. 2729–2735.
17. Chui, M.Y.; Ki, W.-H.; Tsui, C.-Y. A programmable integrated digital controller for switching converters with dual-band switching and complex pole-zero compensation. *IEEE J. Solid-State Circuits* **2005**, *40*, 772–780. [CrossRef]
18. Garcia, O.; de Castro, A.; Soto, A.; Oliver, J.A.; Cobos, J.A.; Cezon, J. Digital control for power supply of a transmitter with variable reference. In Proceedings of the Twenty-First Annual IEEE Applied Power Electronics Conference and Exposition, Dallas, TX, USA, 19–23 March 2006; p. 6.
19. Erickson, R.W.; Maksimovic, D. *Fundamental of Power Electronics*, 2nd ed.; Springer: Berlin, Germany, 2001.
20. Venable, H. The K factor: A new mathematical tool for stability analysis and synthesis. In Proceedings of the Powercon 10, San Diego, CA, USA, 22–24 March 1983; pp. 1–10.
21. Corradini, L.; Mattavelli, P.; Stefanutti, W.; Saggini, S. Simplified Model Reference-Based Autotuningfor Digitally Controlled SMPS. *IEEE Trans. Power Electron.* **2008**, *23*, 1956–1963. [CrossRef]
22. Yang, C.; Liu, C.; Tsai, C. Direct digital compensator design for switching converters. In Proceedings of the 2010 International Symposium on Next Generation Electronics, Kaohsiung, Taiwan, 18–19 November 2010; pp. 143–146.
23. Bjeletić, A.; Corradini, L.; Maksimović, D.; Zane, R. Specifications-driven design space boundaries for point-of-load converters. In Proceedings of the IEEE Applied Power Electronics Conference and Exposition—APEC, Fort Worth, TX, USA, 6–11 March 2011; pp. 1166–1173.
24. Martinez, C.; Valdivia, V.; Roldan, A.M.; Lourido, J.; Quesada, I.; Lucena, C.; Zumel, P.; Barrado, A. Efficient CAD tool for power electronics compensator design. In Proceedings of the 2010 IEEE Energy Conversion Congress and Exposition, Atlanta, GA, USA, 12–16 September 2010.
25. Fernández, C.; Lázaro, A.; Zumel, P.; Valdivia, V.; Martínez, C.; Barrado, A. Design space boundaries of linear compensators applying the k-factor method. In Proceedings of the 2013 Twenty-Eighth Annual IEEE Applied Power Electronics Conference and Exposition (APEC), Long Beach, CA, USA, 17–21 March 2013; pp. 2706–2711.
26. Peterchev, A.V.; Sanders, S.R. Quantization resolution and limit cycling in digitally controlled PWM converters. *IEEE Trans. Power Electron.* **2003**, *18*, 301–308. [CrossRef]
27. Corradini, L.; Maksimovic, D.; Mattavelli, P.; Zane, R. Amplitude Quantization. In *Digital Control of High-Frequency Switched-Mode Power Converters*; Wiley-IEEE Press: Piscataway, NJ, USA, 2015; p. 360.
28. Lin, Y.C.; Chen, D.; Wang, Y.T.; Chang, W.H. A Novel Loop Gain-Adjusting Application Using LSB Tuning for Digitally Controlled DC-DC Power Converters. *IEEE Trans. Ind. Electron.* **2012**, *59*, 904–911. [CrossRef]
29. Al-Atrash, H.; Batarseh, I. Digital controller design for a practicing power electronics engineer. In Proceedings of the APEC 07—Twenty-Second Annual IEEE Applied Power Electronics Conference and Exposition, Anaheim, CA, USA, 25 February–1 March 2007; pp. 34–41.

30. Corradini, L.; Maksimovic, D.; Mattavelli, P.; Zane, R. Discrete-Time Modeling. In *Digital Control of High-Frequency Switched-Mode Power Converters*; Wiley-IEEE Press: Piscataway, NJ, USA, 2015; p. 360.
31. Maksimovic, D.; Zane, R. Small-Signal Discrete-Time Modeling of Digitally Controlled PWM Converters. *IEEE Trans. Power Electron.* **2007**, *22*, 2552–2556. [CrossRef]
32. VandeSype, D.M.; DeGusseme, K.; DeBelie, F.M.L.L.; VandenBossche, A.P.; Melkebeek, J.A. Small-Signal z-Domain Analysis of Digitally Controlled Converters. *IEEE Trans. Power Electron.* **2006**, *21*, 470–478. [CrossRef]
33. Corradini, L.; Maksimovic, D.; Mattavelli, P.; Zane, R. Digital Control. In *Digital Control of High-Frequency Switched-Mode Power Converters*; Wiley-IEEE Press: Piscataway, NJ, USA, 2015; p. 360.
34. Duarte-Mermoud, M.A.; Prieto, R.A. Performance index for quality response of dynamical systems. *ISA Trans.* **2004**, *43*, 133–151. [CrossRef]
35. Zumel, P.; Fernandez, C.; Granda, M.A.; Lazaro, A.; Sanz, M.; Barrado, A. Simple method of direct digital design of compensator for DC-DC converters. In Proceedings of the 2016 IEEE 17th Workshop on Control and Modeling for Power Electronics, COMPEL 2016, Trondheim, Norway, 27–30 June 2016.

6

Power and Voltage Control for Single-Phase Cascaded H-Bridge Multilevel Converters under Unbalanced Loads

Daliang Yang [1,*], Li Yin [1], Shengguang Xu [2] and Ning Wu [3]

1 College of Electrical Engineering, Guangxi University, Nanning 530004, China; 1612305002@st.gxu.edu.cn
2 Liuzhou Power Supply Bureau, Gxuangxi Power Grid Corporation, Liuzhou 545005, China; hu_fm.lzg@gx.csg.cn
3 Electric Power Research Institute, Guangxi Power Grid Corporation, Nanning 530023, China; wu_n.sy@gx.csg.cn
* Correspondence: yangdl@gxu.edu.cn

Abstract: The conventional control method for a single-phase cascaded H-bridge (CHB) multilevel converter is vector (dq) control; however, dq control requires complicated calculations and additional time delays. This paper presents a novel power control strategy for the CHB multilevel converter. A power-based dc-link voltage balance control is also proposed for unbalanced load conditions. The new control method is designed in a virtual $\alpha\beta$ stationary reference frame without coordinate transformation or phase-locked loop (PLL) to avoid the potential issues related to computational complexity. Because only imaginary voltage construction is needed in the proposed control method, the time delay from conventional imaginary current construction can be eliminated. The proposed method can obtain a sinusoidal grid current waveform with unity power factor. Compared with the conventional dq control method, the power control strategy possesses the advantage of a fast dynamic response. The stability of the closed-loop system with the dc-link voltage balance controller is evaluated. Simulation and experimental results are presented to verify the accuracy of the proposed power and voltage control method.

Keywords: cascaded H-bridge (CHB); dc-link voltage balance control; multilevel converter; power control; single-phase system

---

1. Introduction

Multilevel converters have gained increasing attention as of late. Commonly used multilevel converters include cascaded H-bridge (CHB), flying capacitor, and neutral-point-clamped converters [1]. Among these, the CHB converter has advantages of simple implementation, high reliability, and low harmonics [2], hence, it has been widely researched in academia and industry, particularly in relation to static synchronous compensators [3], solid-state transformers [4], and active power filters [5].

Many control schemes have been proposed to enable the pulse width modulation (PWM) converter to achieve a high power factor and near-sinusoidal grid current [6]. These strategies can be broadly classified into two categories: direct power control (DPC) and direct current control (DCC).

DPC was originally proposed by Ohnishi [7], whose method directly controls the active and reactive power of the PWM converter using three-phase instantaneous power theory. Compared with DCC, DPC has advantages of simple structure, fast dynamic response, high efficiency, and wide applicability [8]. Currently, many advanced DPC strategies have been proposed, such as predictive DPC (P-DPC) [9], deadbeat DPC [10], and slide-mode DPC [11]. In the single-phase PWM converter system, instantaneous power estimation is more complicated than in the three-phase system, which limits the development of DPC schemes in single-phase systems [12].

DCC methods have been frequently used with the PWM converter. The most commonly used DCC method utilizes vector control theory (*dq* control) [13–16]. Coordination transformation is needed in *dq* control to change the ac signals into dc signals [17]. The grid voltage phase angle is essential in the coordinate transformation; therefore, a phase-locked loop (PLL) is employed to synchronize the output voltage with the grid voltage vector [18]. To accurately obtain the information on the grid voltage including the amplitude and phase angle, some advanced PLLs have been proposed, such as hybrid filtering technique-based PLL [19], grid sequence separator PLL [20], frequency-fixed SOGI-based PLL [21], and repetitive learning-based PLL [22]. However, the computational complexity of coordinate transformation and PLL increases the calculation burden of the embedded processor like the digital signal processor (DSP) [23]. Therefore, the development of a convenient control method without PLL or coordination transformation is appealing. Recently proposed control methods of the CHB converter are based on model-predictive control (MPC) [24,25]. However, three major problems arise in the MPC scheme used for the CHB converter. First, the calculation burden persists in the MPC method because the number of switching states increases exponentially as more H-bridges are added [26]. The second one is the MPC system sensitiveness problem like sensitive to the parameter [27]. The third problem is that under unbalanced load conditions, the MPC strategy cannot provide a proper control function [28].

It is exceedingly difficult to implement the *dq* control scheme in the single-phase CHB converter system; imaginary orthogonal current and voltage signals must be created for coordinate transformation ($\alpha\beta$ to *dq*) because only one physical axis is accessible [29]. In *dq* control, the imaginary orthogonal current signal is essential in acquiring the dc current signals i_d and i_q for the inner current loop [30]. Thus, an imaginary current construction module is required to produce the imaginary orthogonal current signal [31]. Although the performance of conventional imaginary current construction under the steady state is mostly acceptable, the time delay tends to slow down the system dynamic response and result in further distortion [32,33].

The control scheme performance of the CHB converter in the single-phase system depends on two key factors: sinusoidal grid current waveform with unity power factor and balanced dc-link voltages [34,35]. Therefore, the CHB multilevel converters need a control method to balance the dc-link voltages, especially when the dc-side loads are unbalanced. In [36,37], the dc-link voltages were controlled using individual capacitor voltage controllers. In [38], the presented voltage balance control method uses a weighting factor to adjust the dc-link voltages. In [39], two voltage-balancing techniques were presented, allowing for much more efficient regulation of dc-link voltages. However, these voltage balanced control methods are all based on ac voltage reference duty cycle regulation; a power-based dc-link voltage balance controller has not been analyzed in detail.

In the conventional *dq* control strategy, the current calculation is complicated and additional time delay is required. For optimization purposes, a power control method without coordinate transformation, PLL or conventional imaginary current construction is introduced and applied to control the single-phase CHB multilevel converter system. Moreover, a power-based dc-link voltage balance control scheme is presented to make the proposed power control strategy available under unbalanced load conditions. This power and voltage control method can achieve a sinusoidal grid current waveform and balanced dc-link voltages. Several experimental tests were conducted to compare dynamic responses of the presented control scheme with those of the classical *dq* control method. The proposed voltage balance controller is verified as well.

The remainder of this paper is organized as follows: in Section 2, principles of the proposed control scheme and dc-link voltage balance controller are discussed, and the stability of the closed-loop system with the dc-link voltage balance controller is evaluated. Section 3 presents the simulation results. Experimental results are addressed in Section 4. The proposed control method is discussed in Section 5. Finally, relevant conclusions are drawn in Section 6.

2. Control Method Principles

2.1. Model of the Cascaded H-Bridge Multilevel Converter

Figure 1 presents the topology of a two-cell CHB system in which u_s is the grid voltage, i_{sa} is the grid current, u_c is the ac side voltage, L and R_L denote the filter inductor and line resistance, respectively,. u_{dc1} and u_{dc2} represent the dc-link voltages of the first and second H-bridge, and u_{dc_sum} is the total dc-link voltage (u_{dc1} + u_{dc2}). R_1 and R_2 are the resistive loads connected to the dc side of each H-bridge.

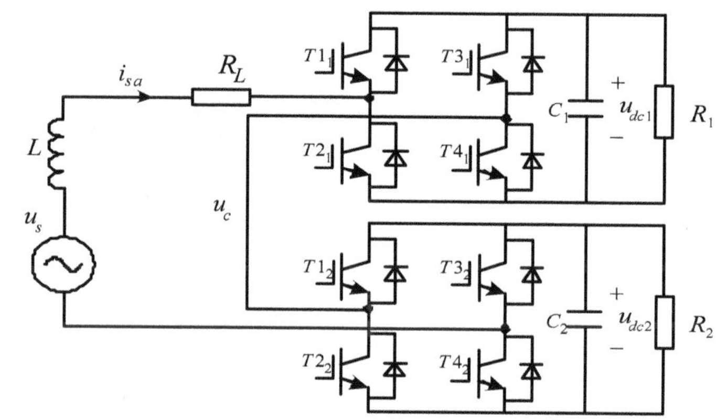

Figure 1. Two-cell single-phase CHB converter.

The mathematical model of the CHB converter is as follows:

$$\begin{cases} L\frac{di_{sa}}{dt} = u_s - u_c - R_L i_{sa} \\ u_{dc\_sum} = u_{dc1} + u_{dc2} \end{cases} \quad (1)$$

To simplify analysis of system stability, according to Equation (1), a small-signal model of the stationary reference frame ($\alpha\beta$ frame) was built, as illustrated in Figure 2. Here, $u_{s\alpha}$ equals u_s, and $u_{s\beta}$ is the orthogonal signal of u_s; $u^{\sim}{}_{s\alpha}$ and $u^{\sim}{}_{s\beta}$ represent the small-signal ac variation of $u_{s\alpha}$ and $u_{s\beta}$; $u^{\sim}{}_{dc1}$ and $u^{\sim}{}_{dc2}$ denote the small-signal ac variation of u_{dc1} and u_{dc2}; $i_{s\alpha}$ equals i_{sa}, and $i_{s\beta}$ is the orthogonal signal of i_{sa}; $i^{\sim}{}_{s\alpha}$ and $i^{\sim}{}_{s\beta}$ represent the small-signal ac variation of $i_{s\alpha}$ and $i_{s\beta}$; $I_{s\alpha}$ and $I_{s\beta}$ stand for the quiescent operating point of $i_{s\alpha}$ and $i_{s\beta}$; $D_{\alpha1}$, $D_{\alpha2}$, $D_{\beta1}$, and $D_{\beta2}$ are the duty cycles of the first and second H-bridge; and $d^{\sim}{}_{\alpha1}$, $d^{\sim}{}_{\alpha2}$, $d^{\sim}{}_{\beta1}$, and $d^{\sim}{}_{\beta2}$ represent the small-signal ac variation of $D_{\alpha2}$, $D_{\beta1}$, and $D_{\beta2}$.

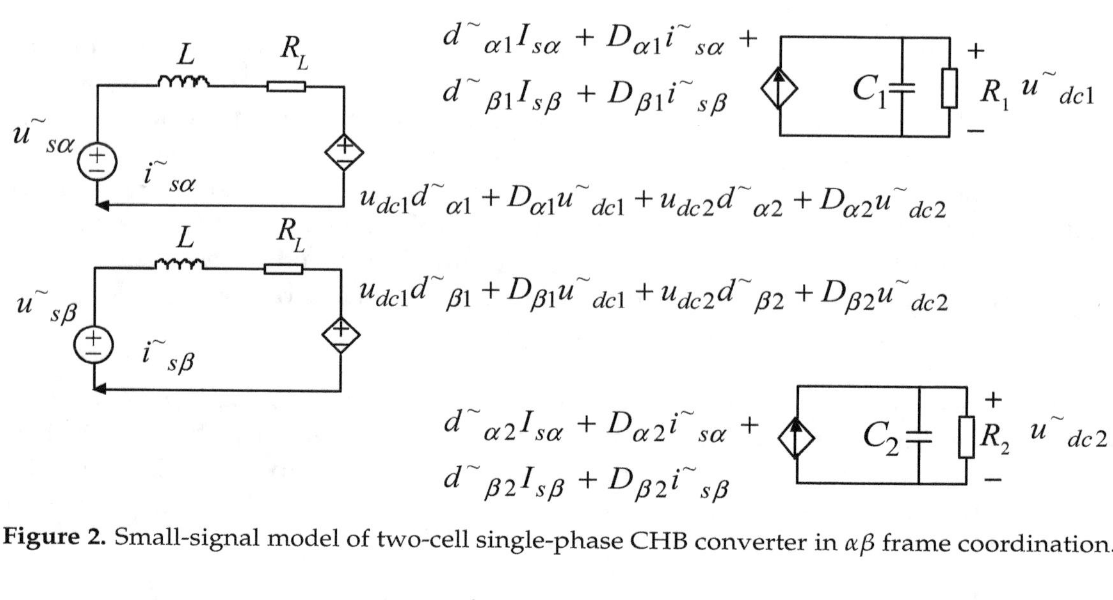

Figure 2. Small-signal model of two-cell single-phase CHB converter in $\alpha\beta$ frame coordination.

2.2. Principle of the Proposed Power Control Method

The main function of CHB multilevel converter is to obtain active power and reactive power from the grid to satisfy load needs. For the CHB converter, the essence of controlling the grid current as well as the dc-link voltage is control of the input and output power. By rapidly and effectively controlling the active and reactive power, the converter can achieve strong dynamic and static characteristics.

In the three-phase CHB converter system, the active power p and reactive power q can be described as:

$$\begin{bmatrix} p \\ q \end{bmatrix} = \begin{bmatrix} u_{s\alpha} & u_{s\beta} \\ u^*_{s\alpha} & u^*_{s\beta} \end{bmatrix} \begin{bmatrix} i_{s\alpha} \\ i_{s\beta} \end{bmatrix} \tag{2}$$

where $u^*_{s\alpha} = u_{s\beta}$, $u^*_{s\beta} = -u_{s\alpha}$.

Only one phase physical variable is available in the single-phase CHB converter system; a second-order generalized integrator (SOGI) block is needed to introduce the imaginary signal. The virtual $\alpha\beta$ stationary reference frame is depicted in Figure 3, the α-axis coincides with the a-axis, and the β-axis is orthogonal to the α-axis.

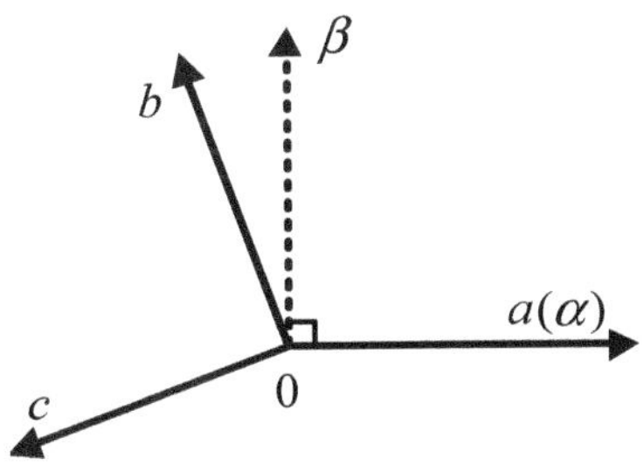

Figure 3. Schematic diagram of virtual $\alpha\beta$ coordinate axis.

In the single-phase CHB converter system, p and q can be described as follows:

$$\begin{bmatrix} p \\ q \end{bmatrix} = \frac{1}{3} \begin{bmatrix} u_{s\alpha} & u_{s\beta} \\ u^*_{s\alpha} & u^*_{s\beta} \end{bmatrix} \begin{bmatrix} i_{s\alpha} \\ i_{s\beta} \end{bmatrix} \tag{3}$$

p^* and q^* are defined as the active power and reactive power reference, respectively; $i^*_{s\alpha}$ and $i^*_{s\beta}$ represent the current reference.

Equations (4) and (5) are shown as follows according to Equation (3), and the coefficient calculation of the proposed power control is shown in Equation (6):

$$\begin{bmatrix} i^*_{s\alpha} \\ i^*_{s\beta} \end{bmatrix} = \frac{3}{u_{s\alpha} * u^*_{s\beta} - u^*_{s\alpha} * u_{s\beta}} \begin{bmatrix} u^*_{s\beta} & -u_{s\beta} \\ -u^*_{s\alpha} & u_{s\alpha} \end{bmatrix} \begin{bmatrix} p^* \\ q^* \end{bmatrix} \tag{4}$$

$$\begin{bmatrix} i^*_{s\alpha} \\ i^*_{s\beta} \end{bmatrix} = \begin{bmatrix} \frac{3u^*_{s\beta}}{u_{s\alpha} * u^*_{s\beta} - u^*_{s\alpha} * u_{s\beta}} p^* + \frac{-3u_{s\beta}}{u_{s\alpha} * u^*_{s\beta} - u^*_{s\alpha} * u_{s\beta}} q^* \\ \frac{-3u^*_{s\alpha}}{u_{s\alpha} * u^*_{s\beta} - u^*_{s\alpha} * u_{s\beta}} p^* + \frac{3u_{s\alpha}}{u_{s\alpha} * u^*_{s\beta} - u^*_{s\alpha} * u_{s\beta}} q^* \end{bmatrix} \tag{5}$$

$$\begin{cases} k_1 = \frac{3u^*_{s\beta}}{u_{s\alpha} * u^*_{s\beta} - u^*_{s\alpha} * u_{s\beta}} \\ k_2 = \frac{-3u_{s\beta}}{u_{s\alpha} * u^*_{s\beta} - u^*_{s\alpha} * u_{s\beta}} \end{cases} \tag{6}$$

If the grid voltage is sinusoidal, then the result of $u_{s\alpha} * u^*_{s\beta} + u^*_{s\alpha} * u^*_{s\beta}$ is constant; p^* is controlled by the outer power loop; q^* is constant because it is given directly; therefore the frequency and phase of the current references $i_{s\alpha}$ and $i_{s\beta}$ are consistent with the instantaneous grid voltage, and the real-time tracking can be achieved. Notably, only $i^*_{s\alpha}$ is used as the current reference, because the CHB multilevel converter is used on a single-phase system.

2.3. Presentation of Proposed Dc-Link Balance Controller

Figure 4 presents the conventional dc-link voltage balance controller. In this approach, the duty cycles of each ac voltage reference u^*_{a1} and u^*_{a2} are modified separately to keep the voltages balanced, and the scheme is suitable for the dq method. A power-based dc-link voltage balance controller adjusting the ac current reference duty cycle is introduced in this paper to regulate the individual H-bridge dc-link voltage.

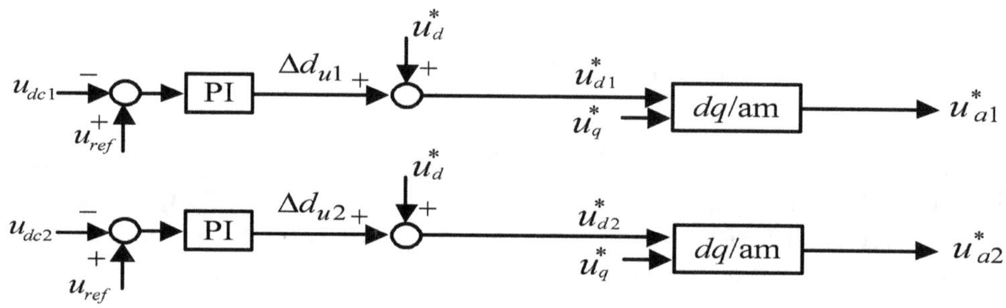

Figure 4. Conventional voltage balance controller [37].

If the dc-link voltage of each H-bridge is unbalanced (mainly via load imbalance), then the power of these H-bridges becomes uneven. To keep the dc-link voltage balanced, the uneven power should be extracted from the grid. Δd_{p1} and Δd_{p2} are used as the compensation values to eliminate uneven power. Δd_{p1} and Δd_{p2} are calculated by the voltage balance controller:

$$\begin{cases} \Delta d_{p1} = (K_p + \frac{K_i}{s})(u_{ref} - u_{dc1}) \cdot u_{dc1} \\ \Delta d_{p2} = (K_p + \frac{K_i}{s})(u_{ref} - u_{dc2}) \cdot u_{dc2} \end{cases} \tag{7}$$

where K_p and K_i represent the proportional and integral coefficient of the proportional-integral (PI) controller, respectively, and u_{ref} is the voltage reference of every H-bridge cell.

The presented voltage balance controller is shown in Figure 5, which modifies the duty cycles of each ac current reference i^*_{sa1} and i^*_{sa2} individually to reduce the dc voltage imbalance. The current reference for each H-bridge can be calculated according to Equations (8)–(10):

$$\begin{cases} i^*{}_{s\alpha 1} = \Delta d_{i^*s\alpha 1} + i^*{}_{s\alpha} \\ i^*{}_{s\alpha 2} = \Delta d_{i^*s\alpha 2} + i^*{}_{s\alpha} \end{cases} \tag{8}$$

$$\begin{cases} i^*{}_{s\alpha 1} = \Delta d_{p1}(k_1 + k_2) + p^* k_1 + q^* k_2 \\ i^*{}_{s\alpha 2} = \Delta d_{p2}(k_1 + k_2) + p^* k_1 + q^* k_2 \end{cases} \tag{9}$$

$$\begin{cases} i^*{}_{s\alpha 1} = k_1(\Delta d_{p1} + p^*) + k_2(\Delta d_{p1} + q^*) \\ i^*{}_{s\alpha 2} = k_1(\Delta d_{p2} + p^*) + k_2(\Delta d_{p2} + q^*) \end{cases} \tag{10}$$

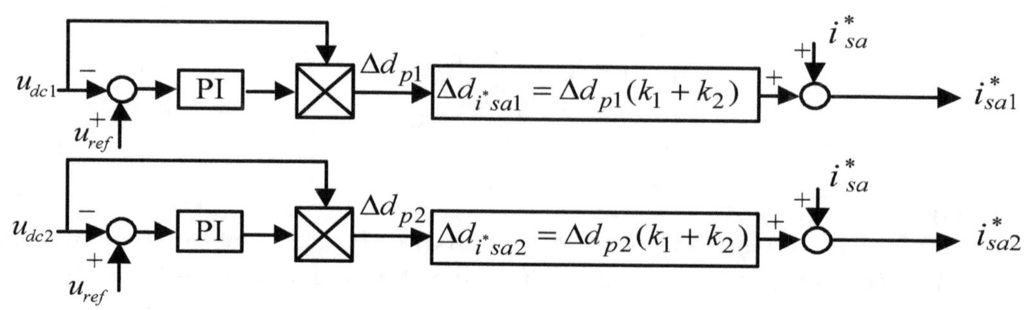

Figure 5. Proposed voltage balance controller.

2.4. Proposed Power Control of the Cascaded H-Bridge Multilevel Converter

The overall control topology of the proposed control method is depicted in Figure 6. The main idea of the proposed control is to obtain command signals in the orthogonal $\alpha\beta$ frame; as such, the dc-link voltages can be balanced.

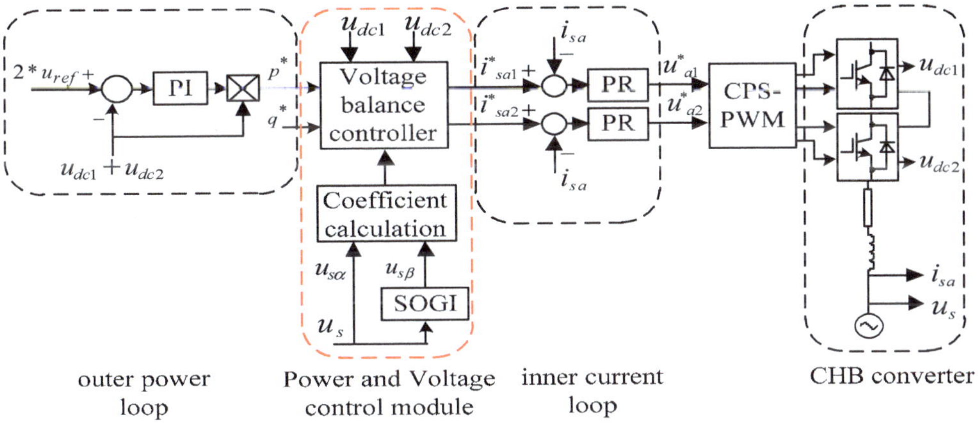

Figure 6. Proposed power and voltage control of CHB converter.

The control system is formed as a double closed-loop control structure using the inner current loop and the outer power loop.

The inner current loop is implemented in the orthogonal $\alpha\beta$ frame. Compared with classical single-phase dq schemes, the current calculation of the proposed control method is simpler because no coordinate transformation, PLL, or conventional imaginary current construction is needed. However, the current references i^*_{sa1} and i^*_{sa2} are ac variables. To track the current references with zero error in the stationary frame and achieve high tracking accuracy when the grid frequency fluctuates, rather than relying on conventional PI controllers or a repetitive controller [40], proportional-plus-resonant (PR) controllers are utilized in the current loop.

As for the outer power loop, a PI controller is used to adjust the total dc-link voltage u_{dc_sum} (u_{dc1} + u_{dc2}). The active power p^* is generated by multiplying the output of the PI controller by the total dc-link voltage u_{dc_sum}. Reactive power reference q^* is given directly.

The detailed control topology of the power and voltage control module is shown in Figure 7. Current reference i^*_{sa1} of the first H-bridge and i^*_{sa2} of the second H-bridge are acquired through the dc-link voltage balance control block; these current references are controlled independently. Deviation values between the grid current i_{sa} and the current references are input into the PR regulators separately, and the modulation voltage signals u^*_{a1} and u^*_{a2} are generated for the first and second H-bridge. Carrier phase-shifted PWM (CPS-PWM) technology is adopted to generate five-level voltage, achieving multi-level modulation. Furthermore, if the reactive power reference q^* is set to zero, then a unity power factor can be achieved.

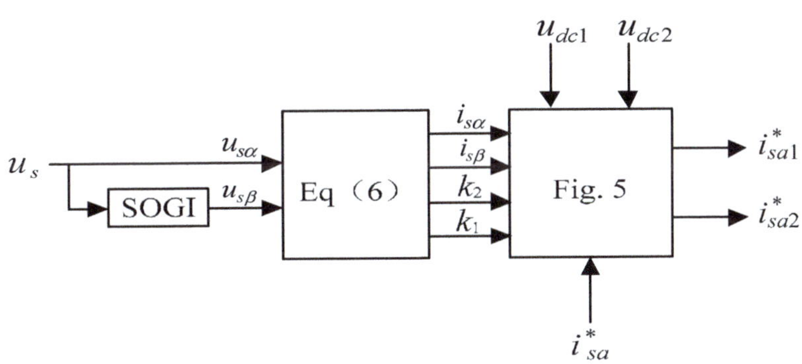

Figure 7. Power and voltage control module.

2.5. Stability Analysis of the Power and Voltage Control Scheme

Assume that the parameters of the H-bridges are identical, such that the dc-link voltage $u_{dc1} = u_{dc2} = u_{dc}$. The double-loop control diagram presented in Figure 6 can thus be transformed as illustrated in Figure 8. In Figure 8, G_V is the voltage gain of CHB, G_{PIV} is the voltage PI regulator, K is the conversion coefficient between p^* and i^*_{sa}, G_{qPR} is the current PR regulator, G_{TPWM} is the transfer function of the PWM modulator, G_{IV} is the transfer function of the voltage to output current i^*_{sa}, and G_{VI} is the transfer function of the current to output voltage u_{dc}. G_{IV} and G_{VI} can be deduced from the small-signal model shown in Figure 2.

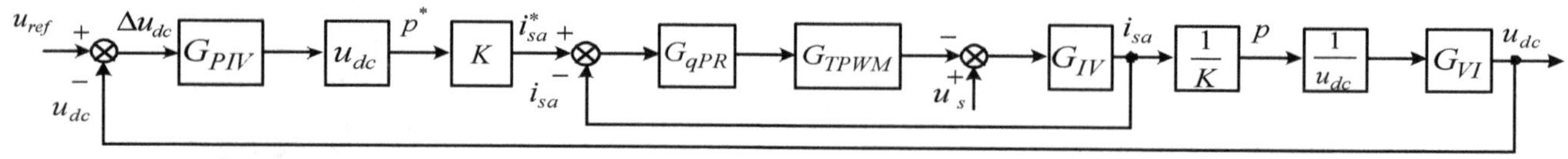

Figure 8. Basic double-loop controller structure diagram of the proposed power control.

The closed-loop transfer function can be expressed as follows:

$$G_V = \frac{G_{PIV} G_{qPR} G_{TPWM} G_{IV} G_{VI}}{1 + G_{PIV} G_{qPR} G_{TPWM} G_{IV} G_{VI}} \tag{11}$$

One bridge of the CHB converter can be simplified according to Figure 9 per Thevenin law. However, the dc-link voltage unbalance problem in CHB remains under the unbalanced load condition; therefore the voltage balance control module should be added. Figure 10 shows the Thevenin equivalent circuit with the dc-link voltage balance module, where Z_1 is the load impedance, Z_{io} is the impedance between the input and output of CHB, M is the gain of the voltage balance control module, G_1 is the conversion coefficient of power to voltage, and i_{out} is the output current. In this case $G_1 = 1/(G_{PIV} * u_{dc})$.

Taking the first H-bridge as an example:

$$G_1[p^* + (u_{ref} - u_{dc1})M]G_V = Z_{io}\frac{u_{dc1}}{Z_1} + u_{dc1} \tag{12}$$

$$u_{dc1} = \frac{G_V G_1}{Z_{io}/Z_1 + 1 + G_V G_1 M} p^* + \frac{G_V G_1 M}{Z_{io}/Z_1 + 1 + G_V G_1 M} u_{ref} \tag{13}$$

Because $Z_1 >> Z_{io}$, GG_1M represents the gain in the basic double-loop controller with the dc-link voltage balance module from Equation (13); the Bode plot is shown in Figure 11. The amplitude margin is 12.7 dB, the cut-off frequency is 272 Hz, the phase margin is 105°, and the cross-over frequency is 78 Hz. All roots of $(Z_{io}/Z_1 + 1 + GG_1M)$ are on the left half-complex plane, ensuring stability of the system with the voltage balance control module.

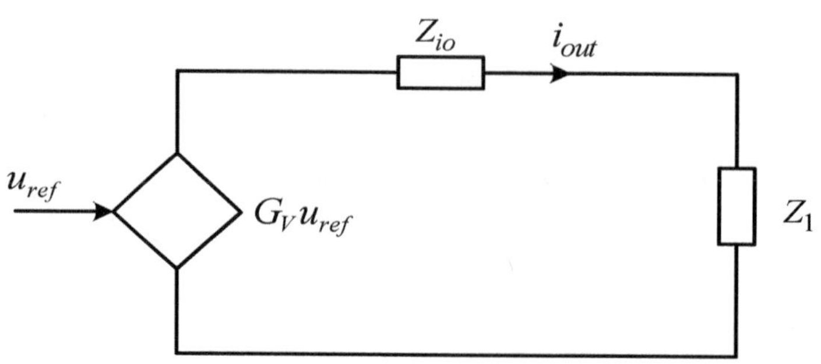

Figure 9. Thevenin equivalent circuit of the basic double-loop.

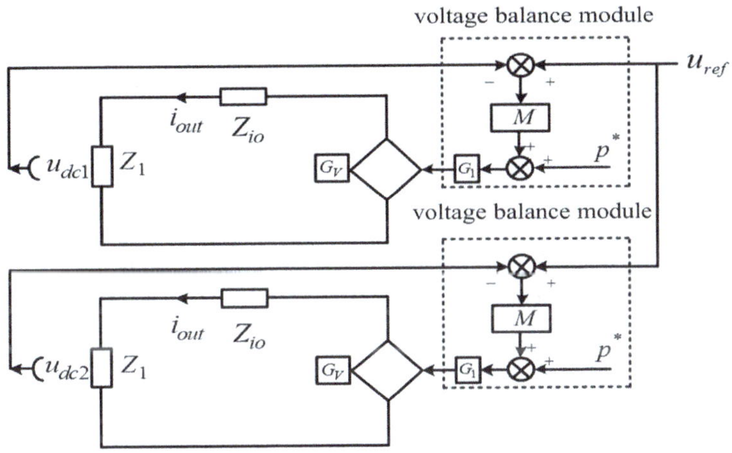

Figure 10. Thevenin equivalent circuit with the voltage balance module controller.

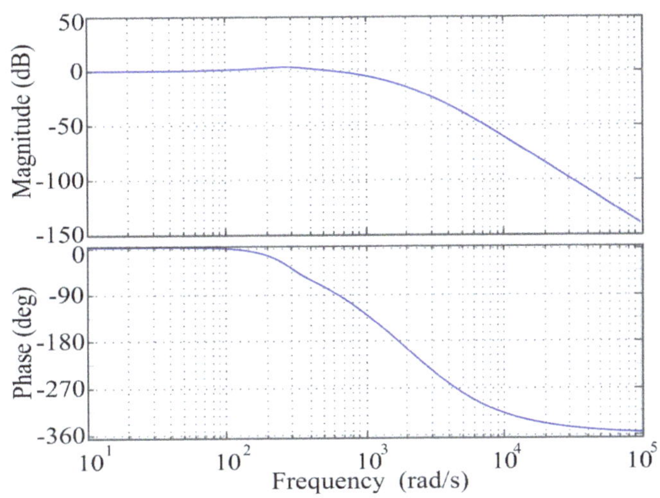

Figure 11. Bode plot of the basic double-loop controller with the dc-link voltage balance module.

3. Simulation Results

The CHB converter system shown in Figure 1 was modelled in the MATLAB/Simulink R2015a (MATLAB/Simulink R2015a, MathWorks, Natick, MA, USA) software environment to verify the power and voltage control scheme. The main system parameters used for simulation are shown in Table 1.

Table 1. Parameters Used For Simulation.

Parameter	Symbol	Simulation Value
Grid voltage *rms* value	u_s	220 V
Grid frequency	f_g	50 Hz
Input inductance	L	3.0 mH
dc-link total voltage	U_{dc}	400 V
dc-link capacitance	C_1, C_2	4700 μF
dc-link load resistance	R_1, R_2	10 Ω, 15 Ω
Switching frequency	f_{sw}	10 kHz
Inner current loop control parameters	P, R, w_c	0.5, 100, 6.28
Outer power loop control parameters	P, I	0.1, 8

3.1. Proposed Power Control Scheme Simulation

Figures 12–14 show the simulation results with unbalanced loads, where $R_1 = 10\ \Omega$ and $R_2 = 15\ \Omega$. Figure 12a displays the grid voltage and grid current waveform in the steady state, revealing that the unity power factor was achieved.

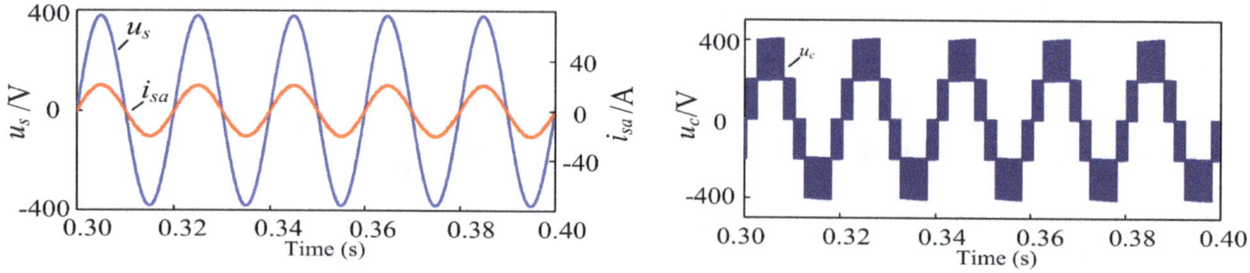

(**a**) (**b**)

Figure 12. Simulation waveforms of CHB converter in steady state: (**a**) Grid voltage and grid current; (**b**) Five-level voltage staircase waveform.

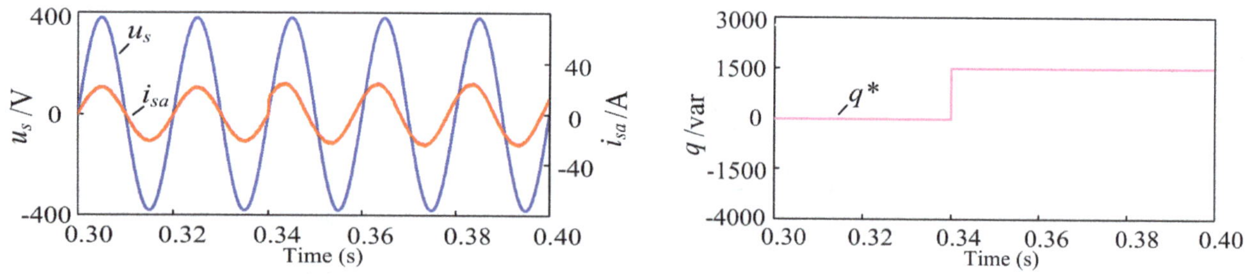

(**a**) (**b**)

Figure 13. q^* changed from 0 var to 1500 var: (**a**) Dynamic response of the grid current; (**b**) Dynamic response of the reactive power.

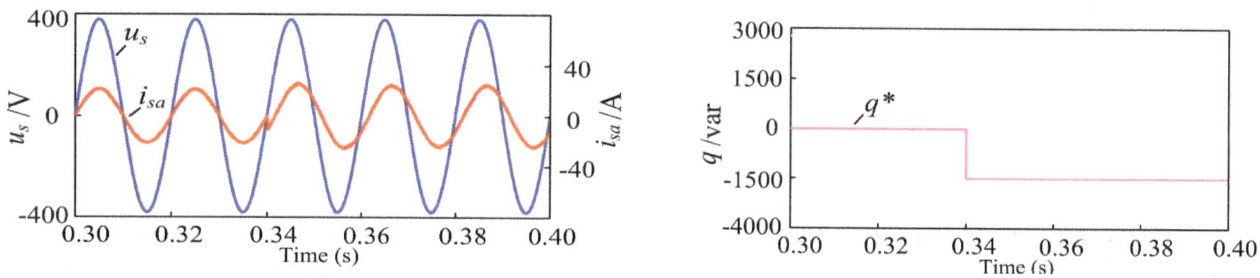

(**a**) (**b**)

Figure 14. q^* changed from 0 var to -1500 var: (**a**) Dynamic response of the grid current; (**b**) Dynamic response of the reactive power.

Figure 12b shows the ac side five-level staircase voltage. Dynamic responses of the proposed method under sudden reactive power change conditions are shown in Figures 13 and 14. In these, the active power reference p^* is governed by the outer power loop, and the reactive power reference q^* is changed from 0 to -1500 var and 0 to 1500 var at 0.34 s.

3.2. Power-Based Voltage Balance Control Simulation

In this simulation, the initial loads of two H-bridges were 15 Ω resistors. At 1.0 s, the resistor of the first H-bridge declined to 10 Ω. Two distinct cases were considered. In Figure 15a, the voltage balance control module was not utilized, and the dc-link voltages became different after the load change. The second H-bridge, whose resistor was unchanged, demonstrated higher dc-link voltage than the first H-bridge. Figure 15b shows the dc-link voltages with the voltage balance control method. Using the proposed voltage balance control module, the dc-link voltages returned to the initial value (200 V).

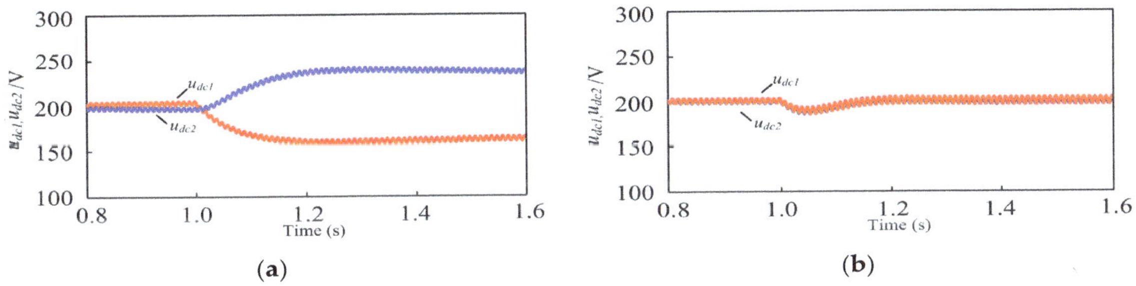

Figure 15. Simulated response of dc-link voltages when load changed: (**a**) Without voltage balance control; (**b**) With voltage balance control.

4. Experimental Results

4.1. Experimental Prototype

A single-phase five-level CHB converter experimental platform was adopted to validate the presented power and voltage control scheme, as shown in Figure 16.

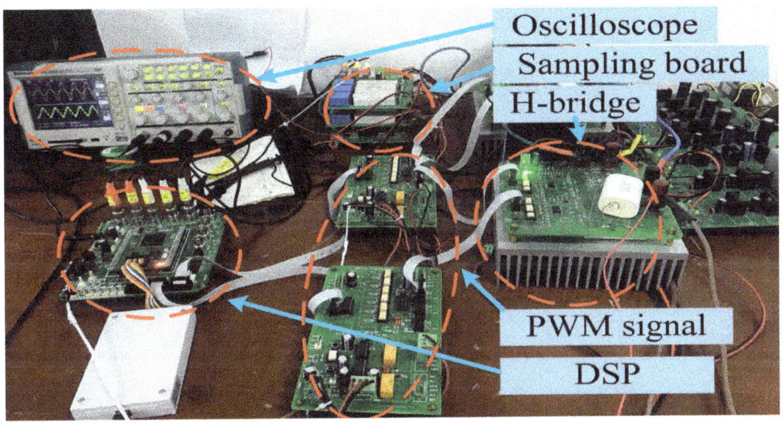

Figure 16. Platform of the experiments.

Two H-bridge converters with serially connected inputs formed the ac side of the experimental setup. The dc output of each converter fed the loads. The converter was interfaced to the grid through a coupling transformer and a voltage regulator to ensure safety during the experiments. The secondary-side voltage of the regulator was set to 80 V (rms), while the dc-link voltage references u_{dc1} and u_{dc2} were set to 50 V. The grid voltage, grid current, ac side voltage, and dc-link voltages were measured using the outer high-performance hall sensors. A TMS320F28377D DSP processor board was utilized to implement the control algorithm. Measured voltage and current signals were fed back to the DSP processor board employing a built-in, 16-bit analog-digital conversion module. The power factor and total harmonic distortion (THD) of the grid current were measured via a Fluke 435 power quality analyzer.

4.2. Steady State

Figure 17 depicts the experimental waveforms of the grid voltage u_s (200 V/division), the grid current i_{sa} (20 A/division), the dc-link voltage of the first H-bridge u_{dc1} (20 V/division), and the ac side voltage u_c (100 V/division) for the proposed scheme in the steady state, where loads R_1 and R_2 in Figure 1 are 10 Ω and 15 Ω, separately. When the reactive power reference q^* was set to zero, the grid current was in phase with the grid voltage, so the unity power factor was achieved. The THD of the grid current in the steady state was 4.3%, indicated in Figure 18. Moreover, this method was found to compensate for the reactive power: CHB converters can generate or absorb reactive power to the power grid when the reactive power reference is positive or negative. In Figure 19, the grid current was leading the grid voltage; thus, the CHB converter was generating reactive power to the system. In Figure 20, the grid current was lagging the voltage, and the CHB converter was absorbing reactive power from the system.

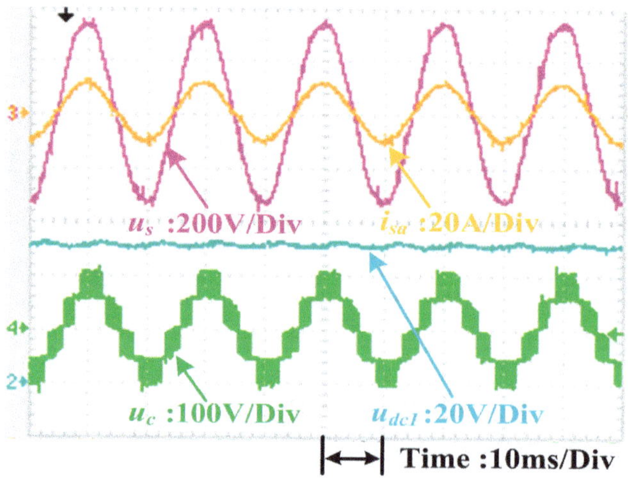

Figure 17. Experimental waveforms of CHB converter in steady state.

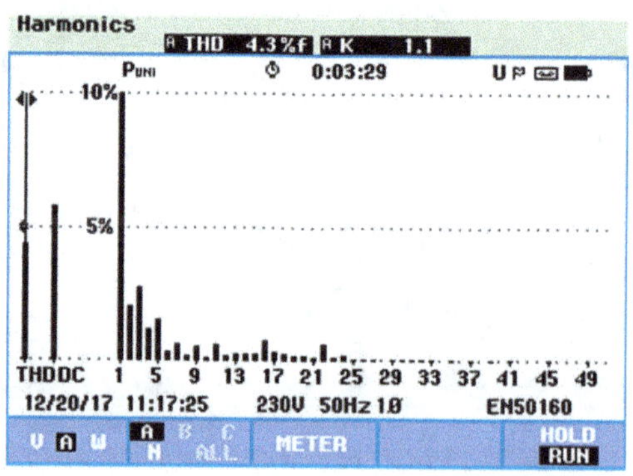

Figure 18. THD of grid current in steady state.

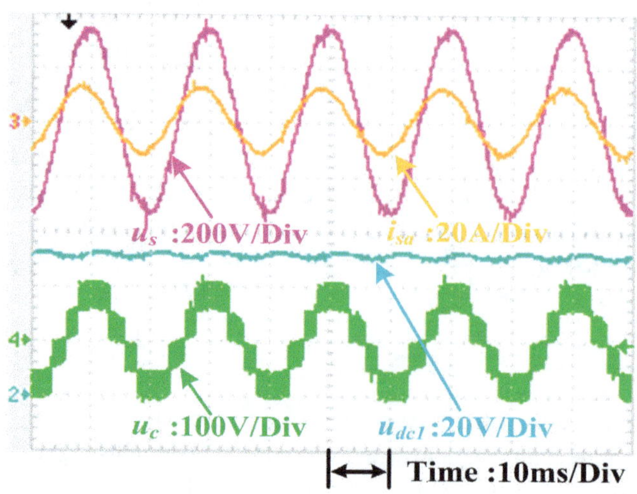

Figure 19. Experimental grid currents of CHB capacitive mode.

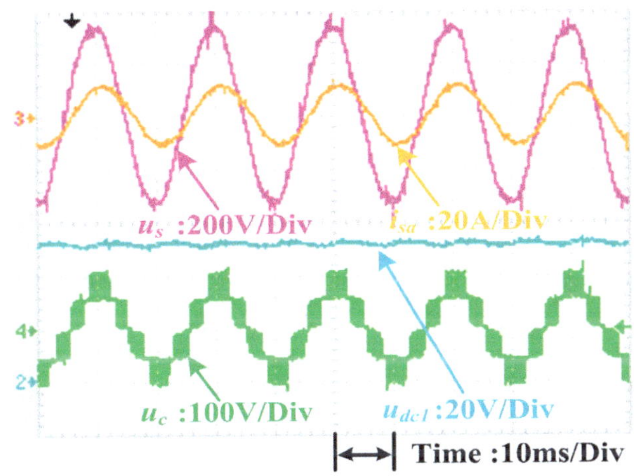

Figure 20. Experimental grid currents of CHB inductive mode.

4.3. Dynamic Response Compared with dq Control

Currently, the most popular control method for the single-phase CHB converter is *dq* control. Several experiments were carried out to compare conventional *dq* control with the proposed control method. These experiments tested the dynamic responses of the systems under sudden power change conditions. In these tests, the total dc-link voltage u_{dc_sum} was kept at 100 V, and the system was connected to two resistive loads where R_1 is 10 Ω and R_2 is 15 Ω, respectively. The dynamic of the d-axis current i^*_d in the *dq* control was managed by the outer voltage loop, and active power p^* in the proposed scheme was controlled by the outer power loop. Therefore, only step changes in the reactive current reference i^*_q in the *dq* control and reactive power reference q^* in the proposed control were applied to evaluate the dynamic control performance of each. Differences in the dynamic performance of the *dq* control scheme and the proposed control method are illustrated in Figures 21–28.

The total dc-link voltage u_{dc_sum} is 100 V; i^*_q and q^* are given directly in each control method. Reactive power reference q^* equals $u_{dc\_sum} * i^*_q$ (100 * i^*_q). For instance, when the step change of i^*_q is 5 A, the step change of q^* equals 100 * i^*_q (500 var); when the step change of i^*_q is −5 A, the step change of q^* equals 100 * i^*_q (−500 var).

As shown, the transient responses of the proposed power control method (less than 1 ms) were faster than the *dq* control method (more than 5 ms). In the *dq* control method, a time delay of 1/4 in the fundamental period arose from classical imaginary current construction.

Grid current and five-level staircase voltage continuous distortions also appeared in the *dq* control method due to conventional imaginary current construction.

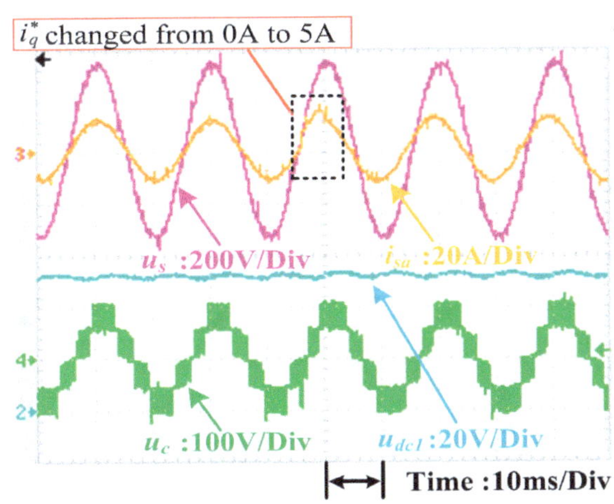

Figure 21. i^*_q changed from 0 A to 5 A (*dq* control method).

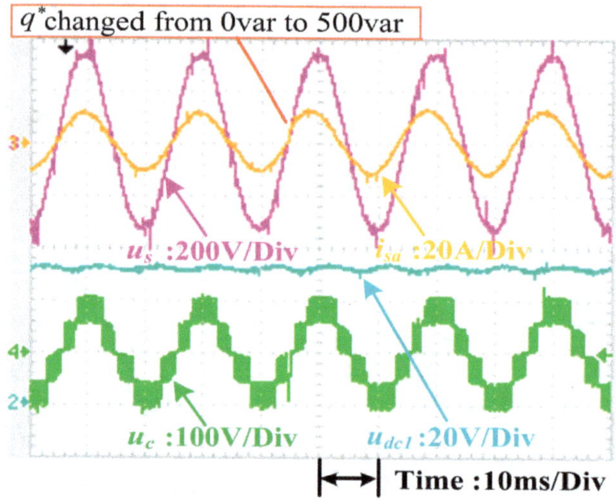

Figure 22. q^* changed from 0 var to 500 var (proposed control method).

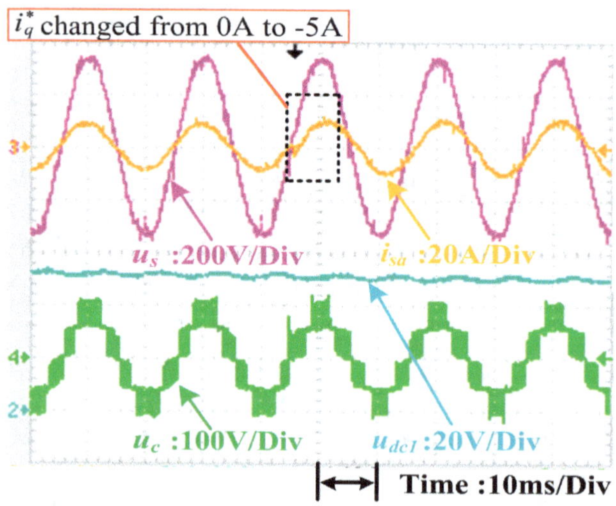

Figure 23. i^*_q changed from 0 A to −5 A (*dq* control method).

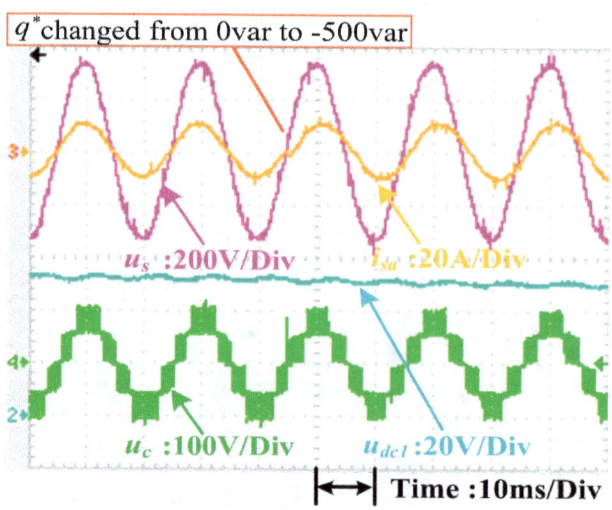

Figure 24. q^* changed from 0 var to −500 var (proposed control method).

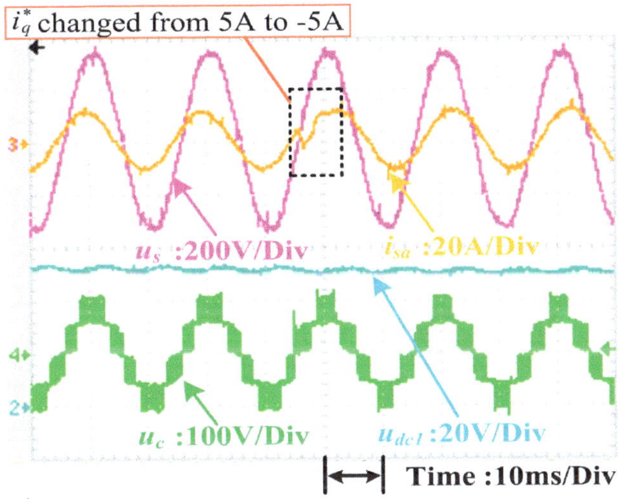

Figure 25. i^*_q changed from 5 A to −5 A (*dq* control method).

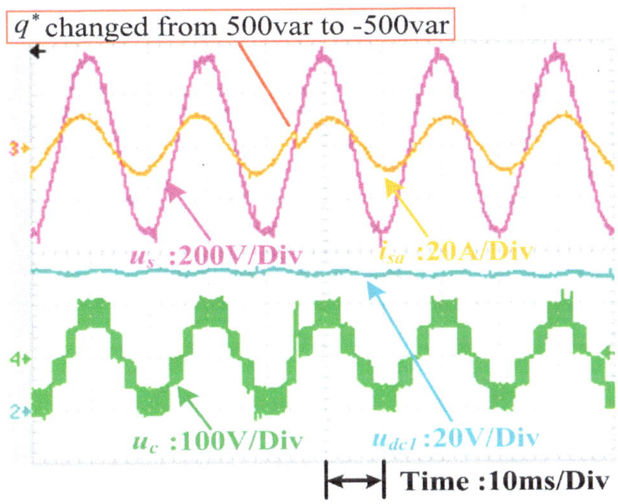

Figure 26. q^* changed from 500 var to −500 var (proposed control method).

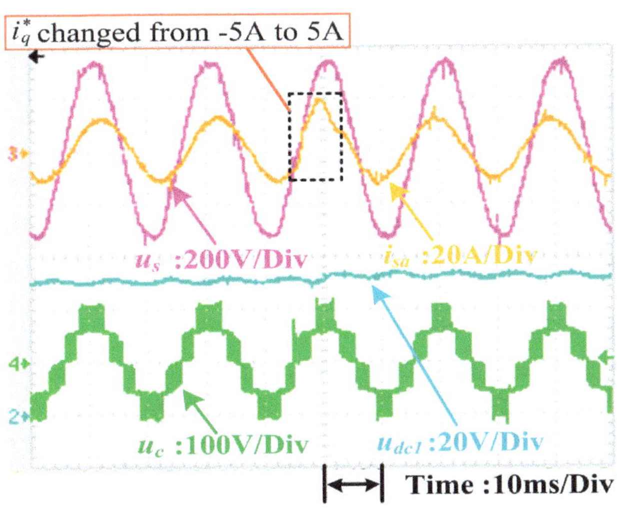

Figure 27. i^*_q changed from −5 A to 5 A (*dq* control method).

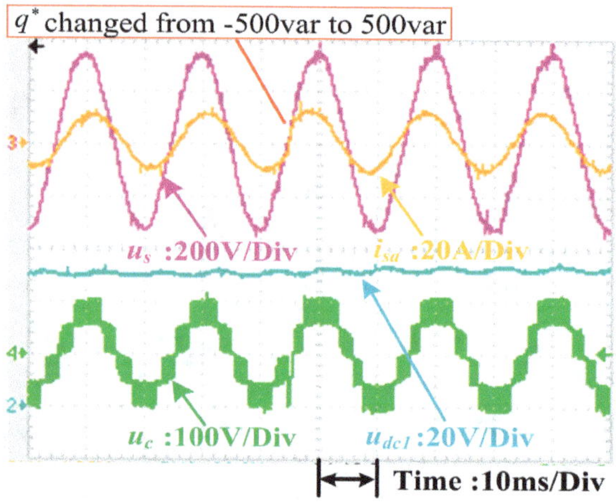

Figure 28. q^* changed from −500 var to 500 var (proposed control method).

4.4. Voltage Balance Control

Besides grid current control, balance control of the dc-link voltages is another important issue. A dc-link voltage balance control method based on power was adopted to render the proposed power control strategy applicable to the CHB converter under unbalanced load conditions.

Unbalanced load experimental tests were conducted to verify the voltage-balancing performance of the presented scheme. In these experiments, the load resistor of the first H-bridge R_1 was stepped from 15 Ω to 10 Ω whereas that of the second H-bridge R_2 was maintained at 15 Ω. In Figure 29a, when the load resistors were changed, the dc-link voltage of the first H-bridge u_{dc1} (10 V/division) declined from 50 V to 45 V, and that of the second H-bridge u_{dc2} (10 V/division) increased from 50 V to 55 V. Figure 29b shows that when the voltage balance controller was adopted, the dc-link voltages could be controlled to 50 V when the load resistors were changed. Thus, the accuracy of the presented dc-link voltage balance control method was confirmed.

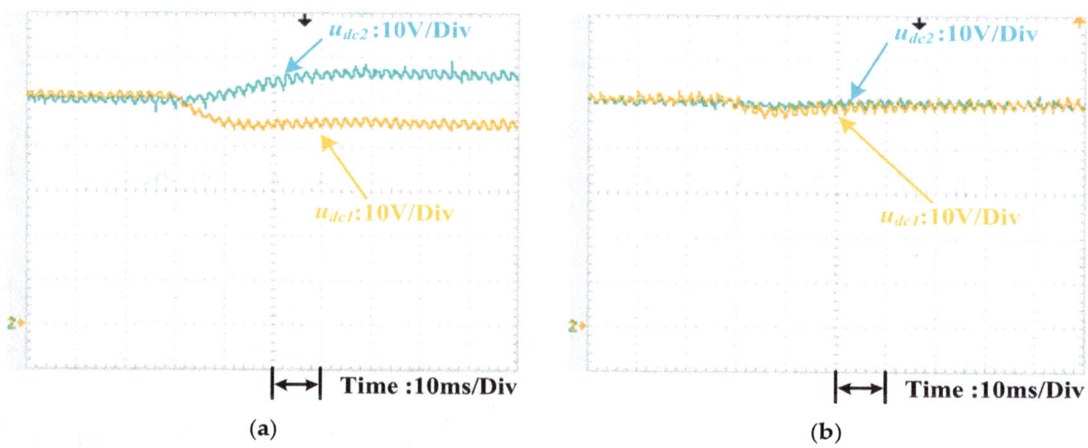

Figure 29. Experimental response of dc-link voltages when load changed: (**a**) Without voltage balance controller, (**b**) With voltage balance controller.

5. Discussion

The proposed power and voltage control method offers many advantages including no coordinate transformation, PLL, or conventional imaginary current construction. Moreover, it was found to exhibit faster transient responses and lower distortions compared with the conventional *dq* control method, as illustrated in Figures 21–28.

The natural frame control scheme presented in [23] also shows the advantages of no coordinate transformation, PLL, or conventional imaginary current construction. Differ from the natural frame control scheme, the proposed scheme connects power control with current control through the instantaneous power theory. Besides, the proposed voltage balance control module is based on power, in which the power compensation values are used to eliminate the uneven power in active power p and reactive power q.

Even so, this control strategy has some limitations: First, although classical imaginary current construction is not necessary, the creation of an imaginary voltage component cannot be avoided. Therefore, an algorithm can be developed to reduce the time delay caused by imaginary voltage construction. Second, only the steady grid voltage condition was considered here, although abnormal operating states (e.g., grid voltage distortion) exist; thus, the control method should be optimized to make it applicable to such states.

6. Conclusions

A power control method is introduced for the single-phase CHB multilevel converter in this paper. A power-based dc-link voltage balance control module is also presented to eliminate the different dc-link voltages caused by the unbalanced loads.

The proposed power and voltage control method is designed in a virtual $\alpha\beta$ stationary reference frame without coordinate transformation or phase-locked loop. So, the complicated calculation issue can be avoided. What is more, conventional imaginary current construction is not necessary. Problems like time delay can also be avoided. The inner loop current calculation is simplified compared with the conventional dq control scheme. Stability of the proposed control scheme can be guaranteed through the analysis based on small-signal model.

Simulation and experimental results were presented to verify the effectiveness of the proposed control method. In the steady state, the proposed method can obtain the sinusoidal grid current and unity power factor under unbalanced load conditions. Upon comparing the dynamic response of the presented control strategy with that of the dq control scheme, the conducted experiments indicate that the salient feature of the proposed scheme is as follows: the proposed scheme maintains a fast dynamic response advantage over the conventional dq control method. This approach has been shown to be useful for the single-phase CHB multilevel converter system. The proposed control scheme can also be utilized with other multilevel converter systems.

Author Contributions: The individual contribution of each co-author with regards to the reported research and writing of the paper is as follows. D.Y. conceived the idea, L.Y. carried out the experiments and analyzed the data, and all authors wrote the paper.

Acknowledgments: This study is partly supported by the National Nature Science Foundation of China (No. 61861003).

Nomenclature

CHB	Cascaded H bridge
PWM	Pulse Width Modulation
DPC	Direct Power Control
DCC	Direct Current Control
P-DPC	Predictive DPC
PLL	Phase-Locked Loop
DSP	Digital Signal Processor
MPC	Model-Predictive Control
SOGI	Second-Order Generalized Integrator
PI	Proportional-Integral
PR	Proportional-plus-Resonant
CPS-PWM	Carrier, Phase-Shifted PWM
THD	Total Harmonic Distortion

References

1. Noman, A.M.; Al-Shamma'a, A.A.; Addoweesh, K.E.; Alabduljabbar, A.A.; Alolah, A.I. Cascaded Multilevel Inverter Topology Based on Cascaded H-Bridge Multilevel Inverter. *Energies* **2018**, *11*, 895. [CrossRef]
2. Aquila, A.D.; Liserre, M.; Monopoli, V.G.; Rotondo, P. Overview of PI-based solutions for the control of DC buses of a single-phase H-bridge multilevel active rectifier. *IEEE Trans. Ind. Appl.* **2008**, *44*, 857–866. [CrossRef]
3. Yu, Y.; Konstantinou, G.; Townsend, C.D.; Agelidis, V.G. Comparison of zero-sequence injection methods in cascaded H-bridge multilevel converters for large-scale photovoltaic integration. *IET Renew. Power Gener.* **2017**, *11*, 603–613. [CrossRef]
4. Wang, L.; Zhang, D.; Wang, Y.; Wu, B.; Athab, H.S. Power and voltage balance control of a novel three-phase solid-state transformer using multilevel cascaded H-bridge inverters for microgrid applications. *IEEE Trans. Power Electron.* **2016**, *31*, 3289–3310. [CrossRef]
5. Yang, Z.; Sun, J.; Li, S.; Huang, M.; Zha, X.; Tang, Y. An adaptive carrier frequency optimization method for harmonic energy unbalance minimization in a cascaded H-bridge-based active power filter. *IEEE Trans. Power Electron.* **2018**, *33*, 1024–1037. [CrossRef]
6. Song, W.; Ma, J.; Zhou, L.; Feng, X. Deadbeat predictive power control of single-phase three-level neutral-point-clamped converters using space-vector modulation for electric railway traction. *IEEE Trans. Power Electron.* **2016**, *31*, 721–732. [CrossRef]
7. Ohnishi, T. Three PWM converter/inverter by means of instantaneous active and reactive power control. In Proceedings of the International Conference on Industrial Electronics, Control and Instrumentation. (IECON 1991), Kobe, Japan, 28 October–1 November 1991; pp. 819–824.
8. Wang, X.; Sun, D. Three-vector-based low-complexity model predictive direct power control strategy for doubly fed induction generators. *IEEE Trans. Power Electron.* **2017**, *32*, 773–782. [CrossRef]
9. Kim, I.; Chan, R.; Kwak, S. A voltage level based predictive direct power control for modular multilevel converter. *IEEE Trans. Power Appl.* **2017**, *11*, 784–792.
10. Cheng, C.; Nian, H.; Wang, X.; Sun, D. Dead-beat predictive direct power control of voltage source inverters with optimised switching patterns. *IET. Power Electron.* **2017**, *10*, 1438–1451. [CrossRef]
11. Ma, H.; Xie, Y.; Shi, Z. Improved direct power control for Vienna-type rectifiers based on sliding mode control. *IET. Power Electron.* **2016**, *9*, 427–434. [CrossRef]
12. Ma, J.; Song, W.; Wang, S.; Feng, X. Model predictive direct power control for single phase three-level rectifier at low switching frequency. *IEEE Trans. Power Electron.* **2018**, *33*, 1050–1062. [CrossRef]
13. Behrouzian, E.; Bongiorno, M. Investigation of negative-sequence injection capability of cascaded H-bridge converters in star and delta configuration. *IEEE Trans. Power Electron.* **2017**, *32*, 1675–1683. [CrossRef]
14. Adam, G.P.; Abdelsalam, I.A.; Ahmed, K.H.; Williams, B.W. Hybrid multilevel converter with cascaded H-bridge cells for HVDC applications: Operating principle and scalability. *IEEE Trans. Power Electron.* **2015**, *30*, 65–77. [CrossRef]
15. Wang, X.; He, Z.; Yang, J. Electric Vehicle Fast-Charging Station Unified Modeling and Stability Analysis in the *dq* Frame. *Energies* **2018**, *11*, 1195. [CrossRef]
16. Liu, J.; Liu, Z. Harmonic Analyzing of the Double PWM Converter in DFIG Based on Mathematical Model. *Energies* **2017**, *10*, 2087. [CrossRef]
17. Gu, Y.; Wang, Y.; Xiang, X.; Li, W.; He, X. Improved virtual vector control of single-phase inverter based on unified model. *IEEE Trans. Energy Convers.* **2014**, *29*, 611–618. [CrossRef]
18. Zheng, X.; Xiao, L.; Wang, Z.; Lei, Y.; Wang, C. Control strategy without phase-locked loop based on coordinate transformation for three-phase AC/DC converter. *IET. Power Electron.* **2015**, *8*, 1701–1709. [CrossRef]
19. Li, Y.; Yang, J.; Wang, H.; Ge, W.; Ma, Y. A Hybrid Filtering Technique-Based PLL Targeting Fast and Robust Tracking Performance under Distorted Grid Conditions. *Energies* **2018**, *11*, 973. [CrossRef]
20. Hamed, H.A.; Abdou, A.F.; Bayoumi, E.; EL-Kholy, E.E. Effective design and implementation of GSS-PLL under voltage dip and phase interruption. *IET. Power Electron.* **2018**, *11*, 1018–1028. [CrossRef]
21. Xiao, F.; Dong, L.; Li, L.; Liao, X. A frequency-fixed SOGI-based PLL for single-phase grid-connected converters. *IEEE Trans. Power Electron.* **2017**, *21*, 1713–1719. [CrossRef]
22. Sahoo, S.; Prakash, S.; Mishra, S. Power quality improvement of grid-connected DC microgrids using repetitive learning-based PLL under abnormal grid conditions. *IEEE Trans. Ind. Appl.* **2018**, *54*, 82–90. [CrossRef]

23. Yang, D.; Wu, N.; Yin, L.; Lu, Z. Natural frame control of single-phase cascaded H-bridge multilevel converter based on fictive-phases construction. *IEEE Trans. Ind. Electron.* **2018**, *54*, 3848–3857. [CrossRef]
24. Chan, R.; Kwak, S. Improved Finite-Control-Set Model Predictive Control for Cascaded H-Bridge Inverters. *Energies.* **2018**, *11*, 355. [CrossRef]
25. Zhang, Y.; Wu, X.; Yuan, X.; Wang, Y.; Dai, P. Fast model predictive control for multilevel cascaded H-bridge STATCOM with polynomial computation time. *IEEE Trans. Ind. Electron.* **2016**, *63*, 5231–5243. [CrossRef]
26. Qi, C.; Chen, X.; Tu, P.; Wang, P. Deadbeat control for a single-phase cascaded H-bridge rectifier with voltage balancing modulation. *IET Power Electron.* **2018**, *11*, 610–617. [CrossRef]
27. Song, W.; Deng, Z.; Wang, S.; Feng, X. A Simple Model Predictive Power Control Strategy for Single-Phase PWM Converters With Modulation Function Optimization. *IEEE Trans. Power Electron.* **2016**, *31*, 5279–5289. [CrossRef]
28. Blahnik, V.; Kosan, T.; Peroutka, Z.; Talla, J. Control of a single-phase cascaded H-bridge active rectifier under unbalanced load. *IEEE Trans. Power Electron.* **2018**, *33*, 5519–5527. [CrossRef]
29. Somkun, S.; Chunkag, V. Unified unbalanced synchronous reference frame current control for single-phase grid-connected voltage source converters. *IEEE Trans. Ind. Electron.* **2016**, *63*, 5425–5436. [CrossRef]
30. Sha, D.; Xu, G.; Xu, Y. Utility direct interfaced charger/discharger employing unified voltage balance control for cascaded H-bridge units and decentralized control for CF-DAB modules. *IEEE Trans. Ind. Electron.* **2017**, *64*, 7831–7841. [CrossRef]
31. Ebrahimi, M.; Khajehoddin, S.; Ghartemani, M.K. Fast and robust single-phase *dq* current controller for smart inverter applications. *IEEE Trans. Power Electron.* **2016**, *31*, 3968–3976. [CrossRef]
32. Shi, J.; Gou, W.; Yuan, H.; Zhao, T.; Huang, A.Q. Research on voltage and power balance control for cascaded modular solid-state transformer. *IEEE Trans. Power Electron.* **2011**, *26*, 1154–1166. [CrossRef]
33. Farivar, G.; Townsend, C.D.; Hredzak, B.; Pou, J.; Agelidis, V.G. Low-capacitance cascaded H-bridge multilevel STATCOM. *IEEE Trans. Power Electron.* **2016**, *32*, 1744–1754. [CrossRef]
34. Qi, C.; Chen, X.; Tu, P.; Wang, P. Cell-by-cell-based finite-control- set model predictive control for a single-phase cascaded H-bridge rectifier. *IEEE Trans. Power Electron.* **2018**, *33*, 1654–1665. [CrossRef]
35. Moeini, A.; Wang, S. A DC Link Sensor-Less Voltage Balancing Technique for Cascaded H-Bridge Multilevel Converters with Asymmetric Selective Harmonic Current Mitigation-PWM. *IEEE Trans. Power Electron.* **2018**, *33*, 7571–7581. [CrossRef]
36. Farivar, G.; Hredzak, B.; Agelidis, V.G. Decoupled control system for cascaded H-bridge multilevel converter based STATCOM. *IEEE Trans. Ind. Electron.* **2016**, *63*, 322–331. [CrossRef]
37. Zhao, T.; Wang, G.; Bhattacharya, S.; Huang, A.Q. Voltage and power balance control for a cascaded H-bridge converter-based solid-state transformer. *IEEE Trans. Power Electron.* **2013**, *28*, 1523–1532. [CrossRef]
38. Chen, H.; Wu, P.; Lee, C.; Wang, C.; Yang, C.; Cheng, P. A flexible DC voltage balancing control based on the power flow management for star-connected cascaded H-bridge converter. *IEEE Trans. Ind. Appl.* **2016**, *2*, 4946–4954. [CrossRef]
39. Moeini, A.; Wang, S. The state of charge balancing techniques for electrical vehicle charging stations with cascaded H-bridge multilevel converters. In Proceedings of the IEEE Applied Power Electronics Conference and Exposition—APEC, San Antonio, TX, USA, 4–8 March 2018; pp. 637–644.
40. Yang, Y.; Zhou, K.; Wang, H.; Blaabjerg, F.; Wang, D.; Zhang, B. Frequency adaptive selective harmonic control for grid-connected inverters. *IEEE Trans. Power Electron.* **2015**, *30*, 3912–3924. [CrossRef]

7

New DC Grid Power Line Communication Technology used in Networked LED Driver

Huipin Lin [1], Jin Hu [2], Xiao Zhou [1], Zhengyu Lu [1] and Lujun Wang [3,*]

1. College of Electrical Engineering, Zhejiang University, Hangzhou 310027, China; linhuipin@zju.edu.cn (H.L.); eezhoux@zju.edu.cn (X.Z.); eeluzy@cee.zju.edu.cn (Z.L.)
2. Zhejiang Dabong Technology Co., Ltd., Hangzhou 311100, China; justinhu@dabong.com.cn
3. Hubei Key Laboratory for High-Efficiency Utilization of Solar Energy and Operation Control of Energy Storage System, Hubei University of Technology, Wuhan 430068, China

* Correspondence: wanglujun@zju.edu.cn

Abstract: In order to reduce the cost and improve the reliability, real-time performance, and installation convenience of remote-controlled light-emitting diode (LED) lighting systems, a networked LED driving technology based on the direct current (DC) grid power line carrier is proposed. In this system, an alternating current (AC)/DC bus converter converts the mains into a DC bus with multiple distributed LED drive powers on the DC bus. The AC/DC bus converter receives the user's control command and modulates it into the DC bus voltage. The DC bus waveform changes to a square wave containing the high and low changes of the address information and command information. The LED drive power of the corresponding address receives energy from the DC bus and demodulates the commands, such as turning the lights on and off, dimming, etc., and performs the action. In order to make the waveform of the AC/DC bus converter have better rising and falling edge, this paper adopts the half-bridge topology with variable modal control. In the modulation process, the circuit works in buck mode and boost mode. Distributed LED drivers have DC/DC circuits and very simple demodulation circuits that dissipate energy and information from the DC bus. Through experiments, the technology not only simplifies the use of communication technology in application, but also reduces the application difficulty.

Keywords: PLC; bus converter; DC bus; LED driver

---

1. Introduction

With the rapid development of light-emitting diode (LED) technology, it is widely predicted that LED will become a new generation of lighting sources. Smart network control is a hot research topic in the field of LED lighting applications. In this application, multiple LED devices generally form an intelligent lighting system through a network, and the LED devices can usually be remotely controlled. In addition to road lighting for large-scale applications [1–4], most of these smart LED lighting systems are often used indoors or outdoors, and the scale is small [5–7].

Figure 1 shows a case for a small-scale LED smart lighting system, such as for commercial lighting, home lighting, landscape lighting, wall washers, high pole lights, etc. In these applications, the number of lamps is small, the distance between two adjacent LED light sources is relatively close, and a computer is typically used to remotely control turning on, turning off, and dimming the lights.

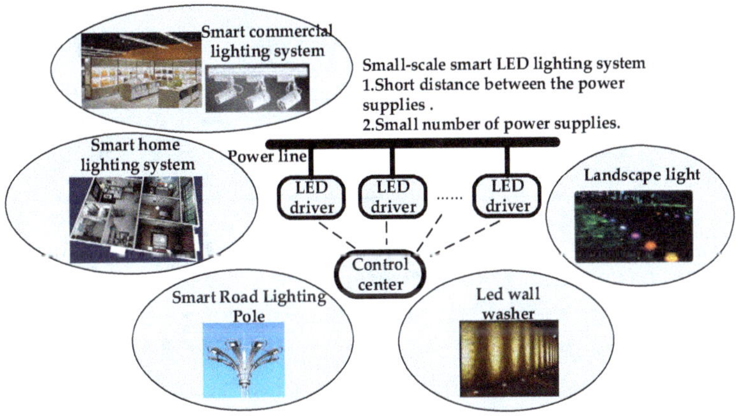

Figure 1. Common small networked control light-emitting diode (LED) smart lighting system.

In this small LED smart lighting system, communication between LED drivers is necessary to achieve control. Which communication technology is used determines the real-time performance and reliability of control in the system, while at the same time it also affects the cost, installation, and maintenance. Figure 2 shows descriptions of several common traditional lighting system communication schemes and their characteristics. These small LED lighting systems require a lower-cost solution, however, for 0–10 V, pulse-width modulation (PWM) control, digital addressable lighting interface (DALI) [8–14] and other remote-control schemes, and additional communication lines must be laid, which is unfavorable for installation and maintenance and costs more. Power line communication (PLC) [15–18] and the remote-control scheme of micropower wireless communication have advantages in practical applications due to the convenient installation. However, the cost of such a solution is still high, which is not conducive to its promotion. Table 1 lists the characteristics of these four options.

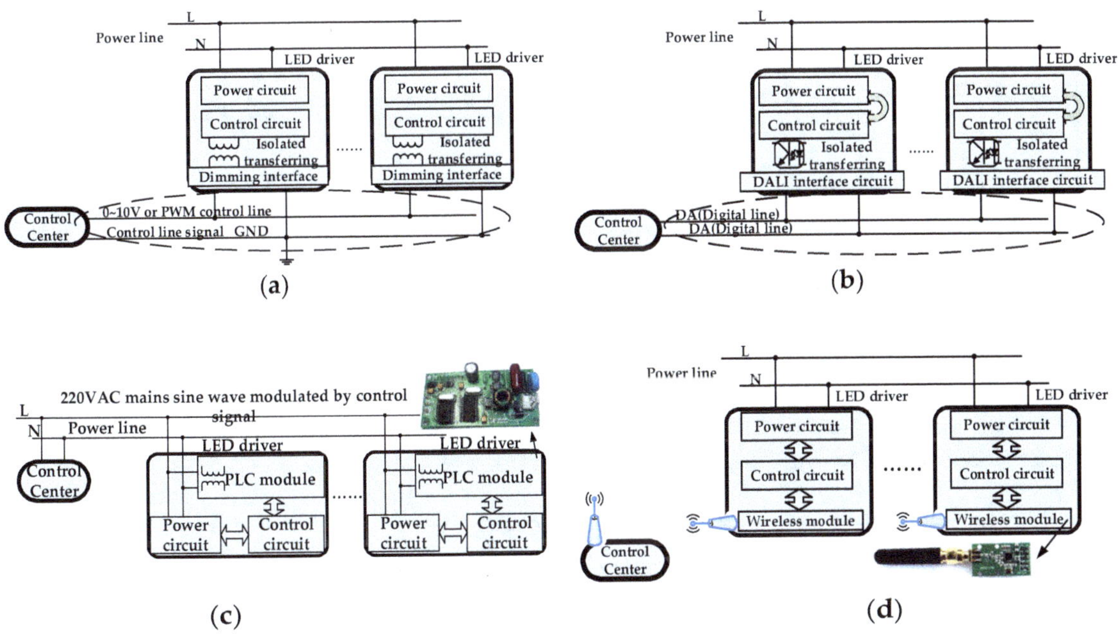

Figure 2. Common traditional remote control schemes: (**a**) 0–10 V or pulse-width modulation (PWM); (**b**) digital addressable lighting interface (DALI); (**c**) traditional alternating current (AC) grid power line communication; (**d**) traditional wireless.

Table 1. Features of common traditional remote control schemes.

Remote Control Scheme	Control Lines	Installation	Cost
0–10 V or PWM	2 Lines	Complex	High
DALI	2 Lines	Complex	High
Power line communication (PLC)	No	Very simple	High
Wireless	No	Simple	High

From the perspective of installation, the micropower wireless communication scheme needs to consider factors such as antenna and electromagnetic interference, which is greatly affected by space, and the PLC scheme is simpler. If the PLC solution can be further simplified, the cost and installation aspects are more advantageous. However, the traditional PLC solution is mainly used for alternating current (AC) mains grid, such as remote meter reading [19–21] and other applications. In the AC mains grid, PLC technology is relatively mature, and it is difficult to have major technological breakthroughs in a short period of time. It is difficult to reduce costs significantly. If power electronics can be used to transmit information while energy is being converted, the PLC module in Figure 2c may no longer be needed and the system cost may be greatly reduced. In the AC mains grid, since the existing transmission and distribution schemes are already very mature, it is difficult to implement communication using power electronics technology. However, if a direct current (DC) grid or an AC microgrid is constructed, it is easy to realize communication while transmitting DC power in the grid.

The LED lighting system based on DC grid has received attention. In [22] an indoor LED smart lighting system was proposed based on DC grid, as shown in Figure 3. The system uses an AC/DC converter to change AC mains to DC bus, and a number of distributed DC/DC LED drivers are connected to the DC bus. The system implements intelligent control and energy saving, but with the DALI communication protocol, a separate control line is still required. A similar DC bus architecture is also proposed in [23].

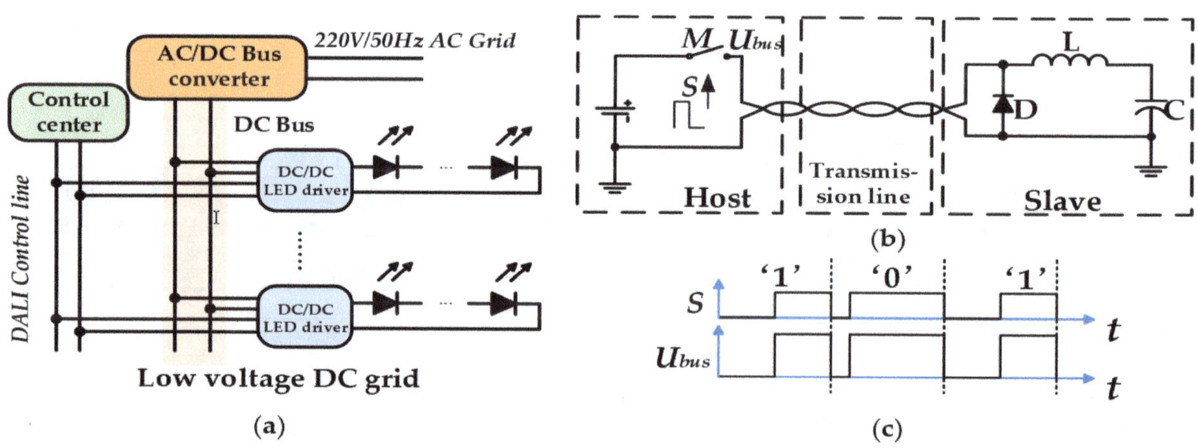

Figure 3. Traditional LED lighting system technology based on direct current (DC) grid: (**a**) Traditional networked LED system based on DC grid; (**b**) Traditional DC grid PLC technology; (**c**) the drive signal waveform of host switch M and the waveform of bus voltage U_{bus}.

PLC in the DC grid has also been studied in recent years, and this technology can eliminate the control line in Figure 3a. A novel DC PLC solution was proposed in [24], which is cheaper for small-scale power line communication applications. Figure 3b,c show the basic principle, as shown in the waveform: when the host switch M is turned on, the voltage U_{bus} goes high, representing logic 1. When switch M is turned off, voltage U_{bus} goes low, representing logic 0. The circuit works in the structure of the buck circuit, and the energy is stored in the output capacitor C of the slave. The change of the voltage U_{bus} during the energy transfer is detected by the slave, and the simultaneous transmission of energy and signals is realized. However, due to the influence of parasitic parameters on the transmission line, the switching device of the circuit is subject to greater voltage stress than the conventional buck circuit.

Some papers have proposed DC PLC schemes [25–29], but they all need additional communication lines. They bring larger parasitic parameters and cause front-stage metal-oxide-semiconductor field effect transistor (MOSFET) high current spikes and damage when the MOSFETs are working at high frequency. So the front-stage MOSFETs need better performance requirements. They are more complicated if used for LEDs.

In order to overcome the problems of the above two types of research, this paper proposes a remotely controlled LED lighting system similar to the DC grid structure of Figure 3a, but with the control line omitted. In this system, an AC/DC bus converter converts the mains into a DC bus, and multiple distributed LED drivers are connected to the DC bus. The AC/DC DC bus converter receives the user's control command and modulates it into the DC bus voltage. The DC bus waveform changes to a square wave containing the high and low changes of the address information and the command information. The LED driving power source corresponding to the address receives energy from the DC bus and demodulates commands such as turning on the light, turning off the light, dimming, etc. In order to make the waveform of AC/DC DC bus converter have better rising and falling edge, this paper adopts the half-bridge topology with variable modal control. In the modulation process, the circuit works in buck mode and boost mode. Distributed LED drivers have DC/DC circuits and very simple demodulation circuits that dissipate energy and information from the DC bus.

With this remote control scheme, the following problems are solved:

1. Compared with 0–10 V, PWM, and DALI controlled systems, the communication line is omitted.
2. The traditional AC mains PLC module is saved, and the signal transceiving function is realized by power electronic technology.
3. Compared with the existing DC PLC technology, the influence of power line on communication is greatly reduced, and communication reliability is improved.

Compared with the conventional schemes described above, this scheme is simpler and costs less. This paper verifies the feasibility of this technology by experiments.

2. Light-Emitting Diode (LED) Networked Drive Technology Based on Direct Current (DC) Grid Power Line Communication (PLC)

2.1. System Architecture

The specific architecture of the LED lighting system proposed in this paper is shown in Figure 2a. It includes a control center, an AC/DC bus converter, a DC bus, and a distributed LED driver. The control center sends control signals to the LEDs to the AC/DC bus converter. This type of technology is relatively mature and can be used for wireless communication such as a General Packet Radio Service (GPRS). The bus converter converts the AC mains into a DC bus voltage v_{bus} with a maximum value of 48 V, and modulates the information to be transmitted into the bus voltage to make it high and low DC levels. The high voltage represents logic 1 and the low voltage represents logic 0. The DC bus is the transmission line, and the distributed LED driver as the load extracts energy from the DC bus drives the LED light source and demodulates the signal from the bus.

2.2. New DC Power Line Carrier Energy and Information Demodulation Principle

The master/slave communication mode is between the AC/DC bus converter and the distributed LED driving power source. The former is the master, the latter is the slave, and the DC bus is used as the medium to provide energy and transfer information to the slave.

The LED driver power supply includes three parts: an energy demodulation circuit, a signal demodulation circuit, and a DC/DC converter. The energy and signal demodulation circuits are both half-wave rectification circuits, as shown in Figure 4a. Taking the jth LED driving power supply in

Figure 4a as an example, the energy demodulating circuit is connected in series with the DC/DC converter, and the output of the signal demodulating circuit is connected to the control circuit of the DC/DC converter. The difference between the two is that the energy demodulation circuit's output capacitance C_{pj} has a large capacitance value. When the v_{bus} voltage is 0, the energy of the DC/DC converter of the latter stage can still be supplied for a period of time, that is, v_{pj} is maintained within the allowable ripple voltage range to keep the circuit working properly.It remains stable for a long time; the output capacitance value C_{sj} of the latter is small. When the v_{bus} voltage is 0, v_{sj} is discharged through the resistor R_{sj} and quickly falls to 0, so that the change of v_{sj} is consistent with the change of v_{bus}. Taking the waveform in Figure 4b as an example, the process of demonstrating a remote dimming operation has the following stages:

t_1: The output of the AC/DC bus converter is 48 V DC, and the load is full power. During this time, the bus converter receives the *j*th LED drive power dimming command from the control center to address A_j.

t_1–t_2: The bus converter modulates the information into the DC bus, and the bus voltage v_{bus} changes to the waveform shown in Figure 4b. All LED drive power supplies receive the command, but only the *j*th addresses A_j. The drive power source executes the dimming command. The total power on the busbar drops from time t_3 when the command is executed. The communication protocol can adopt a mature scheme such as RS232 or a custom scheme.

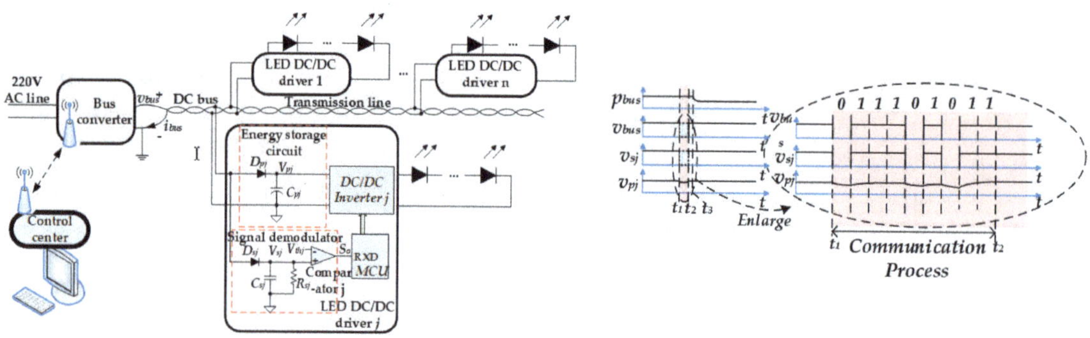

(a) (b)

Figure 4. Networked control LED system based on DC grid PLC: (**a**) system architecture diagram; (**b**) system key waveforms.

2.3. Working Principle of Bus Converter Based on Variable Modal Control

The key device for system energy and information modulation is the AC/DC bus converter. Figure 5a,b shows a schematic diagram and the corresponding buck and boost working modes. The circuit is divided into two stages. The first stage is a power factor correction (PFC) circuit to achieve a high power factor. The output is a fixed DC voltage V_g as the input of the second stage.

The second stage is a half-bridge circuit, the output is a DC bus, and the DC bus is connected to several LED driving power sources. The LED driving power can be regarded as the active load of the second-stage half-bridge circuit, and the total equivalent power is p_{LED}. D_1 and D_2 are body diodes of the half-bridge circuit MOSFETs M_1 and M_2, respectively. The control circuit has a single-chip microcomputer (MCU), which samples the output voltage v_{bus} through the sampling resistors R_1 and R_2, and $R_{Boostcs}$ and R_{Buckcs} are sampling resistors, respectively sampling the inductor currents in boost and buck modes.

In different situations, the half-bridge circuit operates in buck mode or boost mode. Figure 5c shows the main waveforms of these two modes.

t_0–t_1: The AC/DC bus converter does not send a signal to the LED driver, and its output voltage is constant DC 48 V. The upper tube M_1 is operated as a high-frequency switch, the lower tube M_2 is stopped, the half-bridge circuit operates in buck mode as shown in Figure 5a, and the body diode

D_2 of the M_2 acts as a freewheeling diode of the buck circuit. The bus converter charges capacitor C_2, which supplies energy to the distributed LED driver.

t_1–t_{11}: The bus converter receives the command and modulates it into the bus voltage v_{bus}, causing it to change to high and low DC levels. Defining the signal S_{in} passed, when S_{in} is logic 0, the bus converter operates in boost mode, and when S_{in} is logic 1, the bus converter operates in buck mode. When in boost mode, the upper tube M_1 stops working, and M_2 operates as a high-frequency switch. The body diode D_1 of the M_1 operates as a freewheeling diode of the boost circuit, as shown in Figure 5b. At this time, the energy on the bus voltage v_{bus} across the output capacitor C_2 is quickly transferred to the half-bridge input capacitor C_1, and the level rapidly drops to zero. Since the LED driving power supply in Figure 4a has diodes D_{pj} and D_{sj}, the change in bus voltage does not affect the energy and signal demodulating circuits in the LED driving power supply.

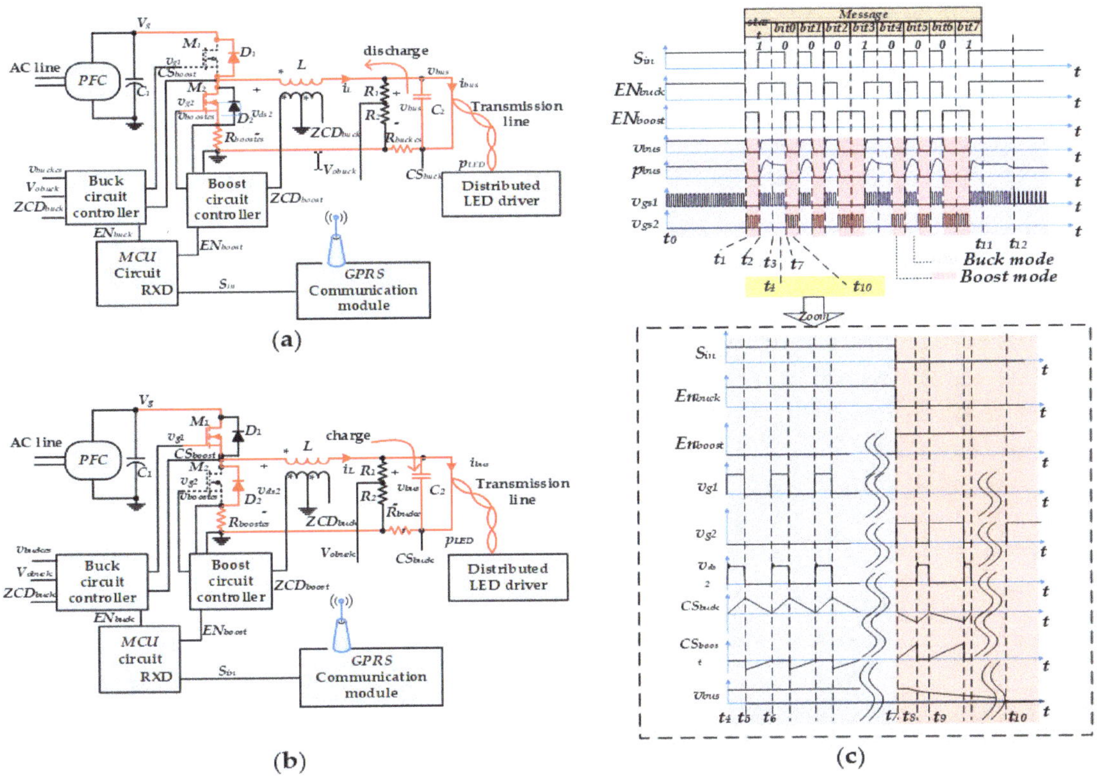

Figure 5. Work principles of bus converter: (**a**) circuit principle of buck mode; (**b**) circuit principle of boost mode; (**c**) working waveform.

The half-bridge circuit adopts different working modes, so that in buck mode, the bus voltage v_{bus} across capacitor C_2 can rise rapidly; in boost mode, the bus voltage v_{bus} across capacitor C_2 can be quickly reduced. There are three advantages to using this method of variable mode control:

(1) The bus voltage v_{bus} changes rapidly, which is beneficial to improve the signal transmission speed and facilitate the operation of the signal demodulation circuit in the LED driver.

(2) If there is no boost mode, the bus converter should make the busbar low voltage and can only rely on the load to discharge naturally. However, in the case of light load of the bus converter, especially when the load is turned off, when the bus converter is equivalent to no-load, the DC bus voltage v_{bus} drops to 0, which takes a long time, and the energy storage of the LED driving power capacitor C_{pj} may not last so long. Boost mode allows the bus to quickly go low under light load conditions.

(3) Since boost mode feeds back the energy of output capacitor C_2 to input capacitor C_1, it can be used again next time, so the energy utilization efficiency in the communication process is higher.

t_{12}–: After the drive power supply executes the dimming command, the total p_{bus} power on the bus is reduced.

As shown in Figure 5a,b, it is divided into two modes, buck and boost. Due to the poor reverse recovery performance of the body diode of the MOSFET, the circuit uses boundary current mode (BCM) regardless of buck or boost mode. The control circuits of both modes are controlled by ST L6562. In buck mode, M_2 is always off and its body diode D_2 operates as a freewheeling diode of the buck circuit. In the main circuit, the signal ZCD_{Buck} that controls M1 turn-on, the peak current signal CS_{Buck} that controls M1 turn-off, and the output voltage signal V_{oBuck} are sampled into the buck circuit controller. In boost mode, M_1 is always off, and its body diode D_1 acts as a freewheeling diode of the boost circuit. In the main circuit, the signal ZCD_{Boost} that controls M_2 turn-on and the peak current signal CS_{Boost} that controls M_2 turn-off are sampled into the boost control circuit.

In the t_4–t_{10} time of Figure 5c, the specific working waveforms of the two working modes are given. At time t_4, the transmitted signal S_{in} is still high, EN_{Buck} is high, and EN_{Boost} is low, indicating that the circuit is in buck mode, M1 is operational, M_2 is disabled, and its drive signal V_{g2} is always low. The body diode D_2 of M_2 operates as a freewheeling diode of the buck circuit. The energy flow of the circuit flows from capacitor C_1 to C_2, as shown in Figure 5a.

$$v_g - v_{bus} = L\frac{i_{Lpk}}{t_5 - t_4} \tag{1}$$

where i_{Lpk} is the peak value of the inductor current during this period.

t_5–t_6: M_1 is turned off, the inductor current drops, the body diode D_2 of M_2 continues to flow, CS_{Boost} is negative voltage, and the half-bridge midpoint voltage V_{ds2} is low. The voltage and current in the circuit satisfy Equation (2), and the circuit gain relationship still satisfies the law of the buck circuit, as in Equation (3).

$$v_{bus} = L\frac{i_{Lpk}}{t_6 - t_5} \tag{2}$$

$$v_{bus} = \frac{t_5 - t_4}{t_6 - t_4} \cdot v_g \tag{3}$$

t_6–t_7: The circuit repeats the working process of the t_4–t_6 switching cycle until time t_7, the transmission signal S_{in} becomes low level, EN_{Buck} is low, and EN_{Boost} is high, indicating that the circuit is in boost mode; M_2 can work, M_1 is disabled, and its drive signal V_{g1} is always low. The body diode D_1 of M_1 operates as a freewheeling diode of the boost circuit. The energy flow of the circuit flows as shown in Figure 5b from capacitor C_2 to C_1.

t_7–t_8: M_2 is turned on, the inductor current rises in the reverse direction, CS_{Buck} is negative voltage, and the half-bridge midpoint voltage V_{ds2} is low. The voltage and current in the circuit satisfy the law of the general boost circuit, as shown in Equation (4).

$$v_{bus} = L\frac{-i_{Lpk}}{t_8 - t_7} \tag{4}$$

t_8–t_9: M_2 is turned off, the inductor current drops, the body diode D_1 of M_1 continues to flow, CS_{Buck} is negative voltage, and the half-bridge midpoint voltage V_{ds2} is high. The voltage and current in the circuit satisfy Equation (5), and the circuit gain relationship still satisfies the law of the boost circuit, as in Equation (6).

$$v_g - v_{bus} = L\frac{-i_{Lpk}}{t_9 - t_8} \tag{5}$$

$$v_g = \frac{t_9 - t_7}{t_9 - t_8} \cdot v_{bus} \tag{6}$$

t_9–t_{10}: The circuit repeats the working process of the t_7–t_9 switching cycle until t_{10}, because the voltage on C_2 is already very low, meaning that most of the energy on C_2 is fed back to C_1.

The working principle of the bus converter described above is that the modal control of the circuit realizes the transformation of the high and low levels on capacitor C_2 to achieve the purpose of simultaneously transmitting energy and information.

2.4. Distributed LED Drive Power DC/DC Converter Working Principle

The distributed LED drive power directly drives the light source and demodulates the communication information from the DC bus. The driving power source is divided into an energy demodulating circuit, an information demodulating circuit, and a DC/DC converter. The principles of the first two have already been introduced. Here, the working principle of the DC/DC converter is introduced.

In theory, there are many solutions for implementing the DC/DC converter of the driving power supply. This paper introduces the principle of demodulation energy and information by taking the buck circuit as an example. Figure 6 shows the schematic and key waveforms of the DC/DC converter in the distributed drive power supply of Figure 4a.

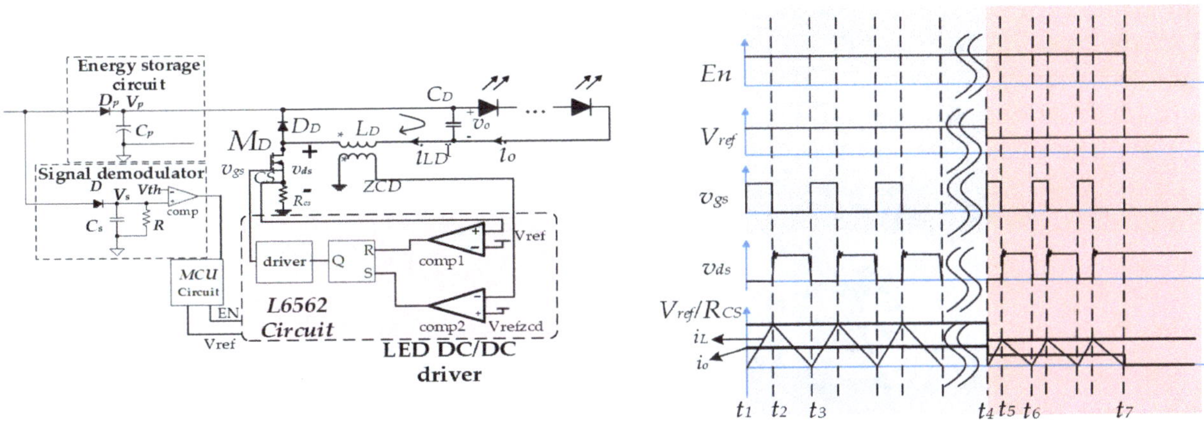

(a) (b)

Figure 6. Principle of distributed drive power DC/DC converter: (**a**) schematic diagram; (**b**) key waveform.

The DC/DC converter uses a buck circuit and operates in a critical mode. Due to the critical mode, the output current i_o and the peak value i_{LDpk} of the inductor current i_{LD} satisfy the following relationship:

$$i_o = \frac{1}{2} i_{LDpk} \tag{7}$$

According to Equation (7), the output current i_o can be made constant by controlling i_{LDpk} to be constant. The control circuit is implemented by chip L6562. The specific working process is as follows:

t_1–t_2: The enable signal EN given by the microcontroller allows the DC/DC converter to operate. The MOSFET MD is turned on and the inductor current rises. The voltage and current in the circuit meet the following equation:

$$V_p - v_o = \frac{L_d \cdot i_{LDpk}}{t_2 - t_1} \tag{8}$$

t_2–t_3: At time t_2, the voltage on the sampling resistor R_{CS} reaches the threshold value V_{ref} of comparator comp1, comp1 gives a positive level, and the output level of the RS flip-flop becomes low, so that the M_D is turned off. At this time, the diode D_D starts freewheeling, and the inductor current decreases. At time t_3, since the inductor current drops to zero, the Z_{CD} voltage drops, and the comparator comp2 output is high, causing the RS flip-flop output to be high, and the drive circuit starts driving the M_D to conduct and begins to enter the next switching cycle. During this period, the voltage and current in the circuit satisfy Equation (9), and the circuit gain relationship still satisfies the law of the buck circuit, as in Equation (10).

$$v_o = \frac{L_d \cdot i_{LDpk}}{t_3 - t_2} \quad (9)$$

$$v_o = \frac{t_2 - t_1}{t_3 - t_1} \cdot V_P \quad (10)$$

The t_3–t_4 circuit repeats the switching cycle state of t_1–t_3. It should be pointed out that V_{ref} is given by the single-chip circuit and V_{ref} changes, according to Equation (7); the output current also changes.

t_4–t_7: The single-chip circuit receives the dimming command from the bus converter, reduces the V_{ref} voltage, and reduces the output current i_o of the DC/DC converter.

t_7: The single-chip circuit receives the shutdown command from the bus converter, turns the enable signal EN to low, the DC/DC converter stops working, and the output current is zero.

Through the working principles described above, the distributed LED driving power supply can drive the LED load according to the command sent by the bus converter.

3. Experimental Verification

The principles of system operation were introduced in the previous section, and the working process of the bus converter and distributed LED driver power supply is analyzed in detail. In order to verify the technology mentioned in this section, this paper establishes a verification system as shown in Figure 4a. The physical diagram is shown in Figure 7. System parameters are shown in Table 2.

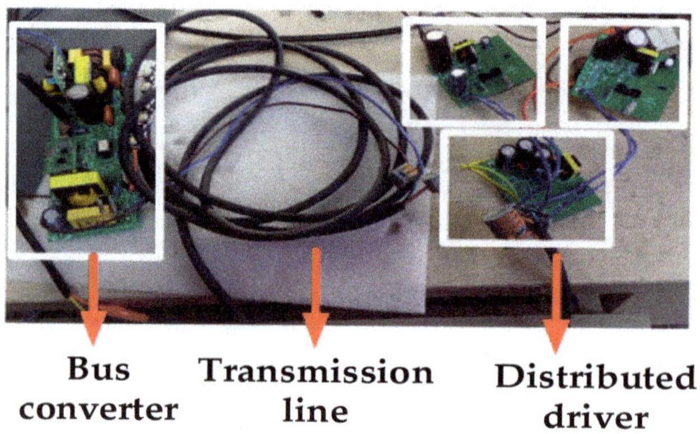

Figure 7. Picture of demo system.

Table 2. System key parameters.

Bus Converter	Quantity	1
	Input voltage	220 V/50 Hz
	PFC voltage	400 V
	Output voltage maximum	48 V
	Total power maximum	30 W
	Communication baud rate	9600
Distributed LED Driver	Quantity	3
	Single rated power	10 W
	Single load structure	3 W LED 3 series 3 parallel
	Single load rated current	700 mA

3.1. Bus Converter Working Principle Verification

The main task of the bus converter is to convert the AC mains into the bus of the DC grid. During the transition, the DC grid output voltage is modulated into a signal that is communicated to the distributed LED drive power source, allowing both signal and energy to be delivered simultaneously. This experiment mainly verifies this function.

Figure 8 shows the input and output waveforms of the bus converter. It can be seen that under the 220 V AC mains, the bus converter converts it into a 48 V DC bus v_{bus}. After the communication signal S_{in} is modulated into the v_{bus}, the S_{in} changes, and the v_{bus} also changes in the same waveform to achieve simultaneous transmission of energy and information. Due to the presence of the PFC circuit, the power factor is approximately 1, and the input voltage and current waveforms are approximately the same.

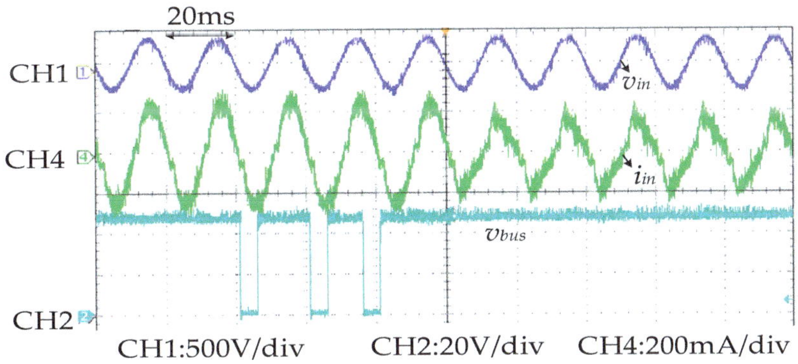

Figure 8. Input and output waveforms of bus converter.

According to the above description, when the DC bus voltage v_{bus} is high, the bus converter is in buck mode, and the output v_{bus} is high. When v_{bus} is low, in order to quickly reduce the voltage of the bus converter output capacitor C_2, the bus converter goes into boost mode, so that the C_2 energy is quickly transferred to the bus converter input capacitor C_1 for continued use in the next cycle. Figure 9 shows the main waveforms of the modal transformation. The pass signal S_{in} is the same as the enable signal of buck mode. It can be seen from the waveform that in steady state, the bus converter operates in buck mode, the signal S_{in} is high, the drive V_{g1} of the upper tube M_1 is working, the drive V_{g2} of the lower tube M_2 is low, and the output voltage v_o is constantly 48 V. When the bus converter receives the command, it modulates the signal S_{in} into the output bus, causing it to become a square wave with the same high and low conversion as S_{in}. At this time, buck and boost modes of the bus converter are switched according to the transmission signal.

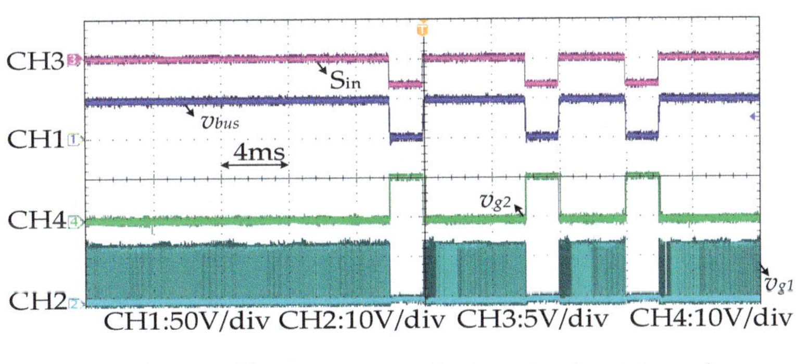

Figure 9. Main waveforms of bus converter during steady state and communication.

The output transformer of the bus converter is continuously charged and discharged during the signal modulation process. It is essentially the energy flow relationship between the input capacitance C_1 of the bus converter and the output capacitor C_2. Figure 10 shows the experimental relationship of energy flow. It can be seen from Figure 10 that in buck mode, the buck sampling resistor voltage v_{Buckcs} is positive, the boost sampling resistor voltage $v_{Boostcs}$ is negative, and the bus converter output capacitor C_2 is charged. In boost mode, the buck sampling resistor voltage v_{Buckcs} is negative, the boost sampling resistor voltage $v_{Boostcs}$ is positive, and the bus converter output capacitor C_2 energy is transferred to the input capacitor C_1.

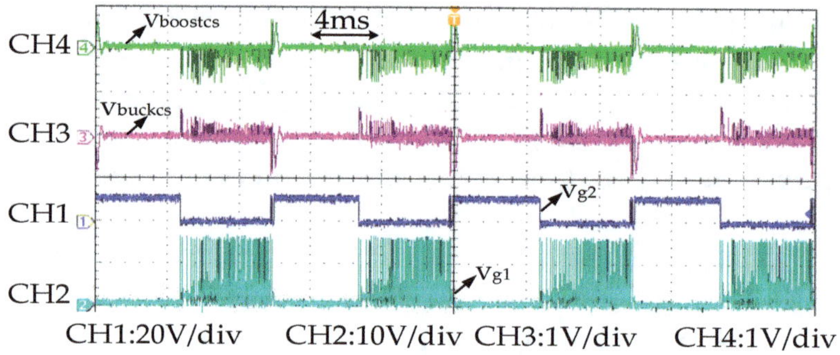

Figure 10. Energy flow relationship in the process of bus converter modal transformation.

As mentioned above, since the bus converter needs to use a body diode of a MOSFET of half-bridge topology as a freewheeling tube in either buck or boost mode, the reverse recovery capability of the MOSFET body diode is poor, in order to reduce this. The effect is that the circuit uses a critical mode. Figure 11a,b shows the key waveforms of the bus converter in buck and boost mode during communication. In buck mode, the sampling resistor voltage v_{Buckcs} and the half-bridge midpoint voltage v_{ds2} are visible, and the circuit operates in a critical mode, avoiding the loss caused by the reverse recovery of the body diode D_2 of the lower tube M_2. In boost mode, the sampling resistor voltage $v_{boostcs}$ and the half-bridge midpoint voltage v_{ds2} are visible, and the circuit operates in a critical conduction mode, avoiding the loss caused by the reverse recovery of the body diode D_1 of the upper MOSFET M_1. This experiment verifies the working principle of the bus converter.

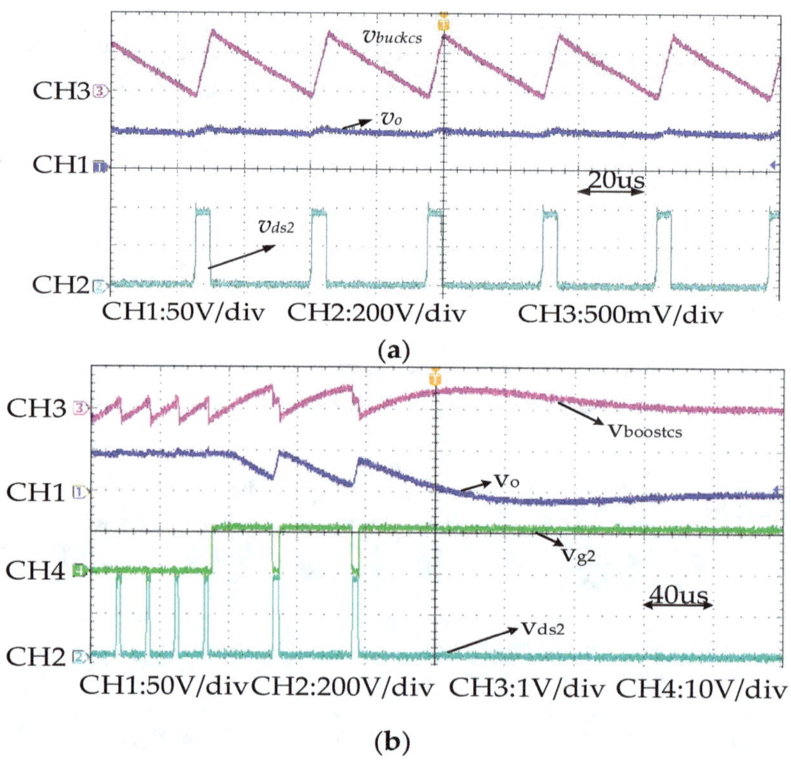

Figure 11. Critical conduction mode waveforms of the bus converter under (**a**) buck mode and (**b**) boost mode.

3.2. Distributed LED Driver Power Supply Working Principle Verification

The main task of the distributed LED drive power source is to receive the energy of the DC bus and demodulate the information. This experiment mainly verifies this function.

Figure 12 shows the main waveforms of the energy and information demodulation of the distributed LED driver power supply during communication. It can be seen from Figure 6 that when the enable signal EN is 1, the MOSFET M_D starts to work. As shown in Figure 12, v_{gs} is a high-low level transition of the driving signal of the MOSFET M_D, enabling the distributed LED driver to operate, so that the bus voltage v_{bus} is high level. At this time through the modulation and demodulation of the signal circuit, so that v_s is also high level, and in the energy circuit, v_p is also high level output. When the enable signal EN is 0, the MOSFET M_D stops working. At this time, the driving signal v_{gs} of M_D is kept low. At this time, in the signal circuit, v_s is also low, and in the energy circuit, due to the capacitance of C_p, as shown in the figure, is slightly dropped and can be maintained within the allowable ripple voltage range to keep the circuit working properly.

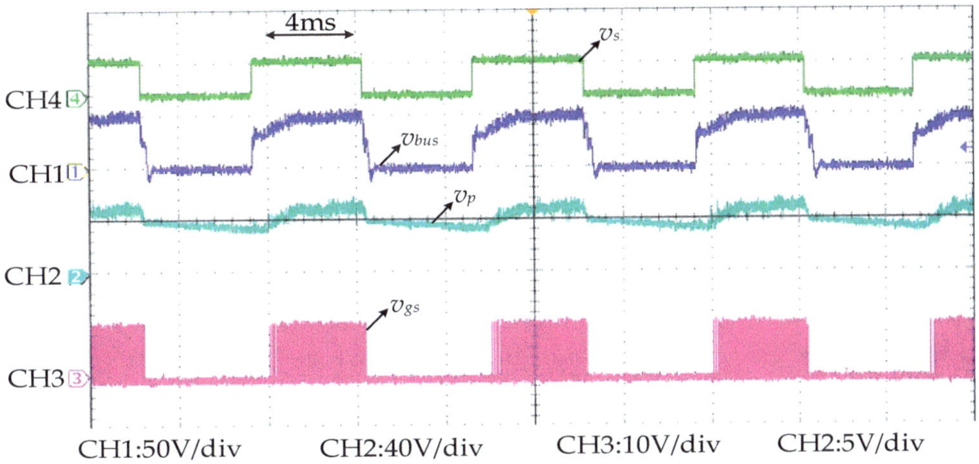

Figure 12. Distributed LED drive power energy and information.

In this section, the distributed LED driver power supply uses a critical conduction mode buck circuit, and the output current i_o is a constant current source. Figure 13 shows the operating waveform of the drive power supply. v_{cs} is the voltage of sampling resistance R_{cs}, R_{cs} is used to sample the current flowing through it. Figure 6a shows, when M_D is turned on, the current of inductor L_D flows through Rcs. v_{gs} is the drive signal of M_D, and v_{ds} is the voltage between drain and source of M_D. The circuit operates in a critical conduction mode, avoiding the loss caused by the reverse recovery of the body diode D_1 of the upper MOSFET M_1, and reducing circuit losses.

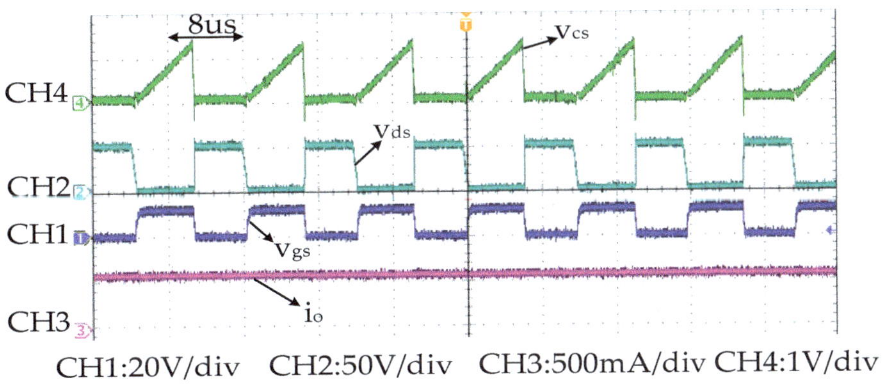

Figure 13. Critical conduction mode waveforms of distributed LED driver.

The LED driving power supply executes the command after demodulating the information on the DC bus. Figure 14 shows the waveform of the action performed after a drive power supply receives a light-off command. When the bus converter receives the command from the control center, it means to turn off the light, so that v_{bus} changes as shown in Figure 14, this signal represents the command

"LED off at a certain address". Through the modulation and demodulation of the circuit, the signal is passed through the signal conditioning circuit to the LED driver at an address as shown in Figure 6a, so that it performs the light-off action, as shown in the CS waveform, and finally becomes a low level.

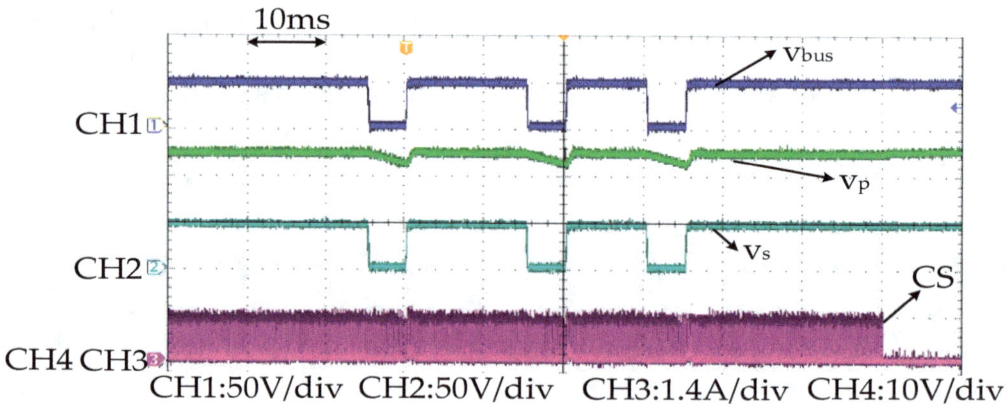

Figure 14. Execution command waveforms of distributed LED driver.

4. Conclusions

With common traditional lighting system remote control schemes like 0–10 V, PWM control, and DALI, two additional communication lines must be used, which brings installation inconvenience and costs more. Traditional PLC or wireless technologies do not need two communication lines. Although they are easy to install, they are not easy to popularize because of their high cost. A DC power grid based on the new power line carrier technology proposed in this paper enables the LED driving power source to have not only energy conversion capability, but also communication capability by using power electronics to realize communication technology. In order to realize the communication function of the driving power source, the proposed system includes a bus converter and a distributed driving power source. The bus converter modulates the information into the DC bus through variable modal control, from which the distributed drive power demodulates energy and information. In contrast with 0–10 V, PWM control, and DALI, this technology does not need communication lines, it is easier to install, and the cost is lower. Compared with the traditional PLC, its communication circuit is simpler, so the cost is lower. Compared with wireless technology, the radio frequency module can be omitted, so the cost is decreased and the installation is simpler. This technology not only simplifies the use of communication technology in application, but also reduces the difficulty of application. The prototype system validated the effectiveness of the technology. It is possible that this technology can also be used in other applications besides LED.

Author Contributions: Conceptualization, H.L. and J.H.; methodology, H.L.; software, H.L.; validation, H.L., J.H. and Z.L.; formal analysis, X.Z.; investigation, H.L.; resources, L.W.; data curation, J.H.; writing—original draft preparation, H.L.; writing—review and editing, H.L. and J.H.; visualization, H.L.; supervision, J.H.; project administration, Z.L.; funding acquisition, L.W.

Nomenclature

V_{bus}	DC bus voltage
i_{bus}	DC bus current
p_{bus}	DC bus input power
v_{pj}	Output capacitance voltage of the energy storage part in the *j*th LED DC/DC driver
C_{pj}	Output capacitance of the energy storage part in the *j*th LED DC/DC driver
v_{sj}	Output capacitance voltage of the signal demodulation part in the *j*th LED DC/DC driver
C_{sj}	Output capacitance of the signal demodulation part in the *j*th LED DC/DC driver
v_{thj}	Comparator reference of the signal demodulation part in the *j*th LED DC/DC driver
R_{sj}	C_{sj} Discharge resistance of the signal demodulation part in the *j*th LED DC/DC driver

S_o	The input signal from the signal demodulation part to MCU in the *j*th LED DC/DC driver
A_j	The address of the *j*th LED DC/DC driver
C_1	Input capacitance of the bus converter
C_2	Output capacitance of the bus converter
V_g	Output voltage of PFC in the bus converter
v_{g1}	Drive signal of the bridge's upper MOSFET in the bus converter
v_{g2}	Drive signal of the bridge's lower MOSFET in the bus converter
CS_{boost}	Peak current signal in Boost mode
CS_{buck}	Peak current signal in Buck mode
$R_{boostcs}$	Sampling resistance of Boost mode
R_{buckcs}	Sampling resistance of Buck mode
ZCD_{buck}	Control signal that M_1 is turn on
ZCD_{boost}	Control signal that M_2 is turn off
V_{obuck}	Feedback signal of sampling output voltage in Buck mode
i_L	Inductor current in the bus converter
R_1	Sampling resistor
R_2	Sampling resistor
S_{in}	Signal transmitted by the bus converter
V_{buckcs}	Voltage across the sampling resistor Rbuckcs in buck mode
$V_{boostcs}$	Voltage across the sampling resistor Rboostcs in boost mode
V_{ds2}	The midpoint voltage of bridge in the bus converter
EN_{buck}	Enable signal of Buck mode
EN_{boost}	Enable signal of Boost mode
i_o	Output current of the LED DC/DC driver
i_{LD}	The current of inductance L_D in the LED DC/DC driver
i_{LDpk}	The current peak of inductance L_D in the LED DC/DC driver
V_p	Output capacitance voltage of the energy storage part in the LED DC/DC driver
C_p	Output capacitance of the energy storage part in the LED DC/DC driver
V_s	Output capacitance voltage of the signal demodulation part in the LED DC/DC driver
C_s	Output capacitance of the signal demodulation part in the LED DC/DC driver
V_{th}	Comparator reference of the signal demodulation part in the LED DC/DC driver
R_s	Cs Discharge resistance of the signal demodulation part in the LED DC/DC driver
L_D	Inductance of the LED DC/DC driver
C_D	Output capacitor of the LED DC/DC driver
v_o	Output voltage of the LED DC/DC driver
R_{cs}	Sampling resistance of the LED DC/DC driver
EN	Enable signal of L6562 in the LED DC/DC driver
V_{ref}	Comparator reference of L6562's comp1 in the LED DC/DC driver
V_{refzcd}	Comparator reference of L6562's comp2 in the LED DC/DC driver
V_{ds}	Voltage between drain and source of M_D

References

1. Yan, W.; Hui, S.Y.R.; Chung, H.S. Energy Saving of Large-Scale High-Intensity-Discharge Lamp Lighting Networks Using a Central Reactive Power Control System. *IEEE Trans. Ind. Electron.* **2009**, *56*, 3069–3078. [CrossRef]
2. Djuretic, A.; Kostic, M. Actual energy savings when replacing high-pressure sodium with LED luminaires in street lighting. *Energy* **2018**, *157*, 367–378. [CrossRef]
3. Beccali, M.; Bonomolo, M.; Leccese, F.; Lista, D.; Salvadori, G. On the impact of safety requirements, energy prices and investment costs in street lighting refurbishment design. *Energy* **2018**, *165*, 739–759. [CrossRef]
4. Jin, H.; Jin, S.; Chen, L.; Cen, S.; Yuan, K. Research on the Lighting Performance of LED Street Lights with Different Color Temperatures. *IEEE Photonics J.* **2015**, *7*, 1–9. [CrossRef]
5. Gentile, N.; Govén, T.; Laike, T. A field study of fluorescent and LED classroom lighting. *Light. Res. Technol.* **2016**, 1–20. [CrossRef]
6. Fantozzi, F.; Leccese, F.; Salvadori, G.; Rocca, M.; Garofalo, M. LED Lighting for Indoor Sports Facilities: Can

Its Use Be Considered as Sustainable Solution from a Techno-Economic Standpoint? *Sustainability* **2016**, *8*, 618. [CrossRef]
7. Perdahci, C.; Akin, H.C.; Cekic, O. A comparative study of fluorescent and LED lighting in industrial facilities. In Proceedings of the 2018 7th International Conference on Clean and Green Energy (ICCGE 2018), Paris, France, 7–9 February 2018.
8. Liang, Y.; Yimin, C.; Baorong, C. Integrated analog dimming controller for 0–10 V dimming system. In Proceedings of the 2013 10th China International Forum on Solid State Lighting (ChinaSSL), Beijing, China, 10–12 November 2013; pp. 147–149.
9. Dai, G.N.; Dai, B.L. Application of DALI Bus in Large High-end Lighting. *Jiangsu Constr.* **2016**, *6*, 113–115. [CrossRef]
10. Wang, S.; Liu, Y.; Yuan-Lin, C.; Jyun-Yan, C. Development of DALI-based electronic ballast with energy saving control for ultraviolet lamps. In Proceedings of the 2010 8th IEEE International Conference on Industrial Informatics, Osaka, Japan, 13–16 July 2010; pp. 214–219.
11. Bellido-Outeirino, F.J.; Flores-Arias, J.M.; Domingo-Perez, F.; Gil-de-Castro, A.; Moreno-Munoz, A. Building lighting automation through the integration of DALI with wireless sensor networks. *IEEE Trans. Consum. Electron.* **2012**, *58*, 47–52. [CrossRef]
12. Liang, T.J.; Huang, J.F.; Yadav, P.K. Design and implementation of dimmable LED control circuit with DALI protocol. In Proceedings of the 2016 IEEE International Conference on Power and Energy (PECon), Nice, France, 28–29 November 2016; pp. 121–126.
13. Wang, P. A Design of DALI Illumination System Based on RS-485. *China Illum. Eng. J.* **2012**, *23*. [CrossRef]
14. Khalilian, H.; Farzanehfard, H.; Adib, E.; Esteki, M. Analysis of a New Single-Stage Soft-Switching Power-Factor-Correction LED Driver With Low DC-Bus Voltage. *IEEE Trans. Ind. Electron.* **2018**, *65*, 3858–3865. [CrossRef]
15. Vallbè, B.; Balcells, J.; Bogónez-Franco, P.; Mata, J.; Gago, X. Immunity of power line communications (PLC) in disturbed networks. In Proceedings of the 2011 IEEE International Symposium on Industrial Electronics, Gdansk, Poland, 27–30 June 2011; pp. 1621–1625.
16. Han, R.; Yang, C.; Chen, X.; Han, X.; Zhang, J.; Li, C. Design of an Intelligent White-OLED Lighting System using power line communication. In Proceedings of the 2012 2nd International Conference on Consumer Electronics, Communications and Networks (CECNet), YiChang, China, 21–23 April 2012; pp. 3191–3194.
17. Beer, A.S.d.; Emleh, A.; Ferreira, H.C.; Vinck, A.J.H. Effects of LED lamps on the power-line communications channel. In Proceedings of the 2013 IEEE 17th International Symposium on Power Line Communications and Its Applications, Johannesburg, South Africa, 24–27 March 2013; pp. 209–213.
18. Amarsingh, A.A.; Latchman, H.A.; Yang, D. Narrowband Power Line Communications: Enabling the Smart Grid. *IEEE Potentials* **2014**, *33*, 16–21. [CrossRef]
19. Kubota, H.; Suzuki, K.; Kawakimi, I.; Sakugawa, M.; Kondo, H. High frequency band dispersed-tone power line communication modem for networked appliances. *IEEE Trans. Consumer Electron.* **2006**, *52*, 44–50. [CrossRef]
20. Mak, S.T. TWACS®, A Power Line Communication Technology for Power Distribution Network Control and Monitoring. *IEEE Trans. Power Deliv.* **1986**, *1*, 66–72. [CrossRef]
21. Ferreira, H.C.; Grove, H.M.; Hooijen, O.; Vinck, A.J.H. Power line communications: An overview. In Proceedings of the IEEE AFRICON'96, Stellenbosch, South Africa, 27–27 September 1996; Volume 552, pp. 558–563.
22. Tan, Y.K.; Huynh, T.P.; Wang, Z. Smart Personal Sensor Network Control for Energy Saving in DC Grid Powered LED Lighting System. *IEEE Trans. Smart Grid* **2013**, *4*, 669–676. [CrossRef]
23. Cardesín, J.; García-Llera, D.; López-Corominas, E.; Calleja, A.J.; Ribas, J.; Gacio, D. Low cost intelligent LED driver for public Lighting Smart Grids. In Proceedings of the 2013 International Conference on New Concepts in Smart Cities: Fostering Public and Private Alliances (SmartMILE), Gijon, Spain, 11–13 December 2013; pp. 1–6.
24. Wu, J.; Li, C.; He, X. A novel power line communication technique based on power electronics circuit topology. In Proceedings of the 2010 Twenty-Fifth Annual IEEE Applied Power Electronics Conference and Exposition (APEC), Palm Springs, CA, USA, 21–25 Feberary 2010; pp. 681–685.
25. Lin, Z.; Du, J.; Wu, J.; He, X. Novel communication method between power converters for DC micro-grid applications. In Proceedings of the 2015 IEEE First International Conference on DC Microgrids (ICDCM), Atlanta, GA, USA, 7–10 June 2015; pp. 92–96.

26. Wang, R.; Du, J.; Hu, S.; Wu, J.; He, X. An embedded power line communication technique for DC-DC Distributed Power System based on the switching ripple. In Proceedings of the 2014 International Power Electronics and Application Conference and Exposition, Shanghai, China, 5–8 November 2014; pp. 811–815.
27. Du, J.; Wu, J.; Wang, R.; Lin, Z.; He, X. DC Power-Line Communication Based on Power/Signal Dual Modulation in Phase Shift Full-Bridge Converters. *IEEE Trans. Power Electron.* **2017**, *32*, 693–702. [CrossRef]
28. Wu, J.; Zong, S.; He, X. Power/signal time division multiplexing technique based on power electronic circuits. In Proceedings of the 2011 Twenty-Sixth Annual IEEE Applied Power Electronics Conference and Exposition (APEC), Fort Worth, TX, USA, 6–11 March 2011; pp. 1710–1714.
29. Li, C.; Wu, J.; He, X. Realization of a general LED lighting system based on a novel Power Line Communication technology. In Proceedings of the 2010 Twenty-Fifth Annual IEEE Applied Power Electronics Conference and Exposition (APEC), Palm Springs, CA, USA, 21–25 Feberary 2010; pp. 2300–2304.

8

Inrush Current Control of High Power Density DC–DC Converter

Ahmed H. Okilly [1], Namhun Kim [2] and Jeihoon Baek [1,*]

[1] Electrical & Electronics and Communication Engineering Department, Koreatech University, Cheonan-si 31253, Korea; ahmed21490@koreatech.ac.kr

[2] Research Center, ESTRA Automotive, Daegu-si 42981, Korea; namhun.kim@estra-automotive.com

* Correspondence: jhbaek@koreatech.ac.kr

Abstract: This paper presents a complete mathematical design of the main components of 2 kW, 54 direct current (DC)–DC converter stage, which can be used as the second stage of the two stages of alternating current (AC)–DC telecom power supply. In this paper, a simple inrush current controlling circuit to eliminate the high inrush current, which is generated due to high input capacitor at the input side of the DC–DC converter, is proposed, designed, and briefly discussed. The proposed circuit is very easy to implement in the lab using a single metal–oxide–semiconductor field-effect transistor (MOSFET) switch and some small passive elements. PSIM simulation has been used to test the power supply performance using the value of the designed components. Furthermore, the experimental setup of the designed power supply with inrush current control is built in the lab to show the practical performance of the designed power supply and to test the reliability of the proposed inrush current mitigation circuit to eliminate the high inrush current at initial power application to the power supply circuit. DC–DC power supply with phase shift zero voltage switching (ZVS) technique is chosen and designed due to its availability to achieve ZVS over the full load range at the primary side of the power supply, which reduces switching losses and offers high conversion efficiency. High power density DC–DC converter stage with smooth current startup operation, full load efficiency over 95%, and better voltage regulation is achieved in this work.

Keywords: DC–DC converter; phase shift PWM; ZVS; inrush current; MOSFET; telecom server

---

1. Introduction

The spread of 5G technology in communications and telecom systems making universal electronic devices makes 5G technology one of the largest sources of electrical energy consumption, including such electronic devices usually operated with DC power, so that the supply AC voltage needs to be rectified. Conventional AC–DC rectifiers can be used to supply these devices, but circuit performance and power efficiency at high power density applications limit the use of such rectifiers; therefore, their energy efficiency must be increased. Recently, highly efficient AC–DC power supply with a high power factor has been modified for this purpose. Figure 1 illustrates the general construction of the AC–DC telecom power supply, where the two stages power supply consists of the power factor correction (PFC) stage and where the DC–DC output converter stage is the best option to get high power performance and good energy quality [1–3].

The design of power supply with a high input power factor requires the modifying of the input AC current waveform to follow the input voltage waveform to reduce the phase shaft between them, and then the reduction of the harmonic contents of the input current [3]. The power supply first stage usually includes an electromagnetic interference (EMI) filter, bridge rectifying circuit, boost converter, and output bulk capacitor; by controlling the boost converter operation, the input power factor can approach to, or near to, unity [4]. Single switch, two switches or full bridge switching topology can be used to implement PFC boost converter, but switching topology proper switches must be used to withstand the output voltage stress of this stage normally (320–400 V) DC [1]. Different techniques and topologies of the analog control PFC stage of the telecom power supply with power factor more than 99% are previously discussed in [5].

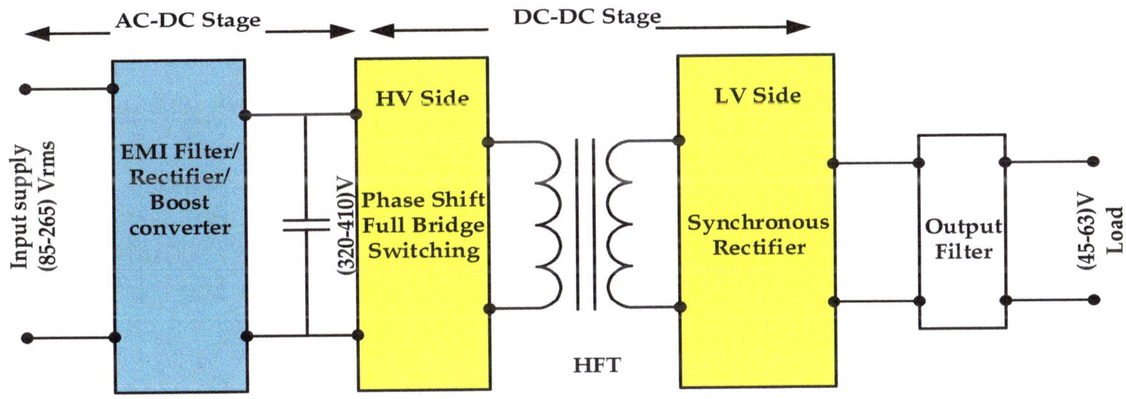

Figure 1. Block diagram of the complete AC–DC power supply for a telecom server.

The power supply second stage is the target of this paper and it is designed to regulate the PFC output voltage to the required load voltage (usually 45–63 V) for the telecom applications. In this stage we must choose an appropriate technique to provide high power density conversion with high efficiency [6], DC–DC converter based on phase shifted ZVS technique is one of the most widely used techniques, because it has features to minimize the switching losses and offer better regulation over a wide load range [7]. Moreover, by using the ZVS technique, it is available to have high power density conversion with low voltage stress and small switching losses [8–10].

Switches on both sides of the high power density DC–DC converter must be designed to withstand the high voltage stress in the primary side and the high current stress in the secondary side [10]. So, one of the most important issue in designing DC–DC converter circuits is to choose the appropriate switching schematics for the converter to reduce the stress in the switches during the converter operation. Based on the phase shifted converter applications, phase shift pulse width modulation (PWM) converter can be implemented using different switching schematics (full bridge, half bridge, dual half bridge) in primary and secondary sides, and analysis of phase shifts isolated DC–DC converter different topologies previously given in [11]. Analysis and design of phase shift controller for a dual half bridge DC–DC converter is provided in [12]. Based on the application of DC–DC converter, it can be designed as an isolated or non-isolated converter; for the isolated DC–DC converter, transformer design and choice are a big challenge to have ZVS for the wide range of the load, which can reduce the switching and conduction losses [13]. Analysis of the phase shifted converter by using series connected transformers for low conduction losses are given in [14,15]. Design and implementation of the non-isolated phase shift ZVS converter is presented in [16].

DC–DC converter control circuits can be implemented using analog or digital control circuits, but analog control chip as compared with digital micro controller unit (MCU) has some demerits such as the temperature drift, fixed control parameters, and slow response speed; on the other hand the price of the MCU and the required analog digital interfacing sensors are expensive as compared with the analog control technique requirement. Optimization analysis and the design of different control techniques for the DC–DC converter have been previously presented in [17–21]. Design of 1 kW efficient phase shift telecom DC–DC converter based on the maximum duty cycle and optimal hold-up time is given in [22]. Controlling techniques to reduce the switching and conduction loss are presented in [23,24].

The bulk capacitor, which is introduced between the two stages of power supply as shown in Figure 1 for adjustment of the input voltage hold-up time and harmonic values of the input waveform, causes high inrush current for a few cycles where a very high "dv/dt" occurred at the initial power application to the power supply; this high dv/dt cause spikes of short-duration and a high peak current, which value may be higher than the circuit component rating current and can seriously damage or destroy these components [25–27]. So appropriate inrush current controlling circuit must be designed to limit such this current. Usually, limiting and reducing of the high inrush current is done by using a large size inductor or resistors in series with the input capacitors [28], but by using these techniques, converter compact design, and weight and power losses cannot be optimally utilized.

In order to reduce the power dissipation on the series resistor, a parallel semiconductor switch or relay can be connected through the resistance. However, based on the converter operating current, size of the relay can be also excessive large. Moreover, by using of a semiconductor switch an appropriate controlling circuit must be designed to control the switch operation. Another technique used to control the inrush current is by handling a soft starter time at the beginning of the DC–DC converter operation; soft starter technique works to limit inrush current based on controlling the duty cycle of the converter switches in order to slowly charge the capacitors, and hence reduce the high dv/dt for limiting the inrush current [29–31]; this technique is usually implemented to limit the inrush current inside the analog controller chips such as PWM controller integrated circuits (IC's) UCC256403 and UCC28950 from Texas Instruments, Dallas, Texas, United States, 2016 [32].

Recently, NTC thermistor is the most famous device used in reducing the inrush current generated in the telecom power supplies. An NTC thermistor has a compact size and a negative temperature coefficient, is connected in series with the input power supply, and when the current flows, its temperature will increase and the device resistance will decrease [33,34]. The shortcoming of this device is that it requires a cool off time after the stopping of the current flowing in order to return back the device high resistance.

Another more applicable and simple technique to limit the converter inrush current can be done using a single MOSFET switch connected with the input side of the DC–DC converter [35–37]. MOSFETs switches are usually considered as ideal devices because they are characterized by fast switching time due to majority carrier, lower switching losses due to fast rise and fall times, as well as very small on-state DC resistance, which helps to reduce the voltage drop through the switch at steady state operation [38]. Control of the inrush current to the required limit can be done by controlling the gate charge transfer characteristics of the MOSFET switch in order to control the slew rate of the input capacitance charging time [39]. This technique offers inrush current control with economic price, compact design, and simple implementation, without using any sophisticated control circuit for the MOSFET operation [38–40]. High reliability using this technique can be obtained with the appropriate choice of the MOSFET switch, as well as design of the biasing circuit schematic and the connection strategy to the input side of the DC–DC converter [41].

In this paper, complete design, analysis, and mathematical calculations of the circuit main components and control systems of 2 kW, 54 V telecom DC–DC converter stage with phase shift ZVS technique, using full bridge at primary and synchronous rectification at secondary side are presented. A sample inrush current controlling circuit based on using MOSFET switch is proposed and is inserted in the input side of the DC–DC converter, proposed controlling circuit schematics as well as design and choice of the circuit components are briefly discussed. The main advantages of the proposed circuit are that the controlling of the gate charge transfer characteristics of the MOSFET occurs without any contribution or connection with the DC–DC converter control circuit, which make the proposed circuit usable with analog and digital control converters, and has fast response and high reliability to control the inrush current to the required value as well as simple implementation for any converter with different input–output operation conditions. The designed main components and control system of the power supply circuit make the power supply output voltage performance follows the standard specifications IEC61000-3-3, which is required by telecom applications.

The result sections of this paper are organized as follow: first, simulation of the complete designed converter is performed using PSIM software to be sure that the reliability of the designed components enhances the required power supply performance. Then, an experimental setup of the designed converter with the proposed inrush current mitigation circuit is inserted between the power supply two stages is performed in the lab, which shows that the designed DC–DC converter with proposed inrush control circuit can achieve smooth startup operation at the input side, where efficiency is more than 95.5%, and with better voltage regulation at the output side of the converter.

2. Design Procedure of the DC–DC Stage of the Telecom Power Supply

Figure 2 shows the schematic circuit of the telecom DC–DC converter stage with the proposed inrush current control circuit connected at the input side of the DC–DC phase shift converter. In designing the high-power density power supply with high conversion efficiency, losses in this stage must be maintained at their lowest value; one of the highest efficiency conversion techniques at high power density is the full bridge phase shift converter, which can offer very small switching losses by means of the ZVS technique at the converter primary side, which leads to increase the conversion efficiency [7,42].

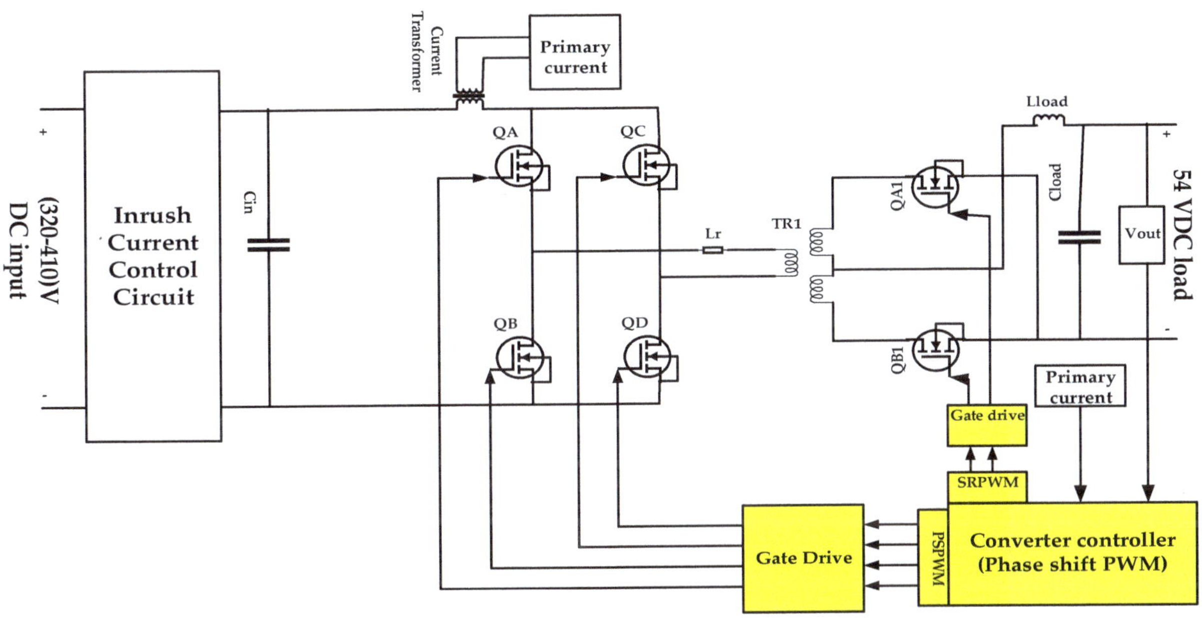

Figure 2. Schematic diagram of DC–DC phase shift ZVS converter with the proposed inrush current control.

The most important components in the DC–DC converter stage to be designed in the next subsections are the input filter capacitor (C_{in}), inrush current control circuit, resonant inductor (L_r), transformer (T_{R1}) turns ratio (a) and magnetizing reactance (L_m), and load inductance and capacitance (L_{load} and C_{load}). Other important factors are the choice of appropriate switches for the full bridge rectifier at the primary side and synchronous rectifier at the secondary side, which can withstand the high voltage stress at transformer primary and high current rated at secondary side [42].

2.1. Input Capacitor Design

Input capacitor of the DC–DC converter is designed to meet the hold-up time (t_{hold}) for the minimum input voltage ($V_{in\ min}$) applied to the converter circuit [15,40], and can be calculated as

$$C_{in} \geq \frac{2P_o\, t_{hold}}{\eta\left(V_{in}^2 - V_{in\ min}^2\right)} \quad (1)$$

where P_o refers to the rated converter power, V_{in} refers to the Converter input voltage, and η is the designed efficiency of the DC–DC converter.

2.2. Proposed Inrush Current Control Circuit

In this subsection the proposed inrush current control circuit based on the controlling of the gate charge characteristic of the MOSFET switch was designed to be used with any DC–DC converter topology (analog or digital); the proposed circuit was tested in PSIM software and also was implemented in the lab to show the power supply two stages (PFC and DC–DC converter) practical current characteristic with, and without, using this circuit.

The input of DC–DC converter stage usually contains filter to adjust the total harmonics distortion (THD) of the input voltage to the DC–DC converter. Input filters usually consist of passive elements such as capacitor and inductor; when power is initially applied to the DC–DC converter circuit, high inrush current will flow due to high dv/dt of the filter capacitor as illustrated in Figure 3, which shows the converter input current curves at initial power application to the converter circuit.

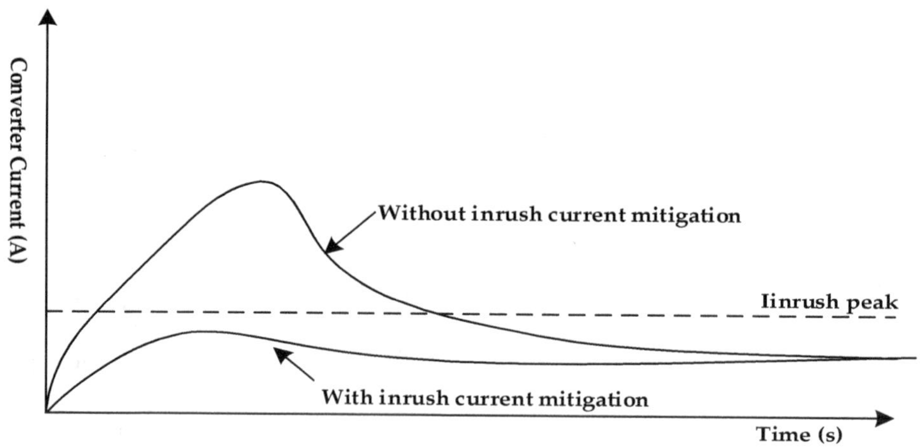

Figure 3. Inrush current at the startup operation of DC–DC converter.

Mathematical analysis, which briefly describes the startup response of the power supply operation and the phenomena of the inrush currents in DC–DC converters, are previously discussed in [26,27]. At the moment of applying the voltage to the DC–DC converter circuit, the current flowing through the input capacitor (C_{in}) can be expressed as

$$I_{Cin} = C_{in}\frac{dV_{in}}{dt} \tag{2}$$

This current has a high amplitude with very small duration and can damage or destroy circuit components if it exceeding the rating of the designed components. Therefore, this current must be effectively managed to ensure system safety operation and stability.

MOSFET switches are charge controlled devices, so if a MOSFET device is connected between two stages of the power supply, as shown in Figure 4, inrush current due to high dV_{in}/dt, which is caused by the input capacitor (C_{in}) at the initial power application, can be controlled by controlling the initially high dV_{in}/dt of the input capacitor by using the ability to control the constant linear slope of the drain voltage transition, which allows accurate control of the inrush current to the capacitive load. This is possible because the current flowing through the capacitor is dependent upon the transition of the voltage as shown in Equation (2). Reliability of this technique depends on the ability to control the MOSFET gate charge transfer curve illustrated in Figure 5 [29].

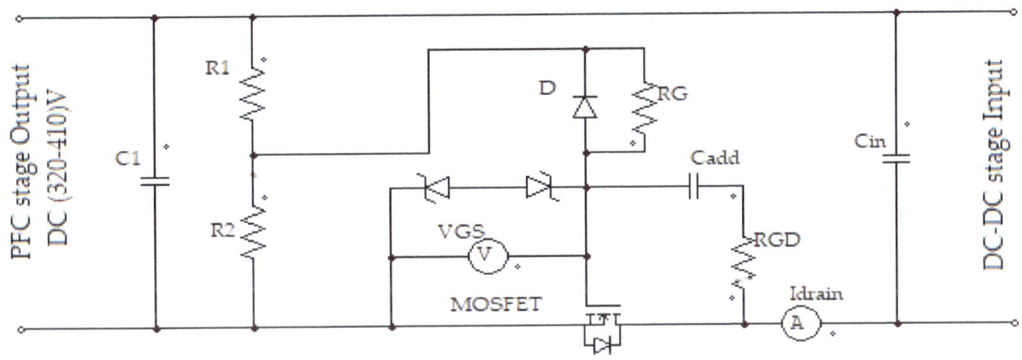

Figure 4. Proposed inrush current controlling circuit.

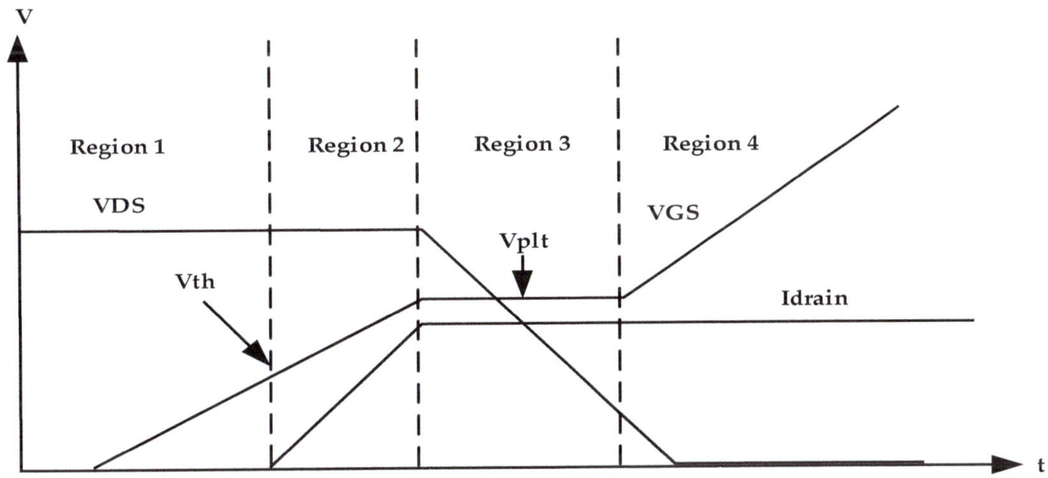

Figure 5. Gate charge transfer characteristics of MOSFET.

Figure 6 shows the equivalent representation of the MOSFET switch, where the gate charge curve is influenced by the MOSFET equivalent input capacitance [29].

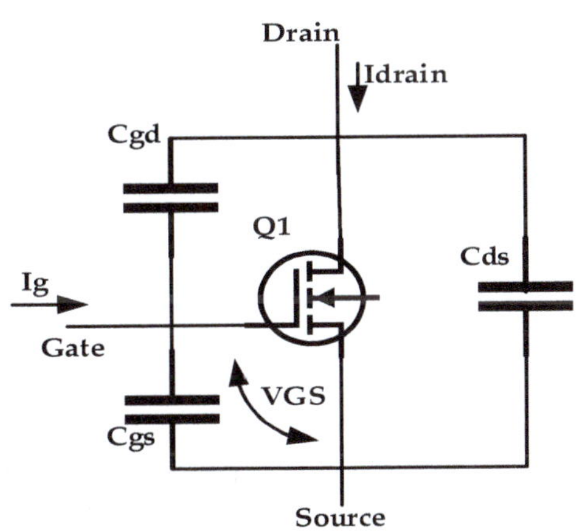

Figure 6. Circuit of N-MOSFET.

Cgd and Cgs represents the gate-drain and gate-source capacitances, respectively; these two capacitance values represent the input capacitance of the MOSFET device. Usually for MOSFET, Cgd is greater than Cgs. So Cgd strongly influences the input capacitance of the MOSFET, this means that MOSFET input capacitance can control the inrush current value, if the value of Cgd is controlled. By inserting the appropriate design capacitance, C_{add}, between the drain and gate terminals of MOSFET, as shown in the proposed circuit in Figure 4, Cgd can be controlled, which then can control the slew rate of the gate-source voltage, V_{GS}, which controls the slew rate of the input capacitance charging ($dVin/dt$), and consequently controls the inrush current of the circuit.

As shown in the characteristics in Figure 5, when MOSFET is turned on, the charging of the equivalent capacitance occurs at Region 1, and the charging time is determined by the equivalent input capacitance of MOSFET. The voltage (V_{GS}) increases until the starting point of Region 2, where it reaches the threshold value (V_{th}). At this time drain current starts to flow and the rate of increase of drain current given by

$$\frac{dI_{drain}}{dt} = g_f \frac{dV_{GS}}{dt} \tag{3}$$

where g_f is the switch forward trans-conductance and can be easily know from the switch datasheet.

At the end of Region 2, charging and discharging of Cgs simultaneously occurs, which causes V_{GS} to be maintained constant at Miller plat voltage (V_{plt}) as shown in Region 3. With V_{plt}, the drain current is saturated at peak constant value dependent of V_{GS} voltage value, and V_{plt} can be calculated as

$$V_{plt} = V_{th} + \frac{I_{drain}}{g_f} \tag{4}$$

The constant voltage of V_{GS} causes the input gate current to flow through the additional capacitance C_{add}, and can be calculated as

$$I_g = \frac{V_{GG} - V_{plt}}{R_G} = C_{add} \frac{dV_{DS}}{dt} \tag{5}$$

$$V_{GG} = V_{in} \times \frac{R_2}{R_1 + R_2} \tag{6}$$

where R_G is designed and connected in a series with the gate for controlling the gate current I_g. Finally, by controlling the value of V_{GS} and I_g, it is possible to control the maximum drain current of the MOSFET and control the inrush current at the initial power application to the power supply circuit, when the drain terminal of the MOSFET is connected to the return path of the DC–DC converter, as shown in Figure 4. R_{GD}, a small value resistor, as compared with R_G, connected with the C_{add} and prevents unwanted high frequency oscillation [29]. Using Equations (5) and (6) inrush peak current value is approximately given by

$$I_{inrush} = -I_{drain} = g_f \left(C_{add} \times R_G \times \frac{dV_{DS}}{dt} + V_{th} - V_{GG} \right) \tag{7}$$

The value of this current must be less than the maximum permissible DC current (inrush peak) of the primary side of the DC–DC converter stage.

Figure 7 shows the flow chart for the complete design procedure for the proposed inrush current control circuit by using the method of the controlling in the gate charge transfer characteristics of the MOSFET.

In Region 4, V_{GS} still increases to higher values. If this voltage reaches a value higher than the gate source breakdown voltage (BV_{GS}), MOSFET may be damaged; therefore, MOSFET must be protected at this region from the higher applied voltage; therefore, PFC output stage voltage (input voltage to the DC–DC converter) is divided using resistors R1 and R2 and only a small voltage is required to be applied to the MOSFET. Additionally, we can use one switch from the family of Zener-protected MOSFETs, such as switches from STMicroelectronics company [43]; in this protected switch, when the voltage applied to the switch is more than the breakdown voltage of the Zener (less than BV_{GS}), the Zener diode breaks down and the voltage is saturated at the safe limit. The design result of the active inrush current control circuit, which used in simulation and experimental modeling is shown in Table 1.

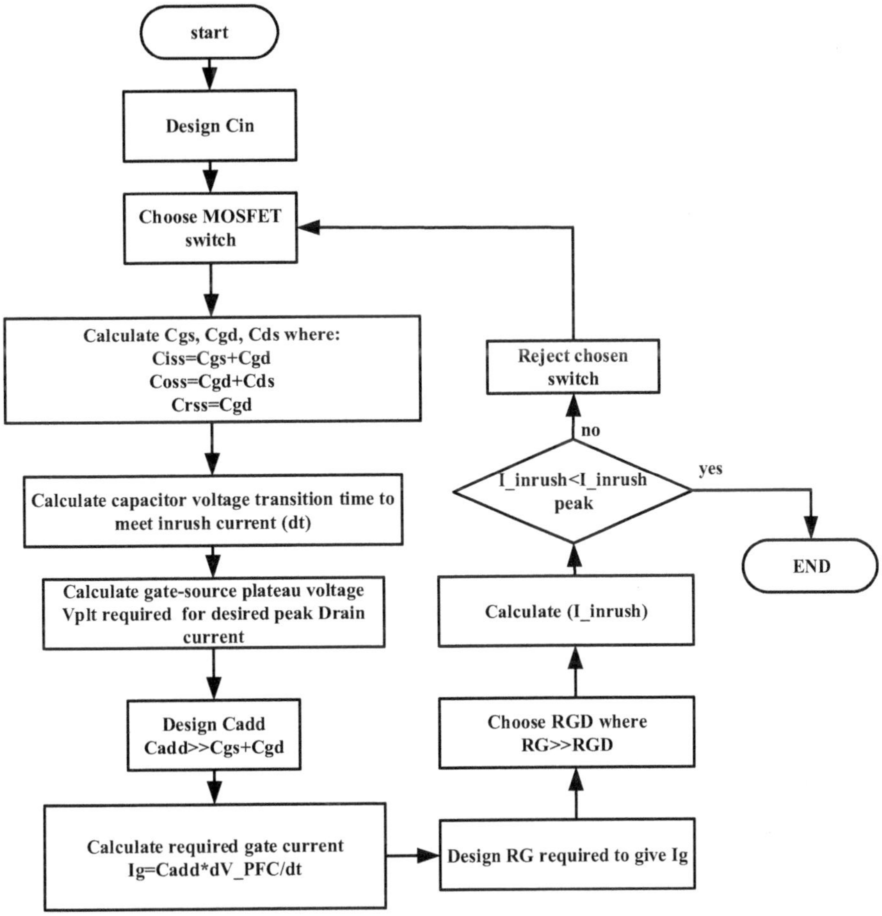

Figure 7. Flow chart of design process of the proposed inrush current control circuit.

Table 1. Design Results for Active Inrush Current Control Circuit.

Parameter	Value	Unit
Input capacitance (C_{in})	350	uF
Chosen MOSFET	STW20NK50Z	-
C_{add}	0.1	uF
R_G	30	kΩ
R_{GD}	500	Ω
R_1	715	kΩ
R_2	84.5	kΩ
Allowable Inrush peak	9	A
Plat voltage (V_{plt})	4.35	V
Gate current (I_g)	3.430	mA

2.3. Transformer (T_{R1}) Turns Ratio (a) and Magentizing Reactance (Lm) Calculation

Transformer turns ratio (a) is calculated based on maximum operating duty cycle (D_{max}) at the minimum input voltage rating of the converter ($V_{in\ min}$) as follows:

$$a = \frac{N_P}{N_S} = \frac{V_P}{V_S} \tag{8}$$

Let D_{max} be about 70% and $V_{in\ min}$ about 320 V, and the transformer turns ratio is calculated as

$$a = \frac{V_P}{V_S} = \frac{\left(V_{in\ min} - 2VD_Q\right)D_{max}}{V_{load} + VD_Q} = 4.1 \tag{9}$$

where VD_Q is the switch voltage drop and it is assumed to be 0.5 V in calculations.

Let $a = 5$, the typical operating duty cycle (D), is calculated as

$$D = \frac{\left(V_{load} + VD_Q\right)a}{\left(V_{in} - 2VD_Q\right)} = 0.68 \tag{10}$$

Transformer magnetizing inductance (L_m) designed based on the maximum magnetizing inductance to realize ZVS as expressed in [13,18], and can be expressed as

$$L_m = \frac{T_{dead}\ a\ V_{load\ min}}{C_{HB}\ V_{in\ min}} \times \left(\frac{T_{s\ min}}{4} - \frac{T_{dead}}{2}\right) \tag{11}$$

where C_{HB} refers to the total equivalent capacitance of the primary H bridge, which can be known from the primary switch data sheet. $T_{s\ min}$ is the minimum switching time depends on the designed minimum switching frequency, also T_{dead} is the PWM dead time which can be calculated according to the previously calculated duty ratio.

One more important issue is the choosing of a transformer with appropriate magnetizing inductance to minimize the output current ripple and to make sure that the converter works in the required control mode, where the smallest value of magnetizing inductance makes the converter work in voltage control mode instead of current control mode [10].

2.4. Resonant Inductor (Lr) Design

Resonant inductor tank is calculated based on the amount of energy required to achieve ZVS condition. The energy absorbed by the inductor values of the resonant inductance (L_r) and the transformer leakage inductance (L_{lk}) must be able to exhaust the energy supplied by the average parasitic capacitance of the primary switches (C_{ossavg}), and also the energy from the transformer winding capacitance (C_w) [42,44].

$$\frac{1}{2}I_P^2(L_r + L_{lk}) \geq \frac{4}{3}C_{ossavg}V_{in}^2 + \frac{1}{2}C_w V_{in}^2 \tag{12}$$

where I_P refers to the converter primary current (A).

2.5. Output Inductance and Capacitance (L_{load} and C_{load})

Output load inductor (L_{load}) is designed based on 10% ripple value in the load DC current (I_{load}), as follows:

$$\Delta I_{load} = \frac{P_{load} \times 0.10}{V_{load}} \tag{13}$$

$$L_{load} = \frac{V_{load} \times (1 - D)}{\Delta I_{load} \times F_s} \tag{14}$$

Output load capacitor (C_{load}) is selected based on hold-up time (t_{hu}) and 20% (200 mV) of the allowable load transient voltage (V_{tran}) as follow:

$$C_{load} \geq \frac{0.9 I_{load} \times t_{hu}}{0.2 V_{tran}} \tag{15}$$

where hold-up time (t_{hu}) is calculated as the time required for the inductor current to reach to 90% of the full load current [41].

$$t_{hu} = \frac{L_{load} \times 0.9 I_{load}}{V_{load}} \tag{16}$$

2.6. DC–DC Converter Controller Design and Implmentation

Phase shift PWM technique is used to control the full-bridge in the primary side of the DC–DC converter by phase shifting the switching pulses of one half-bridge with respect to the other. High power density efficient conversion is available using ZVS technique at high switching frequency in this part. Voltage-mode or current-mode control techniques can be used in this part. Current-mode controlled DC–DC switching is popular and provides a more highly efficient power conversion than voltage mode control. However, the current-mode design can suffer from instability when the duty cycle of the PWM rises above 50% [45]. To overcome this instability, converter primary current slope compensation technique is used to restore reliability over the wide range of duty-cycle [19,46].

Figure 8 shows the schematic of control technique of DC–DC converter, implemented in PSIM software where the primary current and the output voltage are the feedback signals used in this control system. First, primary current is sensed using current transformer with turns ratio (100:1) and then sampled by a resistor to get a primary current signal (VIp); load voltage (V_{load}) signal is also sensed. To overcome the instability in current waveforms at high duty cycles, value slope compensation technique with the ramp signal of 200 kHz is added to the VIp signal to generate the primary current (Iprim); output DC voltage signal is compared with the appropriate reference value and passed to the voltage PI controller to generate the primary current reference value (Iprim-ref), then the primary current (Iprim) is compared with reference value (Iprim-ref) to generate the duty cycle of the PWM generator for the primary and secondary switches QA, QB, QC, QD, QA1, and QB1.

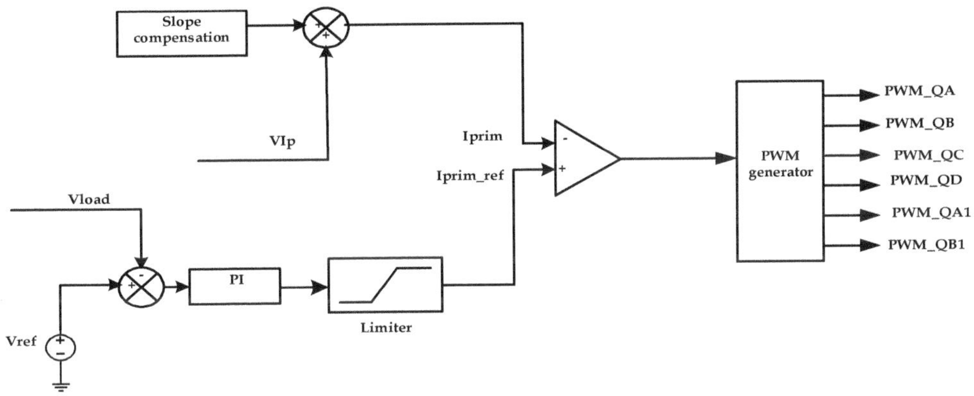

Figure 8. Technique of the phase shift PWM DC–DC converter controller.

Table 2 shows the input/output specifications and the results of the design converter main components of the DC–DC converter.

Table 2. DC–DC Converter Specifications and Design Results.

Parameter	Value	Unit
Input voltage	400 (320–410)	V
Output voltage	54 (45–63)	V
Output voltage transient (V_{tran})	1	Vpp
Rated power	2	kW
Switching frequency	100	kHz
Designed efficiency	95%	-
Input voltage hold up time (t_{hold})	4	ms
Input capacitor (C_{in})	350	uF
Transformer turns ratio	20/4/4	Np/Ns/Nt
Transformer reactance (L_{mag})	2.8	mH
Primary switches	IPW60R070CFD7	-
Secondary switches	IPP110N20N3	-
Output capacitor (C_{load})	2750	uF
Output Inductor (L_{load})	22	uH
Resonant inductor (L_r)	30	uH

3. Simulation Results and Discussions

Phase shift DC–DC converter with power density about 2 kW has been implemented in PSIM software using the designed components in the previous sections to show the system performance under different loading conditions and to test the reliability of inrush controlling circuit to limit the high inrush current at initial power application to the primary side of the power supply circuit.

With full load condition and DC input voltage about 400 V, Figure 9 shows the steady state simulation result of the DC output power curve of the DC–DC converter.

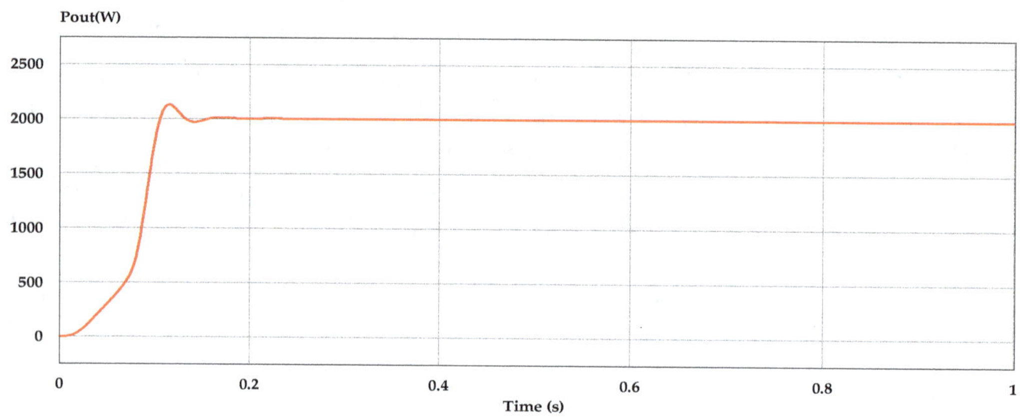

Figure 9. Converter output power curve at full load condition.

The designed power supply efficiency curve at loading condition from 10% to 100% of the rated output power (2 kW) with input voltage 400 V and switching frequency about 100 kHz is illustrated in Figure 10, which shows that the designed power supply offers full-load efficiency at about 95.1%, half-load efficiency at about 95.3%, and the maximum being three-quarter-load efficiency at about 95.6%.

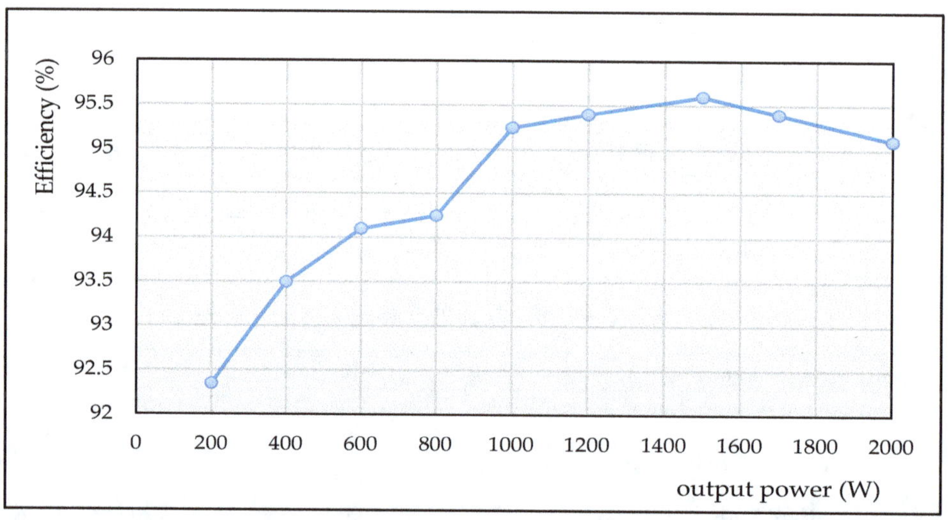

Figure 10. Converter efficiency curve with different loading condition at V_{in} = 400 V.

As mentioned before in Section 2.2, the gate charge transfer characteristics of the MOSFET switch can be used to control the slew rate of the input capacitance charging in order to control and limit the inrush current to the required value. Figure 11 shows the bulk input capacitance charging voltage curves with, and without, using the proposed inrush current control circuit, where the black curve refers to the supply DC input voltage, the red curve is the voltage curve of the input capacitance without using the proposed control circuit, and the blue curve is the voltage curve of the input capacitance using the proposed control circuit. From these curves, it is clearly observed that controlling the slew rate of VGS voltage by controlling the input capacitance of the MOSFET, leads to the reduction of the high dv/dt of the input capacitance, which subsequently reduces the inrush current at initial power application.

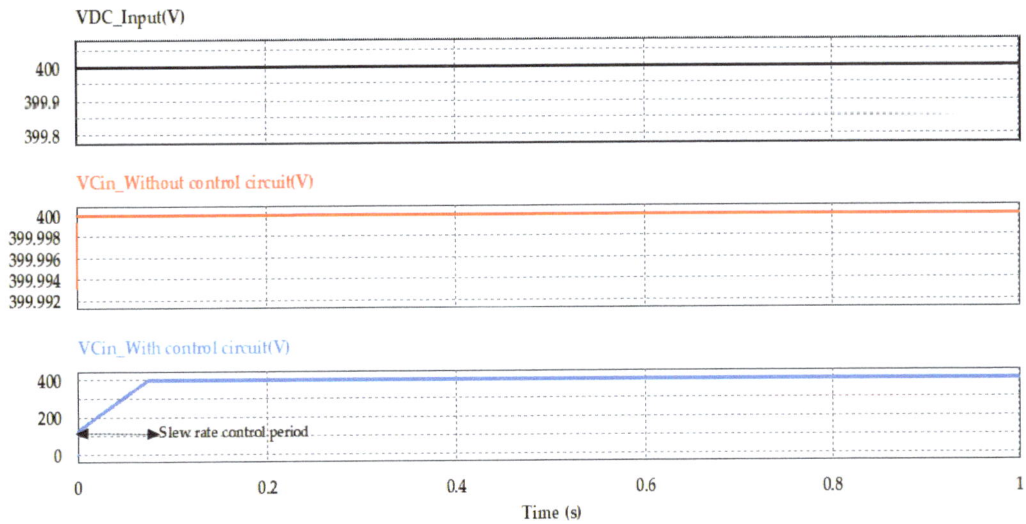

Figure 11. Bulk input capacitance charging voltage with and without using the proposed control.

Waveforms of the average value of the converter input current at switching frequency of 100 kHz with, and without, using the inrush current control circuit are shown in Figures 12 and 13, respectively. From both cases, and as the result of controlling the voltage slew rate at the initial power application to the converter circuit based on the transition time, which is required to meet the required inrush current, we can notice that the proposed controlling circuit reduced the peak inrush current from 16.10 A to about 6.40 A, which is in the allowable range (given in Table 1) of the input DC current to the primary side of DC–DC converter; moreover, it is clear to observe that the peak value of the input current at steady state with full load condition is about 8.20 A in both cases.

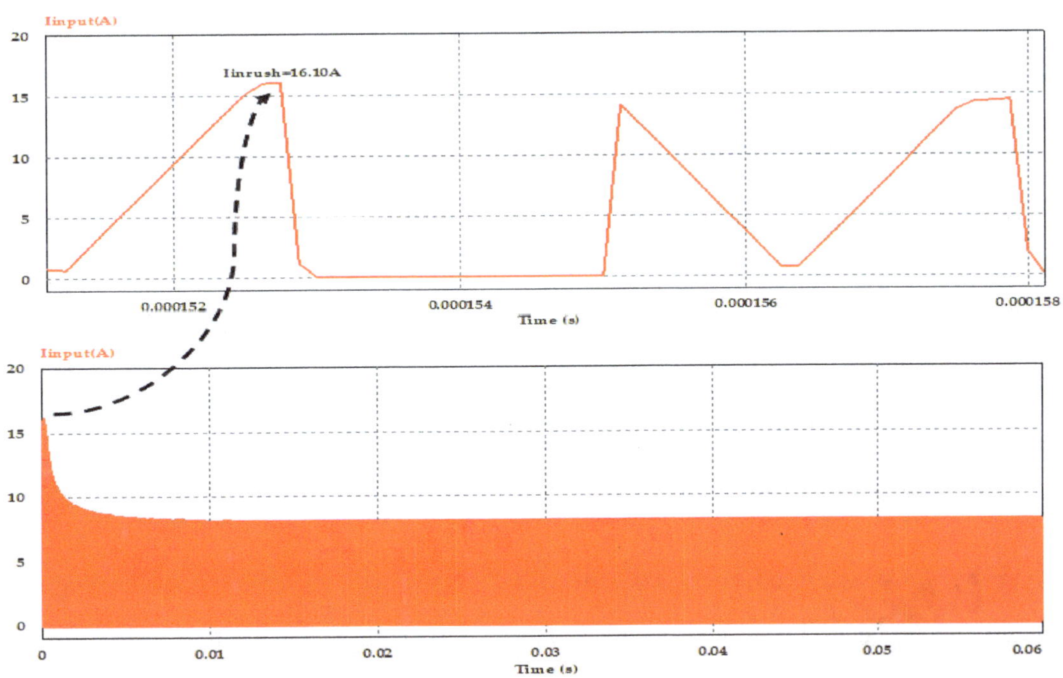

Figure 12. Converter input current without inrush current mitigation circuit at full load and $V_{in} = 400$ V.

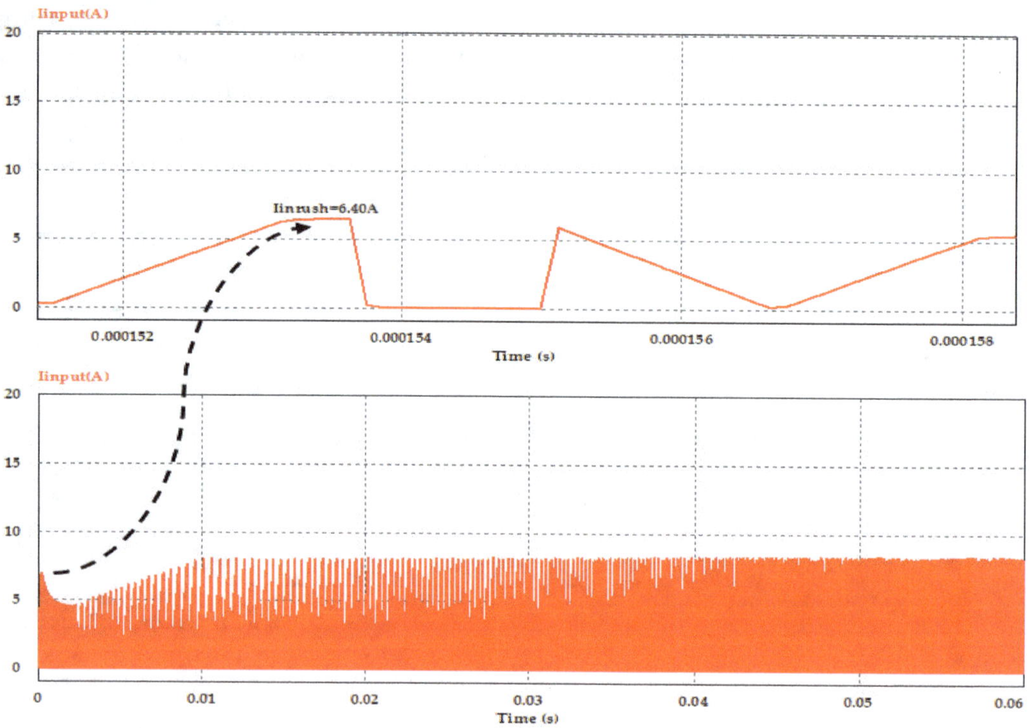

Figure 13. Converter input current with inrush current control circuit at full load and V_{in} = 400 V.

Figure 14 shows the VGS and Idrain characteristics of the MOSFET in the inrush current control circuit. When the VGS voltage reached to Vth (about 3.75 V) of the used MOSFET switch, the drain current starts to increase, at the point when VGS reaches V_{plt}, and circuit inrush current tries to increase but based on the gate charge transfer characteristic depicted in Figure 5, drain current saturated at constant value about 6.40 A as shown in the simulation result. Additionally, from this figure it is clearly observed that the protection of the MOSFET switch, which occurred at Region 4 when the Zener diode circuit broke down before the voltage reached 30 V (BV_{GS} of the used MOSFET) and the voltage was saturated at about 28.5 V.

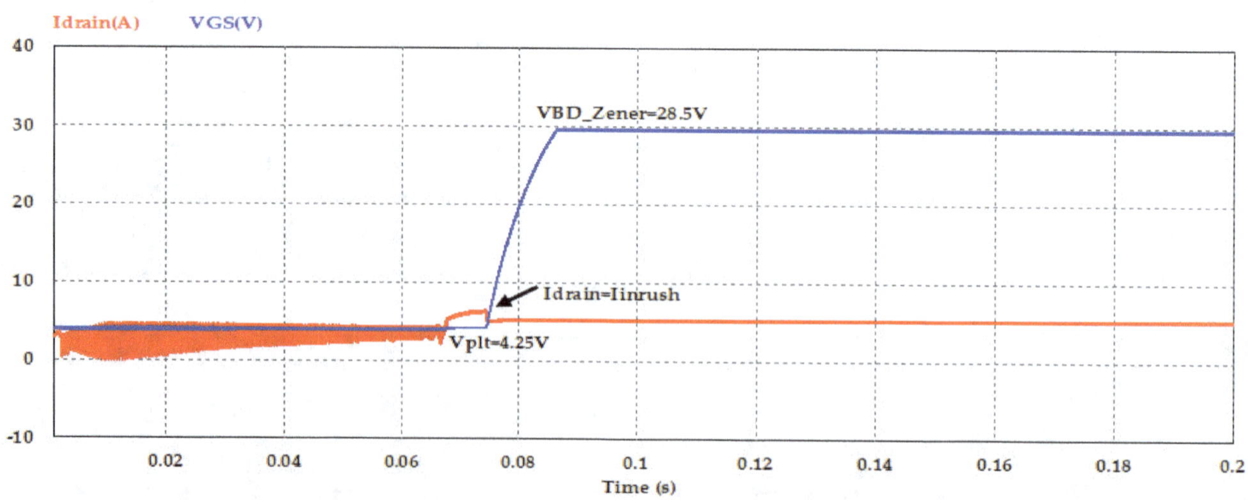

Figure 14. Gate charge waveform of the MOSFET in inrush current mitigation circuit.

Phase shift PWM controlling circuit and the resonant inductor are designed to offer ZVS at the two legs of the bridge at the primary side of the DC–DC converter switches, as shown in current and voltage waveforms of the switches (QA and QC) in Figures 15 and 16.

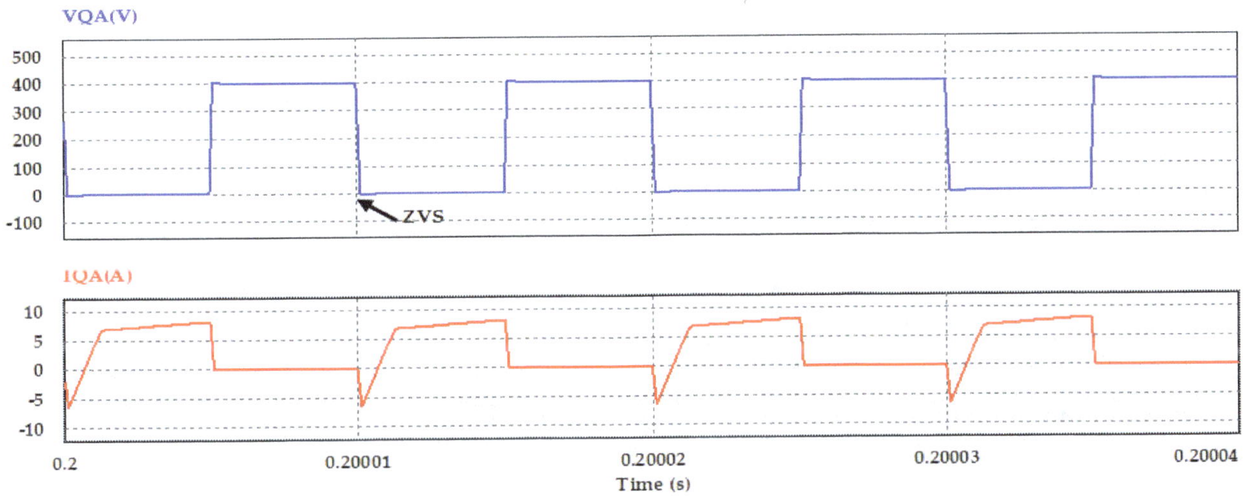

Figure 15. Voltage and current waveforms of switch QA converter primary side at $Po = 2$ kW.

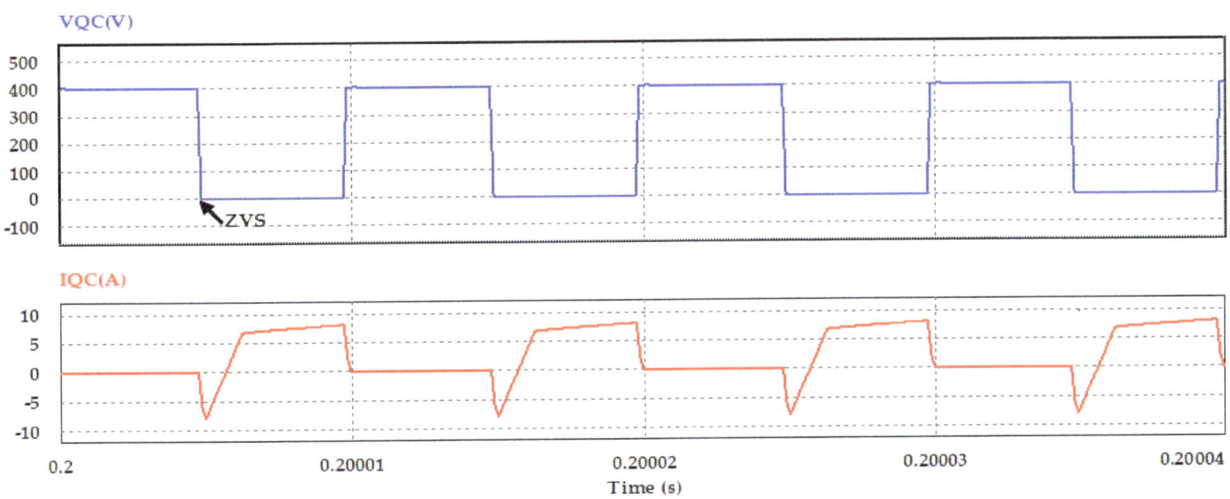

Figure 16. Voltage and current waveforms of switch QC in converter primary side at $Po = 2$ kW.

4. Experimental Setup

The complete designed converter circuit and the proposed inrush current mitigation circuit are experimentally setup and tested with the available maximum DC electronic load in the lab (1000 W), UCC28950 phase shift PWM controller IC from Texas Instruments, Dallas, Texas, United States, 2016 is used to control the DC–DC converter switches, UCC28950 IC provide 4-PWM signals with constant frequency (100 kHz) for the primary side switches, and 2-PWM for the synchronous rectification at the secondary side switches with the availability of the primary current compensation to restore the current stability and the voltage loop control to adjust the output voltage at the specified value.

The proposed inrush current mitigation circuit was tested with the practical case of the telecom two stages AC–DC power supply, where the DC–DC converter was supplied by the PFC converter stage with output voltage of 400 V DC. In order to show the reliability of the proposed inrush current control circuit in mitigation of the high current overshot in the input current to the converter, soft starter of the analog UCC28950 IC is disabled through the (SS/EN) pin and only the proposed inrush current control circuit is connected to the input side of the DC–DC converter circuit.

Figure 17 shows the appearance of the experimental setup of the complete power supply consisting of PFC and DC–DC converter stages and with using the proposed inrush current control circuit connecting in between. 1000W KIKUSI PLZ1004WH, Japan, 2019 DC electronic load was connected at load side, and 2.5 kW PFC converter stage from Infineon with average efficiency more than 95% [47], inrush current control circuit was connected between the two stages and the input AC voltage was applied to the input of the PFC stage so that the output voltage will be 400 V DC, input current to the both stages has been measured using high scale current probe FLUKE i1000s, USA, FLUKE company with scale choice of (10:1) A with, and without, connecting of the proposed inrush current mitigation circuit.

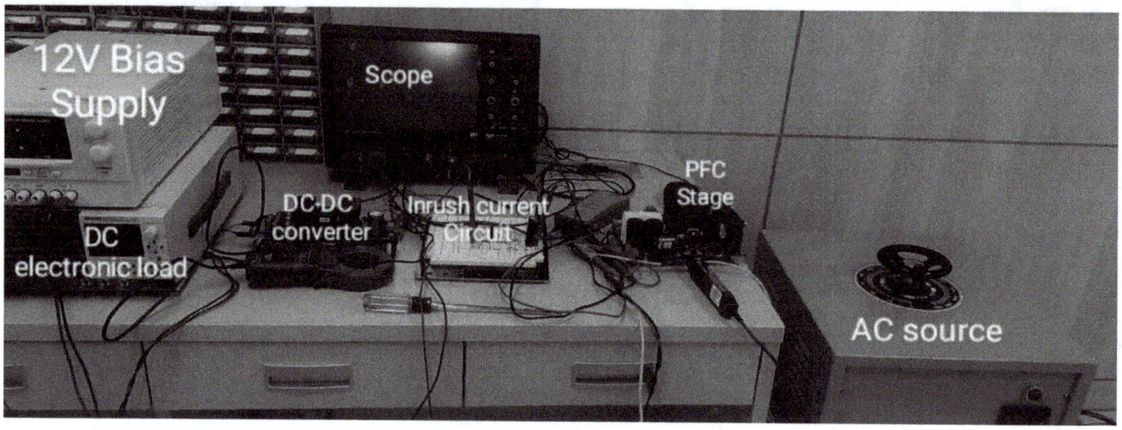

Figure 17. Experimental setup of DC–DC converter circuit with inrush current mitigation circuit.

Figures 18 and 19 show the DC–DC converter stage input current, from which we can investigated that with using inrush current control circuit, the inrush current at the startup of the converter is reduced from 20.32 A to about 4.06 A. Additionally, in both figures, differences in current shape and starting time of each waveform can be observed, this occurred due to changing in the slew rate of the input capacitance voltage, which leads to limit the inrush current at the safe limit using the inrush current controlling circuit, in Figure 19, with the current still increasing gradually until it reached to 4.06 A, at this moment, and as explained before in the controlling circuit characteristics in Figure 14, the drain current is saturated at the safe limit to prevent the inrush current from increasing to higher limits.

Figures 20 and 21 show the PFC converter stage AC input current, from which we can investigate that by connecting the proposed inrush current control circuit between the two stages of the power supply, the peak overshoot of the input current to the PFC stage will reduced from 33.55 A to 15.33 A. Additionally, the difference is noticed between starting time and waveform shapes, due to the controlling of the slew rate of the input voltage by using the proposed inrush current control circuit.

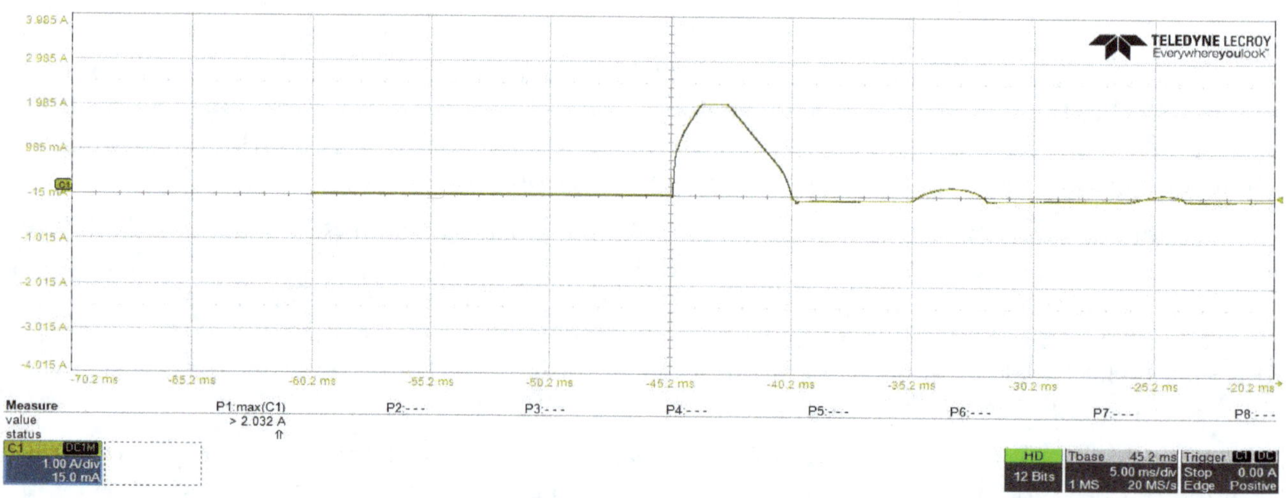

Figure 18. DC–DC converter stage input current without using of the proposed inrush current mitigation circuit.

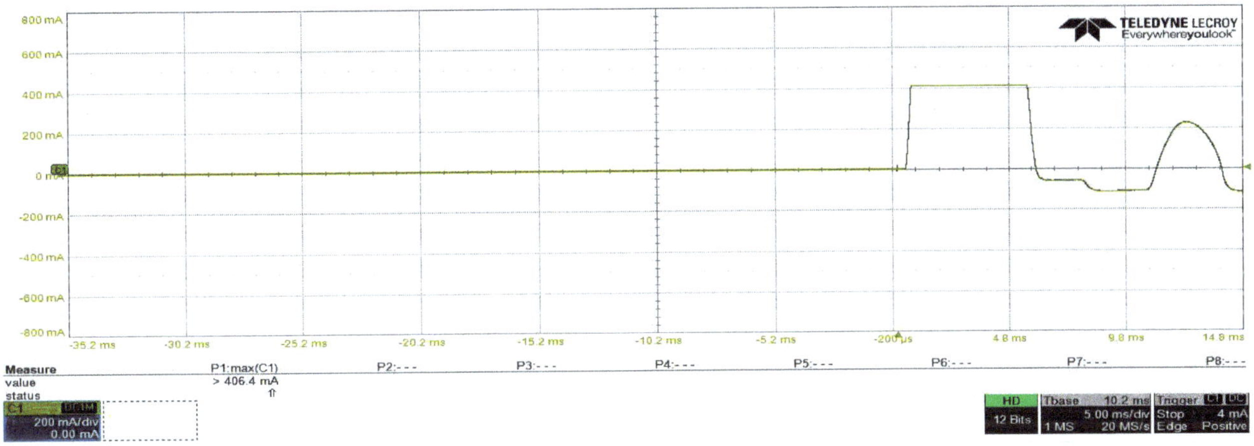

Figure 19. DC–DC converter stage input current with using of the proposed inrush current mitigation circuit.

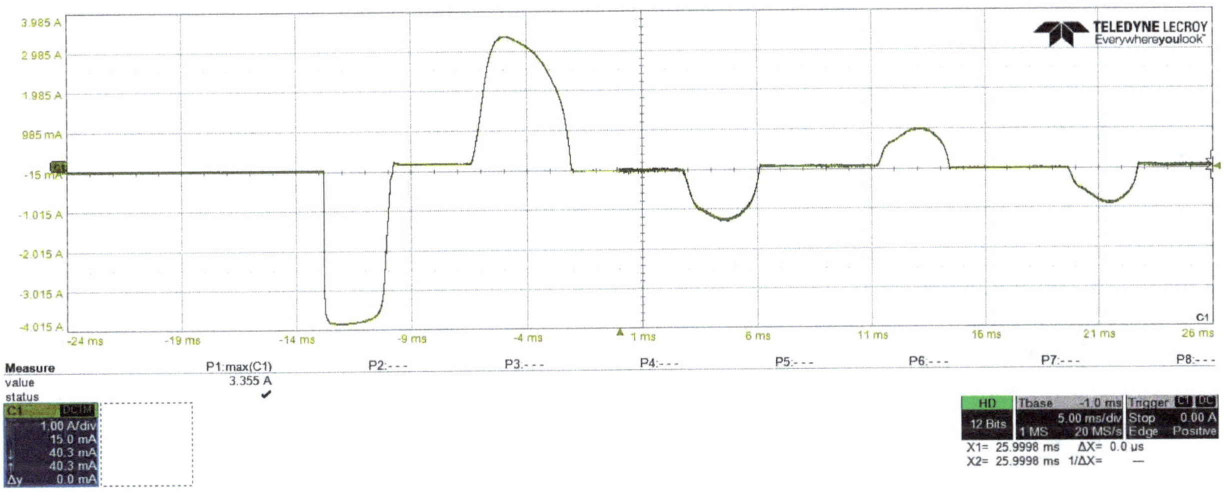

Figure 20. Power factor correction (PFC) converter stage AC input current without using of the proposed inrush current mitigation circuit.

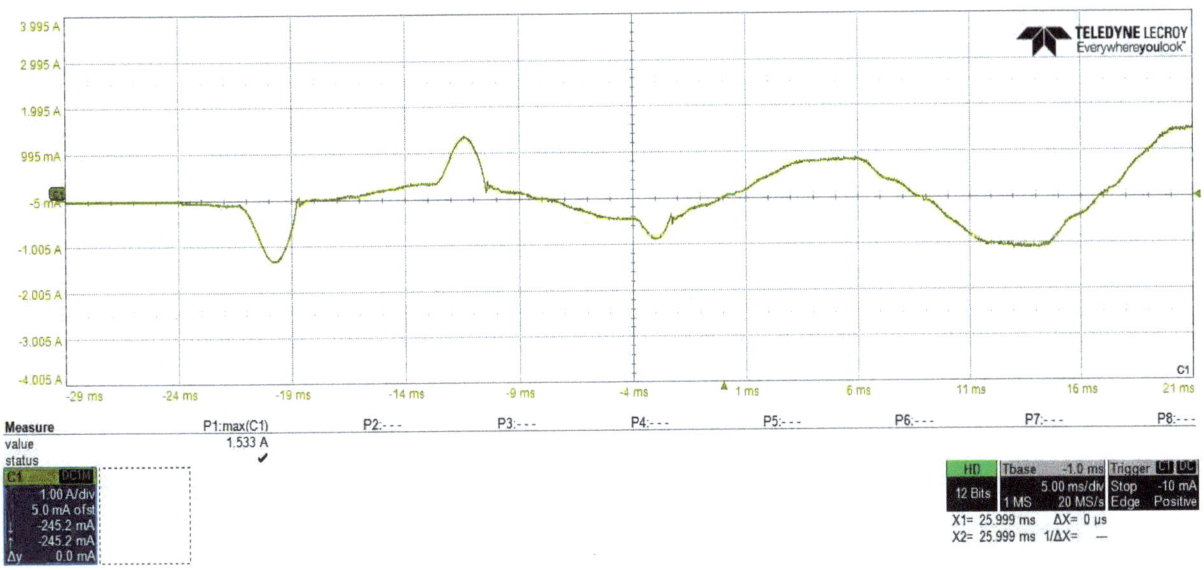

Figure 21. PFC converter stage AC input current with using of the proposed inrush current mitigation circuit.

Figure 22 shows the converter output voltage and current waveforms at steady state operation using the inrush current control circuit, in which the converter DC output power was about 1085.25 W, where the output voltage was constant at about 54.02 V and current value was about 20.090 A.

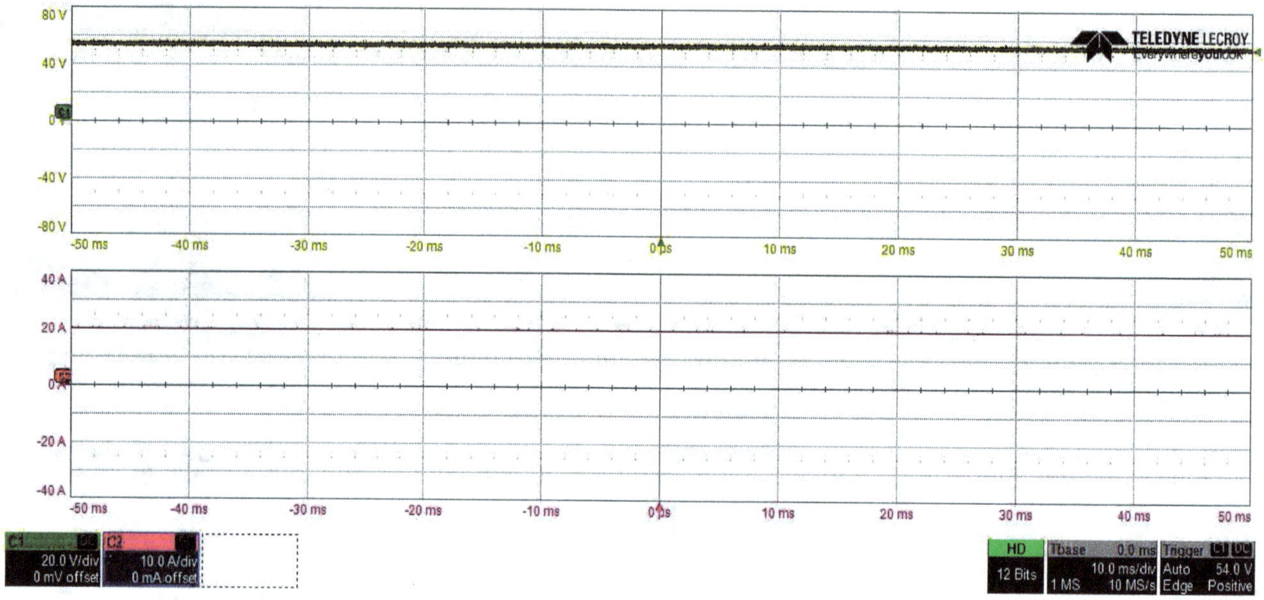

Figure 22. DC–DC converter stage output voltage and current waveforms at V_{in} = 400 V.

DC–DC converter output filter capacitor was designed to maintain the output voltage ripple to the specified value as explained previously in Section 3; in order to clearly observe the ripple component in the output voltage waveform, the vertical axis scale of the voltage was changed and the measurement set in scope was adjusted to measure the ripple peak to peak value and the mean value of the voltage waveform, as shown in Figure 23, which clearly shows that the peak to peak ripple voltage is about 34.4 mV (less than the designed value 200 mV) and the mean output voltage value is about 54.094 V.

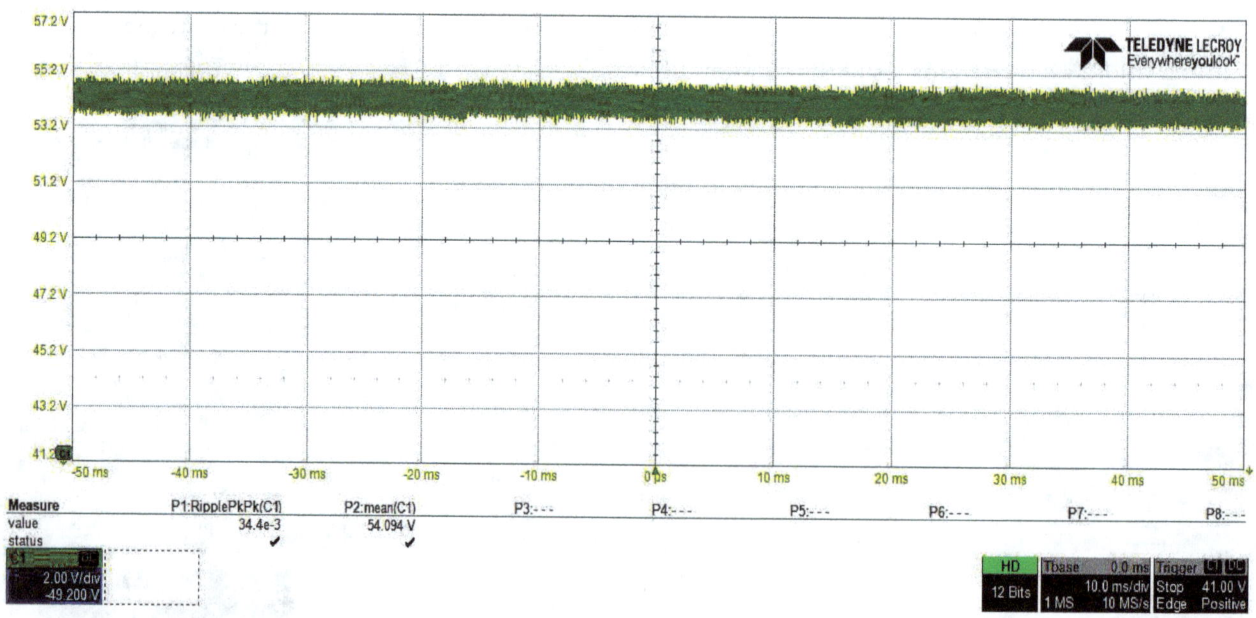

Figure 23. DC–DC converter stage output voltage ripple peak to peak measurement at V_{in} = 400 V.

In case of designing the DC–DC converter stage for the telecom applications, voltage controller must be implemented in order to achieve different loading condition with good voltage regulation.

Figure 24 depicts the output voltage versus output power, which shows that the designed DC–DC converter with the proposed inrush current controlling circuit also provides different loading conditions with good voltage regulation, where the voltage drop is less than 1 V (1.85%) for changing loading conditions from 200 W to 1000 W, which is less than the IEC61000-3-3 standard limit for the limitation of the voltage changes (3.3%) [48].

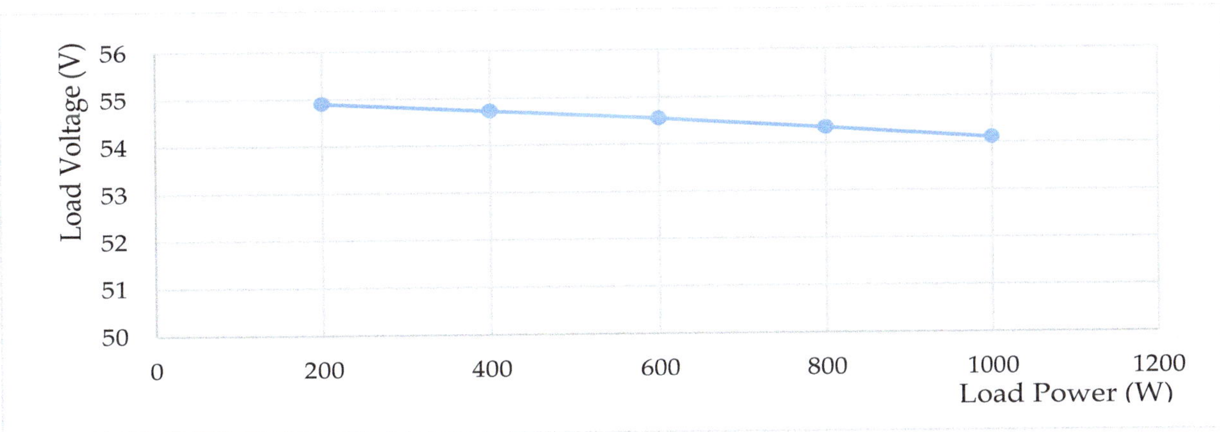

Figure 24. DC–DC converter load voltage-power characteristics at V_{in} = 400 V.

The inrush current control circuit, designed and connected between the two stages of the power supply, to control the inrush current at the initial power application to the power supply, must not effect the performance of the PFC stage at steady state operation. To study the effect of the proposed inrush current control circuit on the PFC converter performance, power analysis of the input side of the PFC converter have been performed with, and without, using the inrush current control circuit with the same loading and input voltage conditions as shown in Figures 25 and 26, which clearly shows that the reduction in the inrush current of the input supply at initial power application leads to a reduction in the supply total reactive power from 103 VA to about 90 VA, and it can also clearly be observed that the power factor value of the power supply was not affected by the connection of the proposed circuit.

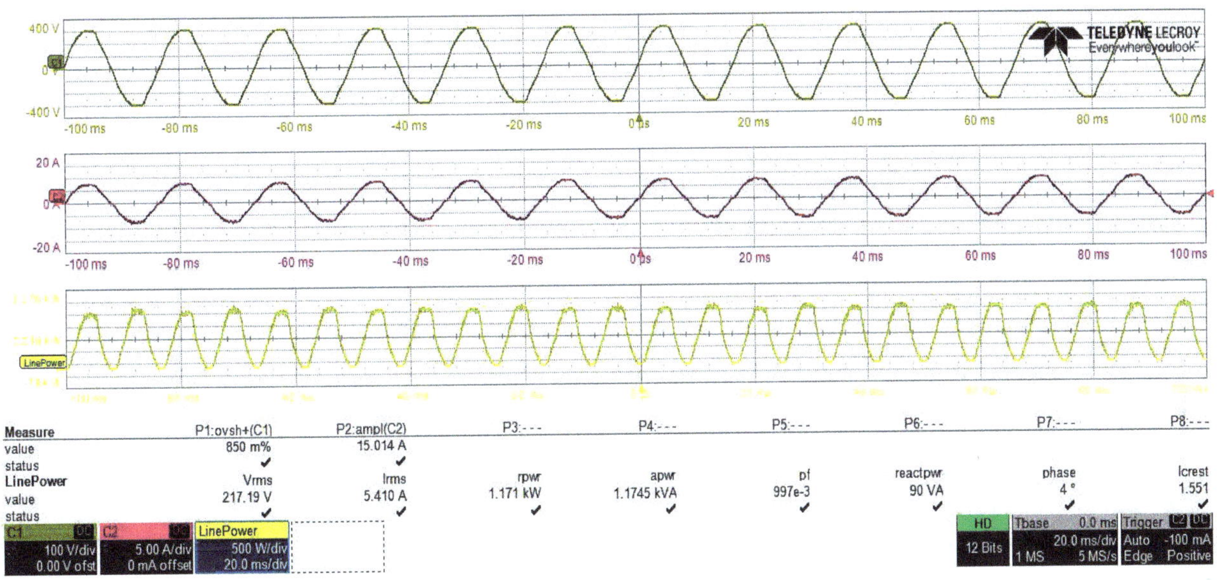

Figure 25. PFC converter stage power analysis with using of the proposed inrush current mitigation circuit.

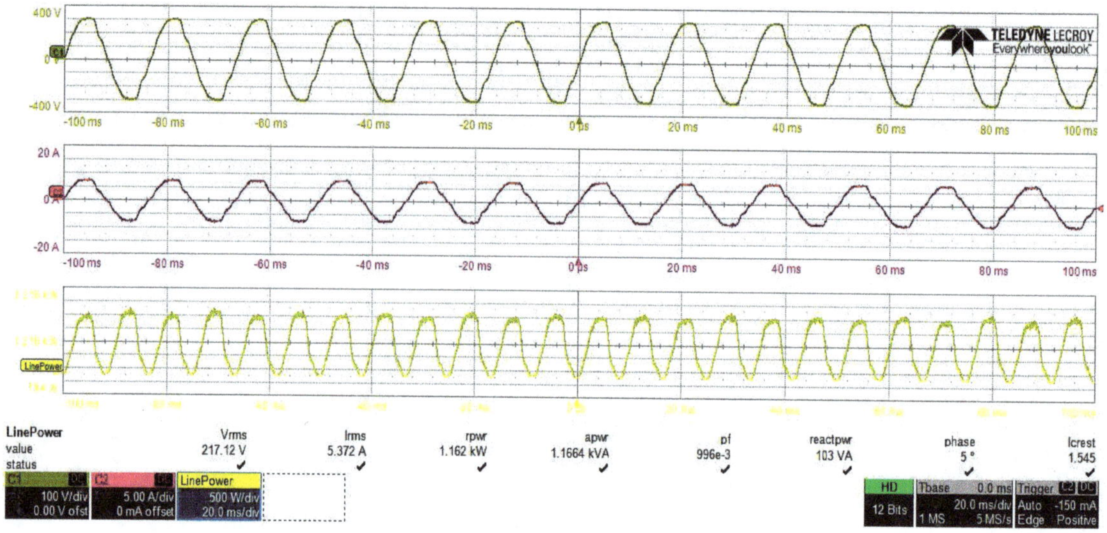

Figure 26. PFC converter stage power analysis without using of the proposed inrush current mitigation circuit.

As shown in Figure 22 the total input power to the two stages with inrush current control circuit is about 1171 W, and as shown in Figure 25, the load power is about 1085.25 W, so the power supply includes about 85.75 W power losses, distributed as 39.5 W in the PFC stage with efficiency of about 96.62% and 46.25 W in DC–DC converter and inrush current control circuit, with efficiency of about 95.73%, which is very close to the simulation results (95.40%) at the same loading. Total losses distribution in the system different parts are performed and the result is depicted in Figure 27.

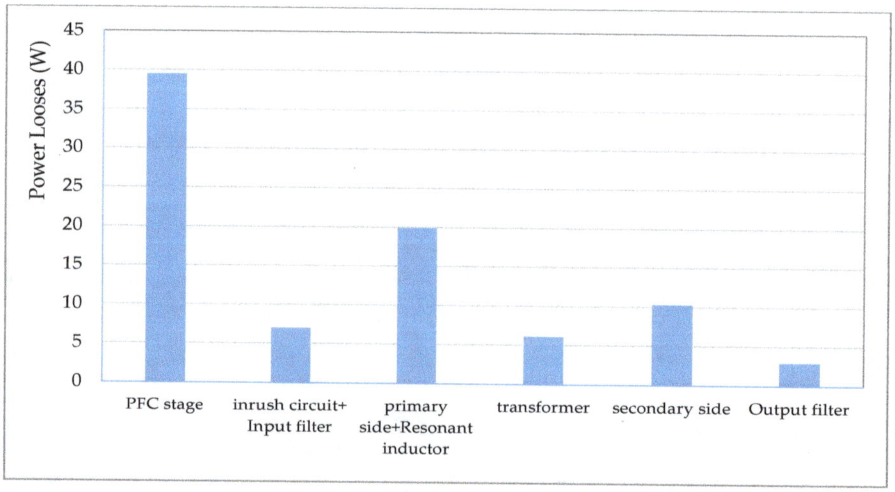

Figure 27. Losses distribution in different parts of the two stages power supply with inrush current control.

From the loss distribution of the power supply, it is noticed that the highest budget of the DC–DC converter losses was accounted by the primary side bridge and the resonant inductor with about 20 W; this loss amount is due to primary side switches internal resistance and forward voltage drop, and also due to the internal DC resistance of the resonant inductor. The second highest budget of the power losses accounted in the secondary side synchronous switches with about 10.5 W. Power loss of about 6 W was dissipated in the converter transformer primary and secondary DC resistances. Internal DC resistance of the output filter inductor and the equivalent series resistance (ESR) of the output electrolytic capacitor introduce power loss about 3 W. The remaining converter power losses budget was accounted by the proposed inrush current control circuit and the ESR of the bulk input electrolytic capacitor. One disadvantage of the proposed inrush current, is that it has an amount of about 5 W of the total supply power losses, but on the other hand, it has many advantages, which can be summarized in the following:

1. Compact design and easy implementation with cheap passive elements and single MOSFET.
2. Controllability and reliability of the circuit to limit the inrush current at the required value.
3. Reduce the current stress on the input side of the two stages of power supply at initial power application.
4. Improve the PFC stage performance by reducing the total reactive power consumption of the circuit.
5. Easy to disconnect and reconnect again between two stages of a faulty case as compared with the IC soft starter (in case where the IC soft starter function is damaged).
6. Ability to use with analog and digital control converter.

5. Conclusions

Complete design of a high power density efficient DC–DC converter stage of the telecom power supply has been discussed in this paper; a high inrush current, which is generated between two stages of power supply where the bulk input capacitor is controlled using the proposed simple inrush current control circuit designed by using single MOSFET switch and some passive elements. Simulation analysis as well as experimental setup of the practical two stages telecom power supply and inrush current control circuit is performed. Phase shifted PWM with ZVS technique is applied in DC–DC converter stage, which provided system experimentally overall efficiency about 95.73% at 50% loading condition. Furthermore, reliability of the converter designed components to achieve the performance of the telecom power supply was established; ability of the proposed inrush current control circuit to control the slew rate of the input capacitance voltage in order to reduce the high dv/dt and to mitigate the inrush current of the input current of the two stages of power supply was also achieved, which ensures a safe and smooth startup operation of the power supply.

Author Contributions: The literature review and manuscript preparation, as well as the simulations, were carried out by A.H.O. Experimental results and implementation of the prototype were carried by A.H.O. and J.B. Final review of manuscript corrections was done by N.K. and J.B. All authors have read and agreed to the published version of the manuscript.

References

1. Garcia, O.; Cobos, J.A.; Prieto, R.; Alou, P.; Uceda, J. Single phase power factor correction: A survey. *IEEE Trans. Power Electron.* **2003**, *18*, 749–755. [CrossRef]
2. Geddam, D.P.; Prakash, D.; Bapu, S.R.; Kumar, Y.R.; Rao, P.S.; Naidu, G.K.M.; Ramanarayanan, V. Design of 1.4 kw telecom rectifier delivering full power from 90Vac to 300Vac. In Proceedings of the IEEE INTELEC 07–29th International Telecommunications Energy Conference, Rome, Italy, 30 September–4 October 2007; pp. 670–676.
3. Kasper, M.; Bortis, D.; Deboy, G.; Kolar, J.W. Design of a Highly Efficient (97.7%) and Very Compact (2.2 kW/dm^3) Isolated AC–DC Telecom Power Supply Module Based on the Multicell ISOP Converter Approach. *IEEE Trans. Power Electron.* **2016**, *32*, 7750–7769. [CrossRef]
4. Sekar, A.; Raghavan, D. Implementation of single phase soft switched PFC converter for plug-in-hybrid electric vehicles. *Energies* **2015**, *8*, 13096–13111. [CrossRef]
5. Kim, Y.-S.; Sung, W.; Lee, B. Comparative performance analysis of high density and efficiency PFC topologies. *IEEE Trans. Power Electron.* **2003**, *6*, 2666–2679. [CrossRef]
6. Biela, J.; Badstuebner, U.; Kolar, J.W. Impact of power density maximization on efficiency of DC–DC converter systems. *IEEE Trans. Power Electron.* **2009**, *24*, 288–300. [CrossRef]
7. Yang, B.; Duarte, J.L.; Li, W.; Yin, K.; He, X.; Deng, Y. Phase-shifted full bridge converter featuring ZVS over

the full load range. In Proceedings of the IECON 36th Annual Conference on IEEE Industrial Electronics Society, Glendale, AZ, USA, 7–10 November 2010; pp. 644–649.

8. Lo, Y.-K.; Lin, C.-H.; Hsieh, M.-I.; Lin, C.-H. Phase-shifted full-bridge series-resonant DC–DC converters for wide load variations. *IEEE Trans. Ind. Electron.* **2010**, *58*, 2572–2575. [CrossRef]
9. Schmidt, J.C.M.; Maragaño, J.C.; Sartori, H.C.; Pinheiro, J.R. Design methodology to achieve high-efficiency isolated ZVS DC/DC converters. In Proceedings of the 2017 IEEE 8th International Symposium on Power Electronics for Distributed Generation Systems (PEDG), Florianopolis, Brazil, 17–20 April 2017; pp. 1–8.
10. Badstuebner, U.; Biela, J.; Kolar, W.J. Design of an 99%-efficient, 5kW, phase-shift PWM DC–DC converter for telecom applications. In Proceedings of the 2010 Twenty-Fifth Annual IEEE Applied Power Electronics Conference and Exposition (APEC), Palm Springs, CA, USA, 21–25 February 2010; pp. 773–780.
11. Bai, H.; Nie, Z.; Mi, C.C. Experimental comparison of traditional phase-shift, dual-phase-shift, and model-based control of isolated bidirectional DC–DC converters. *IEEE Trans. Power Electron.* **2009**, *25*, 1444–1449. [CrossRef]
12. Xiangli, K.; Li, S.; Smedley, K.M. Decoupled PWM plus phase-shift control for a dual-half-bridge bidirectional DC–DC converter. *IEEE Trans. Power Electron.* **2017**, *33*, 7203–7213. [CrossRef]
13. Hurley, W.G.; Wölfle, W.H. *Transformers and Inductors for Power Electronics. Theory, Design and Applications*; Wiley: London, UK, 2013.
14. Koo, G.B.; Moon, G.W.; Youn, M.J. New zero-voltage-switching phase-shift full-bridge converter with low conduction losses. *IEEE Trans. Power Electron.* **2005**, *52*, 228–235. [CrossRef]
15. Kathiresan, R.; Das, P.; Reindl, T.; Panda, S.K. A novel ZVS DC–DC full-bridge converter with hold-up time operation. *IEEE Trans. Ind. Electron.* **2017**, *64*, 4491–4500. [CrossRef]
16. Das, P.; Laan, B.; Mousavi, S.A.; Moschopoulos, G. A non-isolated bidirectional ZVS-PWM active clamped DC–DC converter. *IEEE Trans. Power Electron.* **2009**, *24*, 553–558. [CrossRef]
17. Bojoi, R.; Fusillo, F.; Raciti, A.; Musumeci, S.; Scrimizzi, F.; Rizzo, S. Full-Bridge DC–DC Power Converter for Telecom applications with Advanced Trench Gate MOSFETs. In Proceedings of the 2018 IEEE International Telecommunications Energy Conference (INTELEC), Lingnotto Fiere, Italy, 7–11 October 2018; pp. 1–7.
18. Zhao, Z.; Xu, Q.; Dai, Y.; Yin, H. Analysis, Design, and Implementation of Improved LLC Resonant Transformer for Efficiency Enhancement. *Energies* **2018**, *11*, 3288. [CrossRef]
19. Tao, H.; Zhang, G.; Zheng, Z.; Du, C. Design of Digital Control System for DC/DC Converter of On-Board Charger. *J. Adv. Transp.* **2019**, *2019*, 2467307. [CrossRef]
20. Cho, J.-H.; Seong, H.-Y.; Jung, S.-H.; Park, J.-Y.; Moon, G.-U.; Youn, M.-Y. Implementation of digitally controlled phase shift full bridge converter for server power supply. In Proceedings of the 2010 IEEE Energy Conversion Congress and Exposition, Atlanta, GA, USA, 12–16 September 2010; pp. 802–809.
21. Lee, Y.J.; Bak, Y.; Lee, K.Y. Control Method for Phase-Shift Full-Bridge Center-Tapped Converters Using a Hybrid Fuzzy Sliding Mode Controller. *Electronics* **2019**, *8*, 705. [CrossRef]
22. Cho, I.-H.; Cho, K.-M.; Kim, J.-W.; Moon, G.-W. A new phase-shifted full-bridge converter with maximum duty operation for server power system. *IEEE Trans. Power Electron.* **2011**, *26*, 3491–3500. [CrossRef]
23. Moisseev, S.; Soshin, K.; Sato, S.; Gamage, L.; Nakaoka, M. Novel soft-commutation DC–DC power converter with high-frequency transformer secondary side phase-shifted PWM active rectifier. *IEE Electr. Power Appl.* **2004**, *151*, 260–267. [CrossRef]
24. Kim, Y.-D.; Cho, K.-M.; Kim, D.-Y.; Moon, G.-W. Wide-range ZVS phase-shift full-bridge converter with reduced conduction loss caused by circulating current. *IEEE Trans. Power Electron.* **2012**, *28*, 3308–3316. [CrossRef]
25. Inrush Current in DC–DC Converters. Application Notes, vpt Power. Available online: http://www.vptpower.com/wp-content (accessed on 14 November 2019).
26. Martínez-Salamero, L.; García, G.; Orellana, M.; Lahore, C.; Estibals, B. Start-up control and voltage regulation in a boost converter under sliding-mode operation. *IEEE Trans. Ind. Electron.* **2012**, *60*, 4637–4649. [CrossRef]
27. Aroudi, A.E.; Martínez-Treviño, B.A.; Vidal-Idiarte, E.; Martínez-Salamero, L. Analysis of Start-Up Response in a Digitally Controlled Boost Converter with Constant Power Load and Mitigation of Inrush Current Problems. *IEEE Trans. Circuits Syst. I* **2019**, *67*, 1276–1285. [CrossRef]
28. Mitter, S. *Active Inrush Current Limiting Using MOSFETs*; Motorola Inc.: Phoenix, AZ, USA, 1995; Available online: http://application-notes.digchip.com/010/10-13155.pdf (accessed on 10 January 2020).
29. Kalenteridis; Vasileios; Agorastou, Z.; Siskos, S. A soft start-up technique for inrush current limitation

in DC–DC converters. In Proceedings of the 2019 IEEE Panhellenic Conference on Electronics & Telecommunications (PACET), Volos, Greece, 8–9 November 2019; pp. 1–4.
30. Ma, F.-F.; Chen, W.-Z.; Wu, J.-C. A monolithic current-mode buck converter with advanced control and protection circuits. *IEEE Trans. Power Electron.* **2007**, *22*, 1836–1846. [CrossRef]
31. Sathishkumar, P.; Krishna, T.N.V.; Khan, M.A.; Zeb, K.; Kim, H. Digital soft start implementation for minimizing start up transients in high power DAB-IBDC converter. *Energies* **2018**, *11*, 956. [CrossRef]
32. Managing Inrush Current. Application Report, Texas Instruments. Available online: http://www.ti.com/lit/an/slva670a/slva670a.pdf?&ts=1589053420769 (accessed on 12 January 2020).
33. Jiang, T.; Cairoli, P.; Rodrigues, R.; Du, Y. Inrush current limiting for solid state devices using NTC resistor. In Proceedings of the IEEE Southeast Conference, Charlotte, NC, USA, 30 March–2 April 2017; pp. 1–7.
34. NTC Thermistors Ordering Code. RS Components. Available online: https://docs.rs-online.com/ee83/0900766b80e7f61b.pdf (accessed on 12 January 2020).
35. Dias, V.A.; Pomilio, J.A.; Finco, S. A current limiting switch for applications in space power systems. In Proceedings of the 2017 IEEE Southern Power Electronics Conference (SPEC), Puerto Valas, Chile, 4–7 December 2017; pp. 1–6.
36. Fuengwarodsakul, N.H. Battery management system with active inrush current control for Li-ion battery in light electric vehicles. *Electr. Eng.* **2016**, *98*, 17–27. [CrossRef]
37. Marroqui, D.; Garrigos, A.; Blanes, J.M.; Gutierrez, R. Photovoltaic-Driven SiC MOSFET Circuit Breaker with Latching and Current Limiting Capability. *Energies* **2019**, *12*, 4585. [CrossRef]
38. Lee, E.-J.; Ahn, J.-H.; Shin, S.-M.; Lee, B.-K. Comparative analysis of active inrush current limiter for high-voltage DC power supply system. In Proceedings of the 2012 IEEE Vehicle Power and Propulsion Conference, Seoul, Korea, 9–12 October 2012; pp. 1256–1260.
39. Musumeci, S. Gate charge control of high-voltage Silicon-Carbide (SiC) MOSFET in power converter applications. In Proceedings of the 2015 International Conference on Clean Electrical Power (ICCEP), Taormina, Italy, 16–18 June 2015; pp. 709–715.
40. Manolarou, M.; Kostakis, G.; Manias, S.N. Inrush current limiting technique for low-voltage synchronous DC/DC converters. *IEE Electr. Power Appl.* **2005**, *152*, 1179–1183. [CrossRef]
41. Manousaka, E. DC–DC Buck Converter with Inrush Current Limiter. Master's Thesis, Faculty of Applied Sciences, Delft University of Technology, Delft, Wales, October 2013.
42. UCC28950 600-W, Phase-Shifted, Full-Bridge Application Report. Texas Instruments. Available online: http://www.ti.com/lit/slua560 (accessed on 8 March 2020).
43. N-CHANNEL 500 V –0.105Ω –31A TO-247 Zener-Protected MOSFET. Available online: https://www.st.com/resource/en/datasheet/cd00045267.pdf (accessed on 5 April 2020).
44. *"Power Electronics Handbook" Book*; Available online: http://site.iugaza.edu.ps/malramlawi/files/RASHID_Power_Electronics_Handbook.pdf (accessed on 25 February 2020).
45. Phase-Shifted Full-Bridge, Zero-Voltage Transition Design Considerations. Application Report. Available online: https://www.ti.com/lit/pdf/slua107 (accessed on 8 March 2020).
46. Instruments Texas. *Modeling, Analysis and Compensation of the Current-Mode Converter*; Notes, Application; Texas Instruments Inc.: Dallas, TX, USA, 1999.
47. 2.5kW PFC Evaluation Board with CCM PFC Controller ICE3PCS01G. Design Report, Infineon Company. Available online: https://www.infineon.com/cms/en/product/evaluation-boards/eval_2k5w_ccm_4p_v3/ (accessed on 15 April 2020).
48. IEC/EN61000 Standards for Power Supplies. SL Power Application Note. Available online: https://slpower.com/App-slpower/images/whitepapers/AN-G007_EN61000.pdf (accessed on 12 August 2020).

9

High-Efficiency Design and Control of Zeta Inverter for Single-Phase Grid-Connected Applications

Woo-Young Choi * and Min-Kwon Yang

Division of Electronic Engineering, Chonbuk National University, Jeonju 561-756, Korea; mkyang@jbnu.ac.kr
* Correspondence: wychoi@jbnu.ac.kr

Abstract: The conventional zeta inverter has been used for single-phase grid-connected applications. However, it has high switching losses to operate at high switching frequency in the continuous conduction mode (CCM). To address this drawback, this paper suggests a high-efficiency zeta inverter using active clamp and synchronous rectification techniques. The proposed inverter utilizes the active clamp circuit for reducing switching losses. The non-complementary switching scheme is adopted for not only clamping the switch voltage stresses, but also alleviating the circulating energy. In addition, the synchronous rectification is implemented for reducing the body diode conduction of power switches. By using the silicon carbide (SiC) metal oxide semiconductor field effect transistors (MOSFETs), the switching performance of the proposed inverter is improved. Its operation principle and control strategy are presented. A 220-W prototype has been designed and tested to evaluate the performance of the proposed inverter.

Keywords: zeta inverter; active clamp; synchronous rectification; power efficiency

---

1. Introduction

With the fast growth of renewable energy markets, many single-stage isolated inverters have been developed for single-phase grid-connected applications [1–9]. Among them, the zeta inverter [7–9] has been gaining attention due to the advantages of its low grid current ripple and low circuit component count, compared to other single-stage inverters such as flyback inverters [1–4] and Cuk inverters [5,6]. Of course, although the zeta inverter has large current ripples at the dc side, similar to the flyback inverter, it has been widely adopted for single-phase photovoltaic applications [7–9] due to its simple and flexible circuitry.

As the zeta inverter operates at a constant switching frequency, it has two operation modes: discontinuous conduction mode (DCM) and continuous conduction mode (CCM). The DCM zeta inverter has been used for low-power applications [7]. The grid current is easily controlled, as its control transfer function for the grid current is linear in DCM [8]. However, as the power level increases, the DCM zeta inverter suffers from high conduction losses. The DCM zeta inverters need to be connected in parallel to alleviate conduction losses.

When the zeta inverter operates in CCM, it can withstand a higher power level than the DCM zeta inverter. The CCM zeta inverter has been suggested in [9]. Figure 1 shows its circuit diagram, which is equivalent to the circuit diagram in [9]. It consists of the primary circuit (S_{P1}, L_m, L_{lk}) and the secondary circuit (C_S, $S_{S1} \sim S_{S4}$, L_g). T is a high-frequency transformer, which has the magnetizing inductor L_m and the leakage inductor L_{lk}. The CCM zeta inverter has achieved higher efficiency than the DCM zeta inverter by lowering conduction losses [9]. However, the CCM zeta inverter suffers from high switching losses. In the primary circuit, the switch S_{P1} operates at a high switching frequency to regulate the grid current i_g. When S_{P1} is turned off, a high voltage spike is generated due to the energy stored in the leakage inductor L_{lk}, which increases switching losses [10].

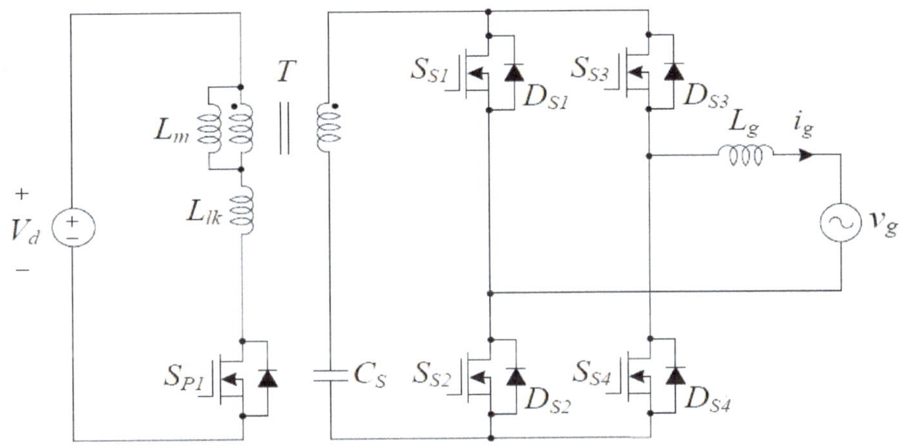

Figure 1. Circuit diagram of the zeta inverter.

The secondary circuit provides the current path to fold the grid voltage v_g and gives the freewheeling path for the grid current i_g. The switches $S_{S1} \sim S_{S4}$ operate at grid frequency according to the polarity of v_g. S_{S2} and S_{S3} are always turned on for the positive grid cycle, while S_{S1} and S_{S4} are always turned on for the negative grid cycle. For the positive grid cycle, when S_{P1} is turned on, i_g flows through S_{S2} and S_{S3}. As S_{P1} is turned off, i_g flows through S_{S2}, S_{S3}, D_{S1}, and D_{S4}. D_{S1} and D_{S4} are the body diodes, which are turned on for the freewheeling of i_g. However, when S_{P1} is turned on again, D_{S1} and D_{S4} are not turned off instantly due to the slow reverse recovery process [11]. The reverse recovery currents of D_{S1} and D_{S4} cause high voltage spikes across D_{S1} and D_{S4}. They are also transferred to the primary circuit, which increases the turn-on switching losses of S_{P1}. It is worse when high-voltage silicon (Si) metal oxide semiconductor field effect transistors (MOSFETs) are adopted, because their body diodes have poor reverse recovery characteristics. These drawbacks limit the practical use of the CCM zeta inverter, despite its advantages.

This paper proposes a high-efficiency zeta inverter to cope with the above-mentioned drawbacks. Figure 2 shows the circuit diagram of the proposed inverter. The active clamp circuit has been used for reducing switching losses in the primary circuit. The active clamp circuit has the auxiliary switch S_{P2} and the clamp capacitor C_P. Conventionally, the auxiliary switch S_{P2} operates complementary to the main switch S_{P1} [12]. As the energy stored in L_{lk} is absorbed in C_P, the switch voltages are clamped to a constant voltage. However, this conventional method increases the circulating current in the primary circuit. It causes additional power losses due to high circulating energy. To address this drawback, the non-complementary switching scheme [13] has been adopted. In the non-complementary method, S_{P2} is turned on for a short time before S_{P1} is turned on. As the leakage energy is recycled with reduced circulating current, power losses associated with the circulating energy can be minimized in the proposed inverter.

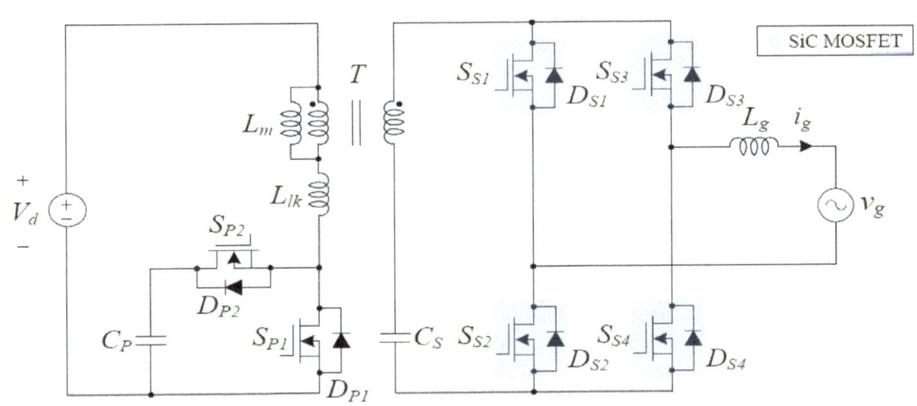

Figure 2. Circuit diagram of the proposed inverter.

In the secondary circuit, the synchronous rectification [14] is used for reducing the body diode conduction of MOSFETs. With synchronous rectification, S_{S1} (S_{S2}) and S_{S4} (S_{S3}) are turned on for the positive (negative) grid cycle when S_{P1} is turned off. It allows i_g to freewheel through the MOSFET channels instead of the body diodes. It avoids the body diode conduction, increasing the power efficiency, because the voltage drop across the on-state resistance of the MOSFET is lower than the forward voltage drop of the body diode [15]. The silicon carbide (SiC) MOSFET has been utilized for the synchronous rectification of $S_{S1} \sim S_{S4}$. Due to the advantages over Si MOSFETs such as wider bandgap and higher electric field capacity [16], it improves the switching performance of the proposed inverter.

This paper is organized as follows. Section 2 describes the operation principle of the proposed inverter in the steady-state condition, along with the analysis of active clamp and synchronous rectification techniques. It also describes the control strategy for the CCM operation of the proposed inverter. Section 3 discusses the experimental results to verify the performance of the proposed inverter. Section 4 presents the conclusion of this paper.

2. Proposed Inverter

2.1. Operation Principle

Figure 3 shows the circuit diagram of the proposed inverter, which shows the reference directions of currents and voltages. The proposed inverter consists of the primary circuit (S_{P1}, S_{P2}, C_P, L_m, L_{lk}) and the secondary circuit (C_S, $S_{S1} \sim S_{S4}$, L_g). The active clamp circuit consists of S_{P2} and C_P. T is a high-frequency transformer, which has the magnetizing inductor L_m and the leakage inductor L_{lk}. It is assumed that $L_m >> L_{lk}$. Its turns ratio n is N_S / N_P. N_P is the primary winding turns. N_S is the secondary winding turns. All of the power switches are considered ideal except for their body diodes. V_d is the dc input voltage. v_g is the grid voltage. L_m and L_g are large enough so that the currents i_{Lm} and i_g are continuous during one switching period T_s, respectively. C_P and C_S are large enough so that the voltages V_{Cp} and V_{Cs} are constant during T_s, respectively. V_{Cs} is considered as $|v_g|$ because the secondary circuit provides the current path to fold v_g.

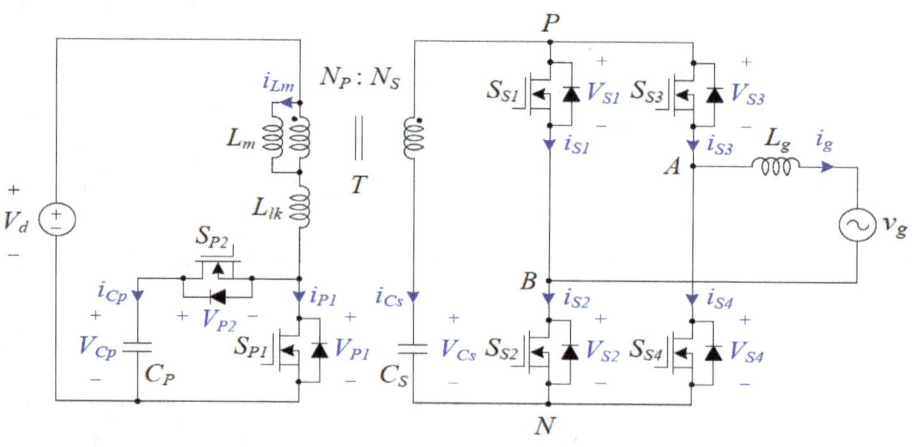

Figure 3. Circuit diagram of the proposed inverter with reference directions of currents and voltages.

Figure 4 shows the switching circuit diagrams of the proposed inverter during T_s for the positive grid cycle. Figure 5 shows the switching waveform diagrams of the proposed inverter during T_s for the positive grid cycle. $V_{gate,P1}$ and $V_{gate,P2}$ are the gate signals for S_{P1} and S_{P2}, respectively. $V_{gate,S1} \sim V_{gate,S4}$ are the gate signals for $S_{S1} \sim S_{S4}$, respectively. S_{P1} and S_{P2} operate with a constant high switching frequency f_s (= $1/T_s$). S_{P2} is turned on for a short time before S_{P1} is turned on. For the positive grid cycle, S_{S2} and S_{S3} are always turned on. S_{S1} and S_{S4} operate complementary to S_{P1}. The proposed inverter has four switching modes during T_s for the positive grid cycle, which are outlined below.

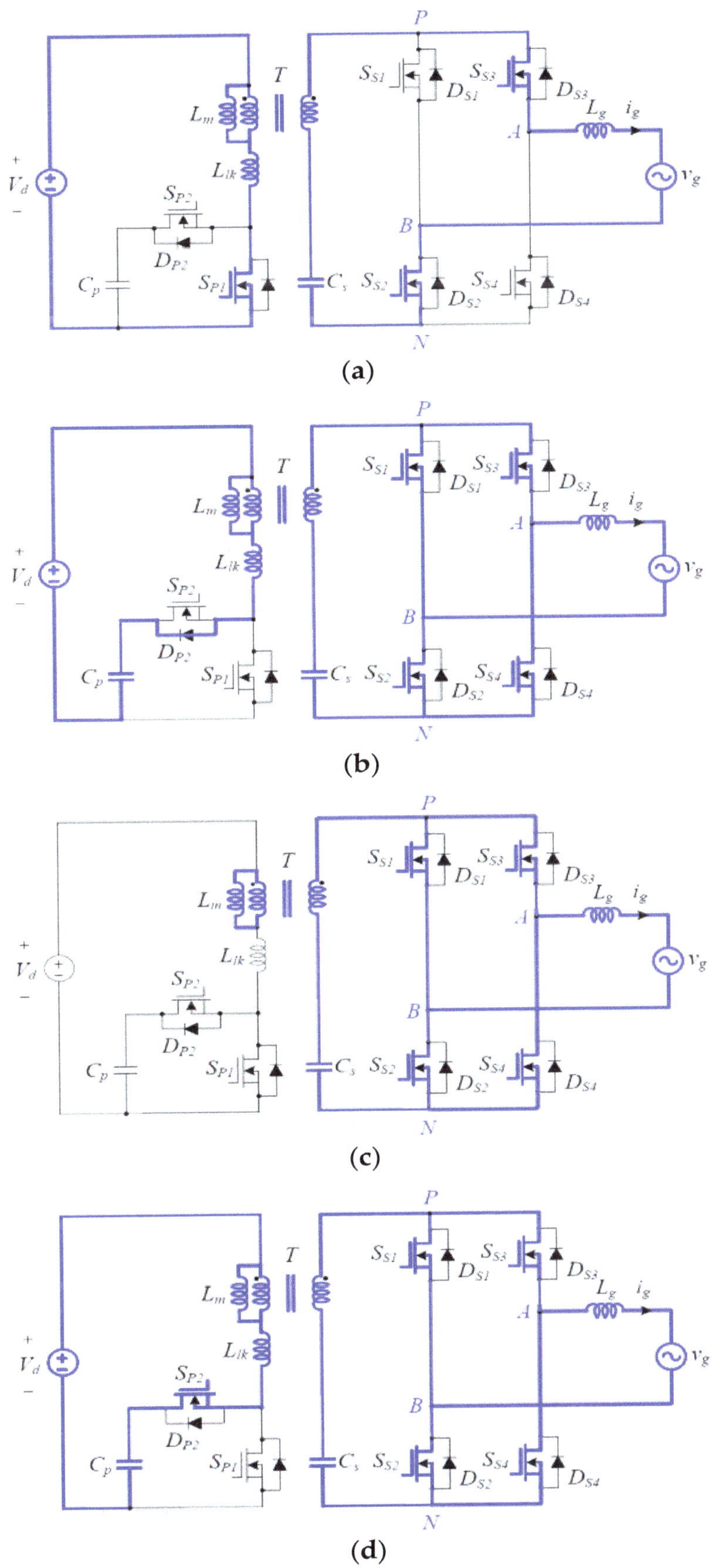

Figure 4. Switching circuit diagrams of the proposed inverter during T_s for the positive grid cycle: (**a**) Mode 1; (**b**) Mode 2; (**c**) Mode 3; and (**d**) Mode 4.

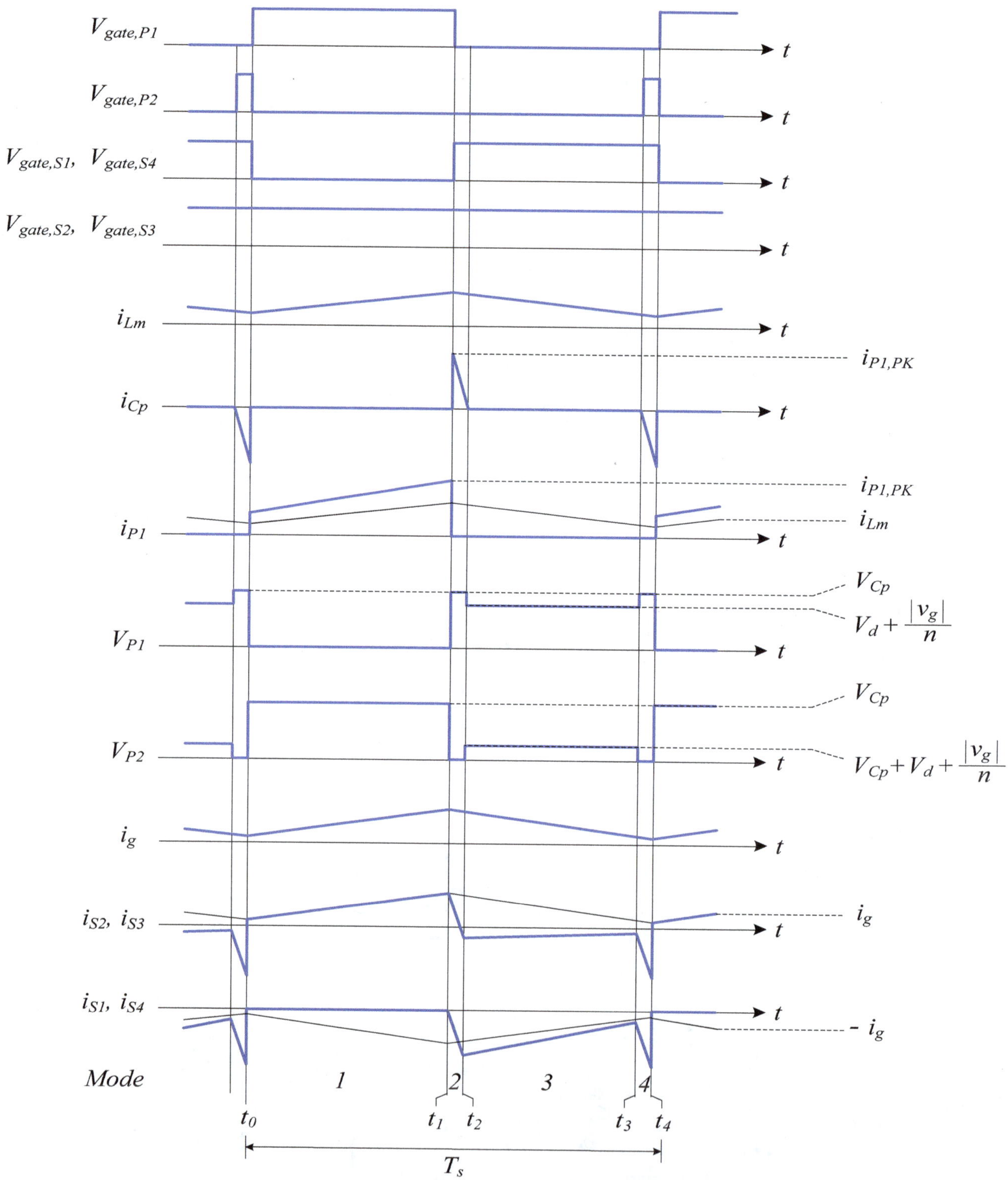

Figure 5. Switching waveform diagrams of the proposed inverter during T_s for the positive grid cycle.

Mode 1 [t_o–t_1]: At $t = t_o$, S_{P1} is turned on. L_m and L_{lk} store the energy from V_d. i_{Lm} increases linearly at the rate of $di_{Lm}/dt = V_d/(L_m + L_{lk})$. The voltage across the secondary winding of T is $nL_mV_d/(L_m + L_{lk})$. i_g increases linearly at the rate of $di_g/dt = nL_mV_d/(L_m + L_{lk})/L_g$.

Mode 2 [t_1–t_2]: At $t = t_1$, D_{P2} is turned on as S_{P1} is turned off. The energy stored in L_{lk} is stored in C_P. The switch voltage V_{P1} is clamped to V_{Cp}. S_{S1} and S_{S4} are turned on in this mode. As the voltage across the secondary winding of T is v_g, i_{Lm} decreases linearly at the rate of $di_{Lm}/dt = -v_g/nL_m$. As i_g freewheels through $S_{S1} \sim S_{S4}$, i_g decreases linearly at the rate of $di_g/dt = -v_g/L_g$.

Mode 3 [t_2–t_3]: At $t = t_2$, D_{P2} is turned off as i_{Cp} becomes zero. The switch voltage V_{P1} is clamped to $V_d + v_g/n$. The switch voltage V_{P2} is clamped to $V_{Cp} + V_d + v_g/n$. i_{Lm} and i_g keep decreasing linearly as in Mode 2.

Mode 4 [t_3–t_4]: At $t = t_3$, S_{P2} is turned on for transferring the energy stored in C_P to the grid. The absorbed leakage energy is recycled to the grid. The switch voltage V_{P1} is clamped to V_{Cp}. i_{Lm} and i_g keep decreasing linearly as in Mode 2.

From Mode 2 to Mode 4, the following current relations are obtained as $i_{Cs} = -i_{S1} - i_{S3} = -i_{S2} - i_{S4}$ and $i_g = i_{S2} - i_{S1} = i_{S4} - i_{S3}$ with respect to the nodes P, N, A, and B. Next, the switching cycle begins as S_{P1} is turned on, and S_{P2}, S_{S1}, and S_{S4} are turned off. For the negative grid cycle, S_{S1} and S_{S4} are always turned on. S_{S2} and S_{S3} operate complementary to S_{P1}. The operation principle for the negative grid cycle is not described here, because it can be explained analogously as the operation principle for the positive grid cycle.

2.2. Active Clamp Circuit

In the active clamp circuit, S_{P2} is turned on for a short period before S_{P1} is turned on. The absorbed leakage energy is recycled as the energy stored in C_P is transferred to the grid. Figure 6 shows the detailed timing diagrams for $V_{gate,P1}$, $V_{gate,P2}$, and i_{Cp}. T_{ON} is the turn-on time of S_{P1}. T_{OFF} is the turn-off time of S_{P1}. T_{DP2} is the turn-on time of D_{P2}. T_{SP2} is the turn-on time of S_{P2}. As S_{P2} is turned on, the capacitor current i_{Cp} can be represented as:

$$i_{Cp}(t) = -\frac{V_{Cp} - V_d - |v_g|/n}{L_{lk}} t. \quad (1)$$

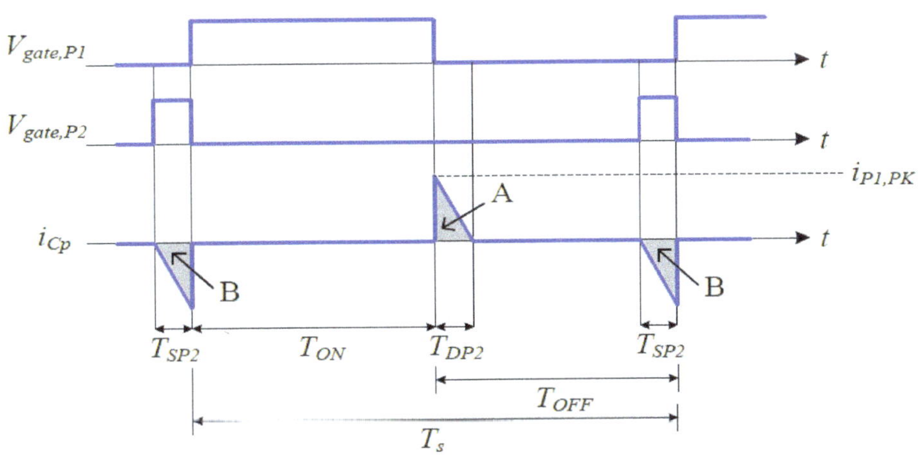

Figure 6. Detailed timing diagrams for $V_{gate,P1}$, $V_{gate,P2}$, and i_{Cp}.

By the charge balance for the capacitor C_P during T_s, the areas A and B for two current waveforms should be equal to:

$$\frac{1}{2} i_{P1,PK} T_{DP2} = \frac{V_{Cp} - V_d - |v_g|/n}{2L_{lk}} T_{SP2}^2 \quad (2)$$

where $i_{P1,PK}$ is the peak value of i_{P1}. Supposing that $T_{DP2} = T_{SP2}$, the capacitor voltage V_{Cp} is represented as:

$$V_{Cp} = V_d + \frac{|v_g|}{n} + \frac{L_{lk} i_{P1,PK}}{T_{SP2}}. \quad (3)$$

Also, the circulating energy E_{Cir} during T_{SP2} is represented as:

$$E_{Cir} = \frac{(V_{Cp} - V_d - |v_g|/n)^2}{2L_{lk}} T_{SP2}^2. \quad (4)$$

The circulating energy E_{Cir} will be smaller as T_{SP2} is getting shorter. However, in the active clamp method using the complementary switching scheme [12], the circulating energy is related with the turn-off time T_{OFF}, which is much longer than T_{SP2}. Thus, the non-complementary switching scheme results in higher power efficiency by reducing the circulating energy.

2.3. Synchronous Rectification with SiC MOSFETs

Figure 7 shows the detailed switching circuit diagrams for the operation principle of the secondary circuit for the positive grid cycle. Figure 7a shows the switching circuit diagram of the secondary circuit in the previous inverter in Figure 1. i_g freewheels through the body diodes D_{S1} and D_{S4} when S_{P1} is turned off. When S_{P1} is turned on, D_{S1} and D_{S4} are not turned off immediately because of the slow reverse recovery process [11]. Then, the diode reverse recovery currents cause high voltage spikes across D_{S1} and D_{S4}. Also, the diode reverse recovery currents are transferred to the primary circuit through the transformer, which causes high current spikes for S_{P1}. On the other hand, Figure 7b shows the switching circuit diagram of the secondary circuit in the proposed inverter. When S_{P1} is turned off, S_{S1} and S_{S4} are turned on. i_g freewheels through the MOSFET channels instead of the body diodes, which minimizes the conduction of the body diodes. When S_{P1} is turned on again, S_{S1} and S_{S4} are turned off without the reverse recovery process of the body diodes.

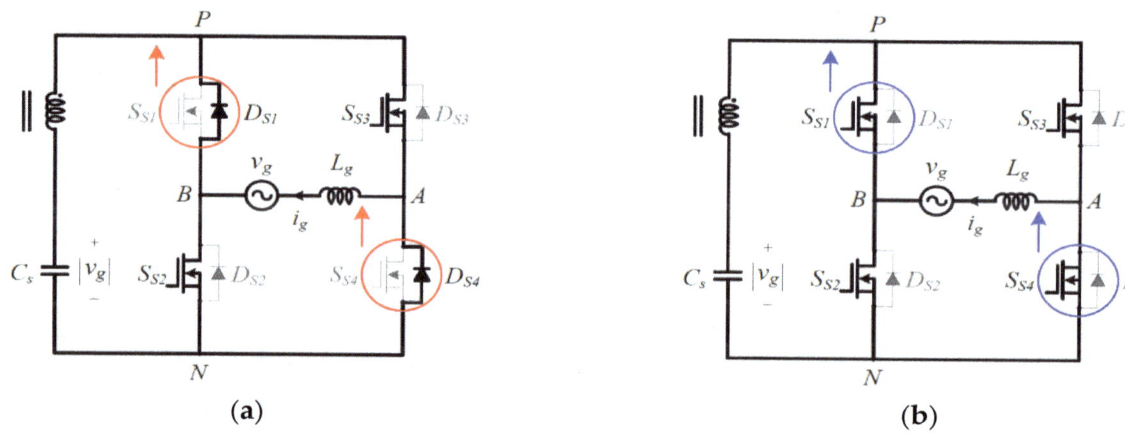

Figure 7. Detailed switching circuit diagrams for the operation principle of the secondary circuit for the positive grid cycle: (**a**) previous inverter; and (**b**) proposed inverter.

When S_{S1} and S_{S4} are turned off, the voltage across the secondary winding of T is $nL_mV_d/(L_m + L_{lk})$. The switch voltage stresses V_{S1} and V_{S4} are represented as:

$$V_{S1} = V_{S4} = |v_g| + \frac{nL_mV_d}{L_m + L_{lk}}. \quad (5)$$

The switch voltage stresses V_{S2} and V_{S3} are identical to V_{S1} and V_{S4} in Equation (5). Suppose that the system parameters are given as V_d = 48 V, n = 4.5, and v_g = 220 V_{rms} with negligible leakage inductor. The switch voltage stress is approximately calculated as 527 V without considering any voltage oscillations. Practically, the use of high voltage (>700 V) MOSFETs is inevitable to withstand the switch voltage stresses for a high switching frequency. However, when Si MOSFETs are adopted for $S_{S1} \sim S_{S4}$, the proposed inverter will suffer from high conduction losses due to the high on-state resistances of Si MOSFETs. Thus, the SiC MOSFET is an attractive alternative to the Si MOSFET for high-frequency and high-voltage applications. Due to its advantages over Si MOSFETs such as its wider bandgap and higher electric field capacity, it improves the switching performance of the proposed inverter, increasing the power efficiency.

2.4. Control Strategy

Supposing that the leakage inductor is negligible, the volt-second balance for L_g during T_s gives the following voltage equation as:

$$nV_dDT_s - v_g(1-D)\,T_s = 0 \tag{6}$$

where D is the duty cycle of S_{P1}. By rearranging Equation (6), we have the relation between v_g and V_d as:

$$\frac{v_g}{V_d} = \frac{nD}{1-D}. \tag{7}$$

Assuming no power losses in the inverter circuit, we have the relation between i_d and i_g as:

$$\frac{i_d}{i_g} = \frac{nD}{1-D} \tag{8}$$

where i_d is the input current. Suppose that C_P keeps the charge balance, i_d is considered as i_{P1} during T_s. Then, the following current relations are represented as:

$$i_d = i_{P1} = kDi_{Lm} \tag{9}$$

where k is a proportional factor. From equations (8) and (9), we have the relation between i_{Lm} and i_g as:

$$i_{Lm} = \frac{n}{k(1-D)}i_g. \tag{10}$$

The average voltage for L_m during T_s can be represented with respect to the deviation Δi_{Lm} of i_{Lm} as:

$$V_dD - \frac{|v_g|}{n}(1-D) = L_m\frac{\Delta i_{Lm}}{T_s}. \tag{11}$$

From Equation (11), D can be represented as:

$$D = \frac{|v_g|}{|v_g| + nV_d} + \frac{n}{|v_g| + nV_d}\left(\frac{L_m\Delta i_{Lm}}{T_s}\right). \tag{12}$$

By using equations (7) and (10), D in Equation (12) can be written as:

$$D = D_n + D_c = \frac{|v_g|}{|v_g| + nV_d} + \frac{nL_m}{kV_dT_s}\Delta i_g \tag{13}$$

where D_n is the nominal duty cycle, and D_c is the control duty cycle as:

$$\begin{aligned} D_n &= \frac{|v_g|}{|v_g|+nV_d} = \frac{V_g|\sin\omega t|}{V_g|\sin\omega t|+nV_d}, \\ D_c &= \frac{nL_m}{kV_dT_s}\Delta i_g. \end{aligned} \tag{14}$$

V_g is the absolute peak value of v_g. ω is the angular frequency of v_g. Supposing that v_g is measured exactly with a phase-locked loop (PLL) [17], D_n plays the role of providing the nominal voltage compensation. By using D_n, the non-linear system in Equation (11) is transformed to the first-order linear system, which can be controlled by the control duty cycle D_c. In order to regulate the grid current i_g with low harmonic currents, D_c is implemented by a proportional-resonant (PR) control [18] whose ideal transfer function C_{PR} (s) is:

$$C_{PR}(s) = k_p + \frac{k_r s}{s^2 + \omega^2} \tag{15}$$

where k_p and k_r are the PR control gains, respectively. However, it is unable to realize the PR controller in Equation (15) with an infinite gain. Thus, the following non-ideal transfer function is adopted in practice as:

$$C_{PR}(s) = k_p + \frac{2k_r \omega_c s}{s^2 + 2\omega_c s + \omega^2} \tag{16}$$

where ω_c is the angular frequency at the cutoff frequency of the controller. In addition, the harmonic compensators can be added to the PR controller to minimize the harmonic currents for the selective harmonic frequencies [19]. Its transfer function C_{HC} (s) is expressed as:

$$C_{HC}(s) = \sum_{h=3,5,7} \frac{2k_{rh} \omega_c s}{s^2 + 2\omega_c s + (h\omega)^2} \tag{17}$$

where h is the harmonic order and k_{rh} is the resonant control gain for each harmonic frequency. Since the third, fifth, and seventh harmonics are significant under the grid environment, the third to seventh harmonic compensators are implemented. The harmonic compensators provide high gains at the selected harmonic frequencies, helping minimize the steady-state error and the disturbance by the selected frequency components. Figure 8 shows the control block diagram of the proposed inverter. I^*_g is the peak magnitude of the current reference i^*_g. The duty cycle D is generated by summing D_n and D_c.

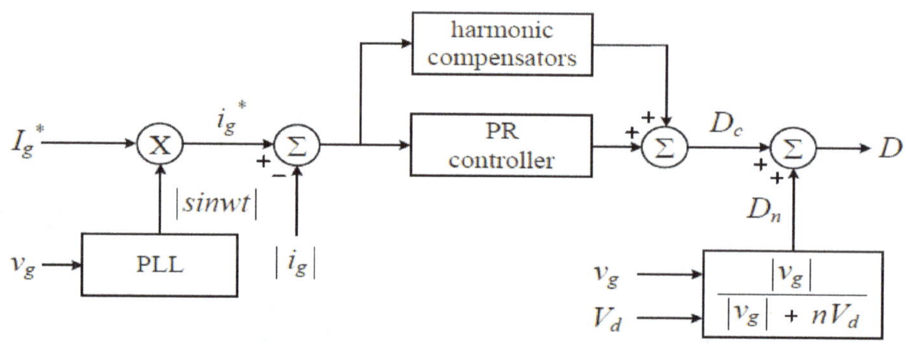

Figure 8. Control block diagram of the proposed inverter.

3. Experimental Results

A 220-W prototype system has been designed and tested to evaluate the performance of the proposed inverter. Table 1 shows the system parameters and the circuit components. As a SiC MOSFET, UJC1206k (UnitedSiC) has been used for $S_{S1} \sim S_{S4}$. As a digital signal controller, dsPIC30F6015 (Microchip) has been used for implementing the current controller and generating the duty cycle signals for all of the power switches. Figure 9 shows the picture of the prototype system. The prototype system includes the power circuit and the control circuit. Even though it is not optimized for the commercialized level, it is expected that the proposed inverter could achieve high power density if it is implemented with advanced devices such as gallium nitride (GaN) devices [20] and planar transformers [21].

Table 1. System parameters and circuit components.

Symbol	Quantity	Value
V_d	dc input voltage	48 V
v_g	grid voltage	60 Hz/220 V_{rms}
f_s	switching frequency	50 kHz
L_m	magnetizing inductor	60 μH
L_{lk}	leakage inductor	0.5 μH
N_P	primary winding turns	14
N_S	secondary winding turns	63
C_P	clamp capacitor	1.0 μF
S_{P1}, S_{P2}	primary swiches	IXFK150N30P3
C_S	secondary capacitor	1.0 μF
$S_{S1} \sim S_{S4}$	secondary switches	UJC1206K
L_g	filter inductor	2.0 mH

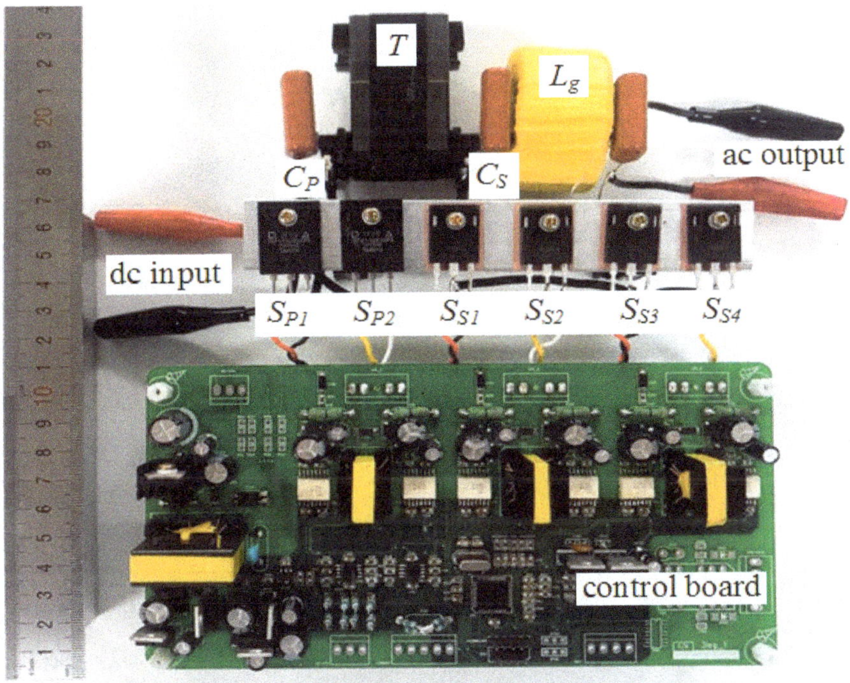

Figure 9. Picture of the prototype system.

The proposed inverter has been simulated to verify its operation principle. It has been simulated by the physical security information management (PSIM) software for the system parameters in Table 1. Figure 10 shows the simulation waveforms of the proposed inverter. The steady state operation of the proposed inverter can be verified from Figure 10a–d. Figure 10e,f show the simulation waveforms of the proposed inverter in the transient state condition. Figure 10e shows v_g and i_g as the output power changes from 110 W to 220 W. Figure 10f shows v_g and i_g as the output power changes from 220 W to 110 W.

Figure 11 shows the experimental waveforms of the previous inverter in Figure 1. The previous inverter has been designed and tested for the same system parameters as the proposed inverter. It uses the same circuit components in Table 1, except that STW40N95K5 (STMicroelectronics), as a Si MOSFET, has been adopted for $S_{S1} \sim S_{S4}$. Figure 11a shows the gate signal $V_{gate,P1}$ and the switch voltage V_{P1} for S_{P1}. When S_{P1} is turned off, V_{P1} has a high voltage spike, which results from the energy stored in the transformer leakage inductor. Figure 11b shows the gate signal $V_{gate,P1}$ and the switch current i_{P1} for S_{P1}. When S_{P1} is turned on, i_{P1} has a high current spike, which results from the reverse recovery current of the body diodes in the secondary circuit.

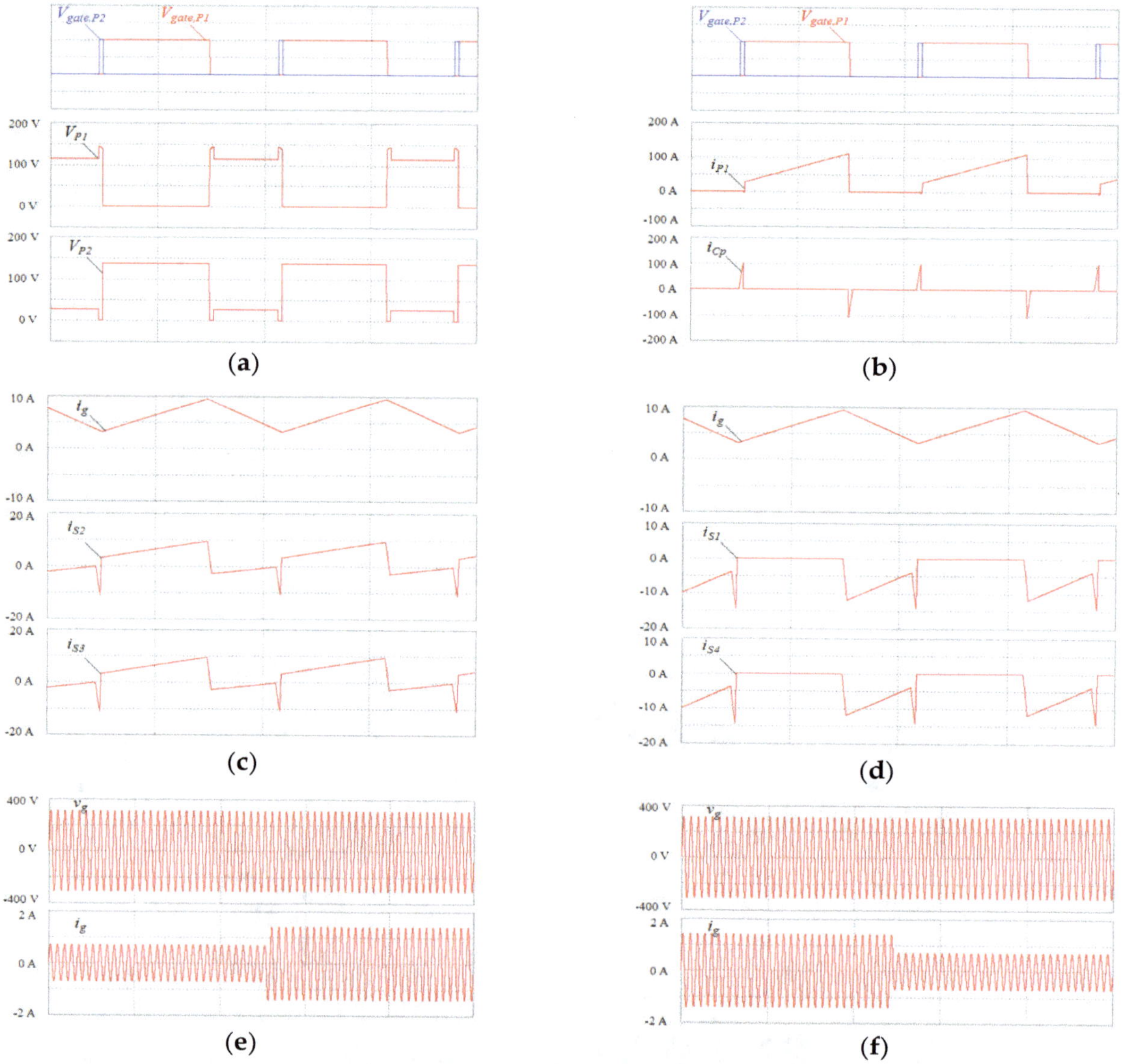

Figure 10. Simulation waveforms: (**a**) $V_{gate,P1}$, $V_{gate,P2}$, V_{P1}, and V_{P2}; (**b**) $V_{gate,P1}$, $V_{gate,P2}$, i_{P1}, and i_{Cp}; (**c**) i_g, i_{S2}, and i_{S3}; (**d**) i_g, i_{S1}, and i_{S4}; (**e**) v_g and i_g from 110 W to 220 W; and (**f**) v_g and i_g from 220 W to 110 W.

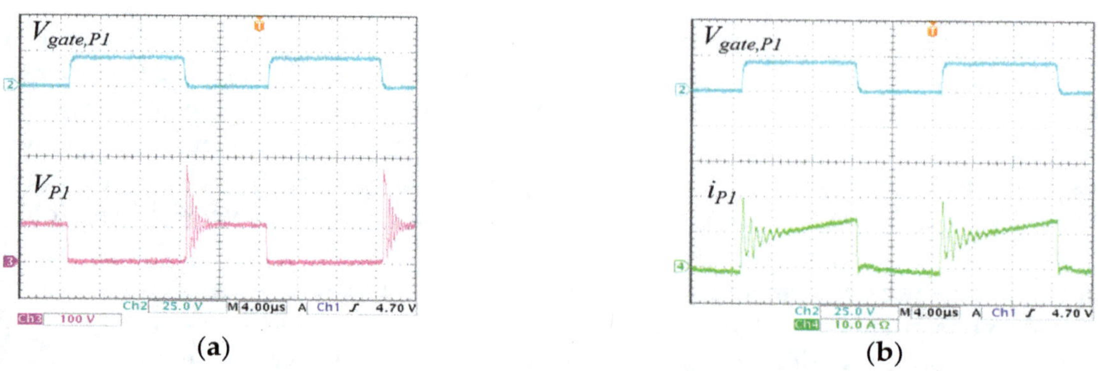

Figure 11. Experimental waveforms of the previous inverter: (**a**) $V_{gate,P1}$ and V_{P1}; and (**b**) $V_{gate,P1}$ and i_{P1}.

Figure 12 shows the experimental waveforms of the proposed inverter. Figure 12a shows $V_{gate,P1}$, $V_{gate,P2}$, and V_{P1}. Figure 12b shows $V_{gate,P1}$, $V_{gate,P2}$, and V_{P2}. As shown in Figure 12a,b, V_{P1} and

V_{P2} are maximally clamped to the capacitor voltage V_{Cp}. Figure 12c shows $V_{gate,P1}$, $V_{gate,P2}$, and i_{P1}. The switch current i_{P1} in Figure 12c has lower current spike than the switch current i_{P1} in Figure 11b because of the synchronous rectification of the secondary circuit. Figure 12d shows $V_{gate,P1}$, $V_{gate,P2}$, and i_{Cp}. It is observed that the capacitor charging and discharging currents are well balanced in the proposed inverter.

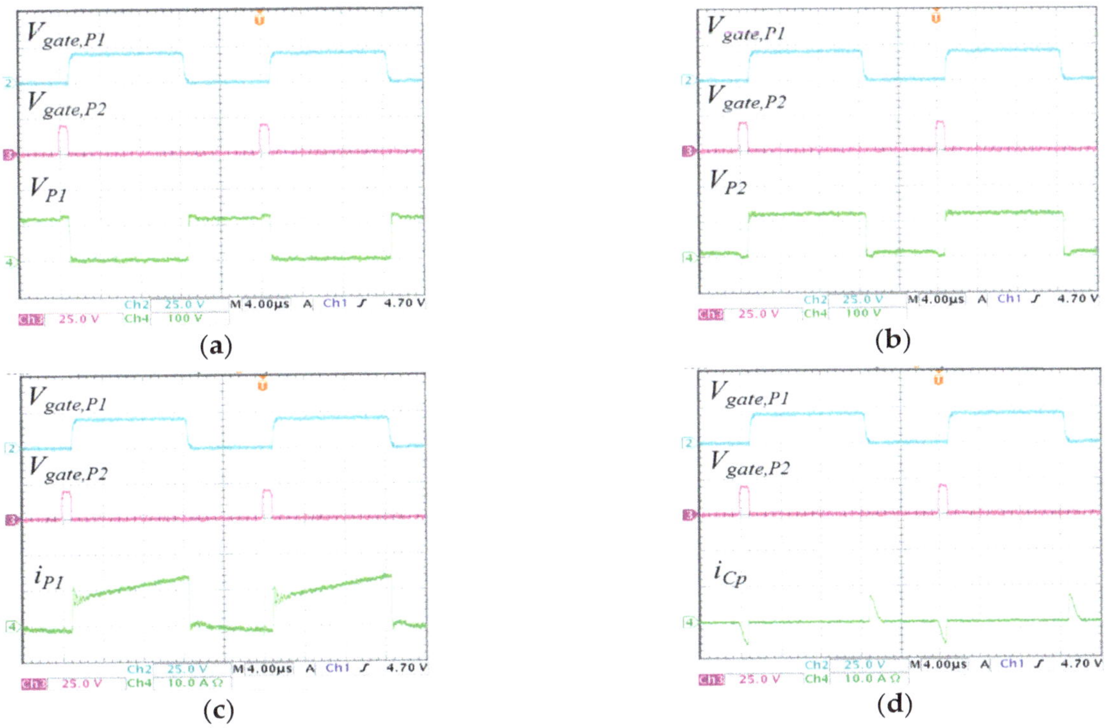

Figure 12. Experimental waveforms of the proposed inverter: (**a**) $V_{gate,P1}$, $V_{gate,P2}$, and V_{P1}; (**b**) $V_{gate,P1}$, $V_{gate,P2}$, and V_{P2}; (**c**) $V_{gate,P1}$, $V_{gate,P2}$, and i_{P1}; and (**d**) $V_{gate,P1}$, $V_{gate,P2}$, and i_{Cp}.

Figure 13 shows the experimental waveforms in the secondary circuit for the positive grid cycle. Figure 13a shows V_{S1} in the previous inverter. It is observed that there is high voltage oscillation across S_{S1}, which results from the slow reverse recovery process of the body diode of the Si MOSFET. Figure 13b shows $V_{gate,S1}$ and V_{S1} in the proposed inverter. With synchronous rectification, the SiC MOSFET channel has been used for the rectification. It is shown that voltage oscillation across S_{S1} has been much alleviated due to the fast switching operation of the SiC MOSFET. The switch currents i_{S1}, i_{S2}, i_{S3}, and i_{S4} are shown from Figure 13c–f with respect to i_g in the proposed inverter.

Figure 14 shows the experimental waveforms of the proposed inverter. Figure 14a shows v_g and i_g as the proposed inverter supplies 110 W into the grid. Figure 14b shows v_g and i_g as the proposed inverter supplies 220 W into the grid. As i_g is in phase with v_g, the power factor is measured as 0.99 in Figure 14.

Figure 15 shows the measured power efficiency curves. The curve A shows the measured power efficiency of the previous inverter. It has achieved the peak efficiency of 93.5% at the rated power. The curve B shows the measured power efficiency of the proposed inverter when the synchronous rectification has been implemented without the active clamp circuit. The proposed inverter has improved the power efficiency, achieving the peak efficiency of 94.2% at the rated power. The curve C shows the measured power efficiency of the proposed inverter when both active clamp and synchronous rectification techniques have been implemented. The proposed inverter has achieved the peak efficiency of 95.0% at the rated power by using active clamp and synchronous rectification techniques.

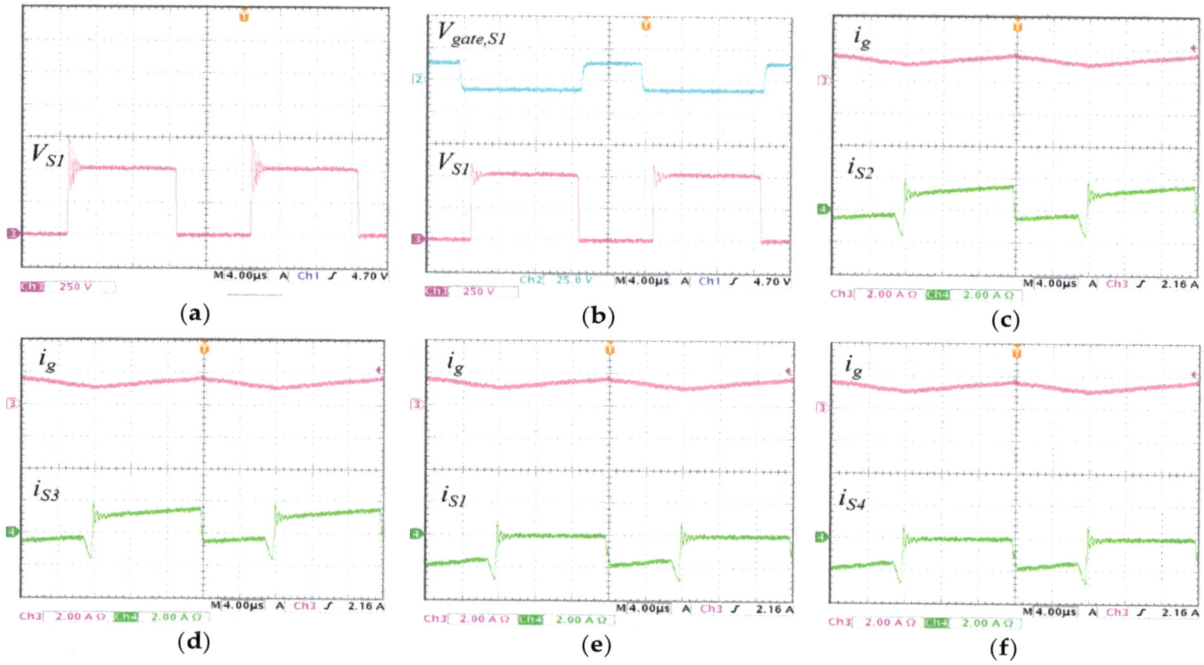

Figure 13. Experimental waveforms in the secondary circuit for the positive grid cycle: (**a**) V_{S1} in the previous inverter; (**b**) $V_{gate,S1}$ and V_{S1} in the proposed inverter; (**c**) i_g and i_{S2} in the proposed inverter; (**d**) i_g and i_{S3} in the proposed inverter; (**e**) i_g and i_{S1} in the proposed inverter; and (**f**) i_g and i_{S4} in the proposed inverter.

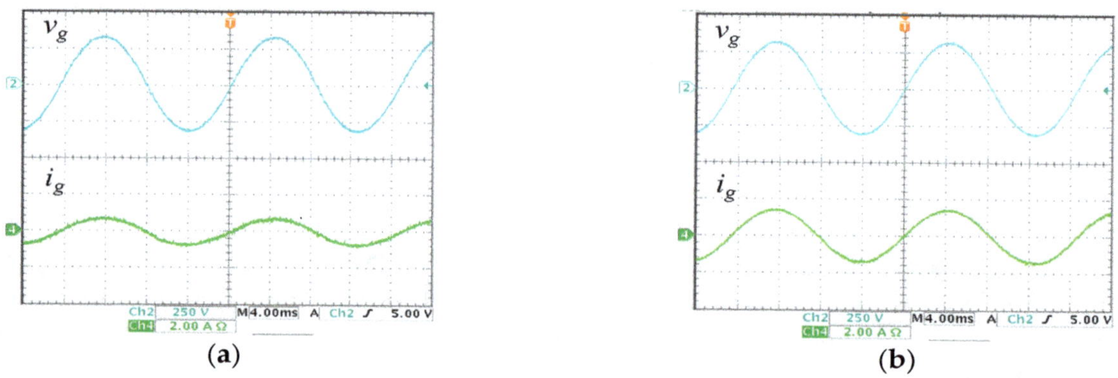

Figure 14. Experimental waveforms of the proposed inverter: (**a**) v_g and i_g as the proposed inverter supplies 110 W into the grid; and (**b**) v_g and i_g as the proposed inverter supplies 220 W into the grid.

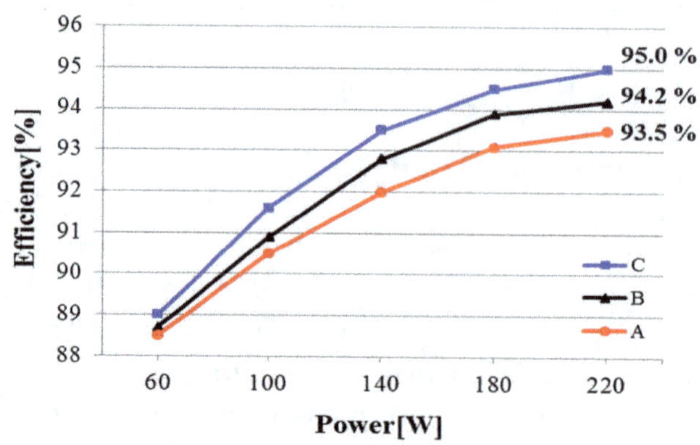

Figure 15. Measured power efficiency curves.

Figure 16a shows the measured power efficiency curves up to 440 W. The prototype system has shown the peak efficiency of 95.0% at 220 W. The efficiency decreases gradually as the output power level goes up from 300 W to 440 W. The prototype system has achieved the efficiency of 94.3% at 440 W. Figure 16b shows the measured power efficiency curves for 50 kHz and 100 kHz, respectively. As the switching frequency increases, the power efficiency decreases because of the switching losses. In order to improve the power efficiency and the power density further, the circuit design scheme should be advanced by considering the GaN devices [20] and the planar transformers [21] for megahertz operations.

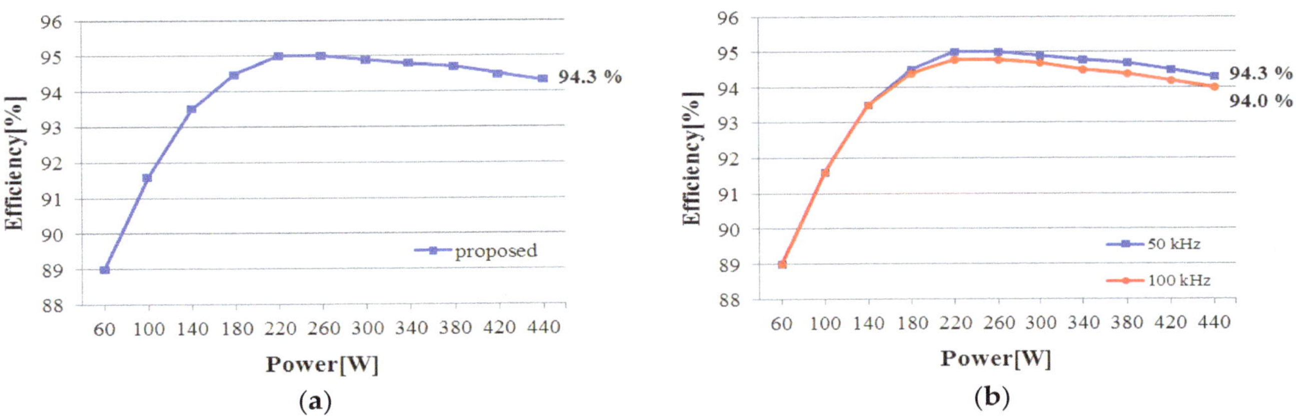

Figure 16. Measured power efficiency curves: (**a**) power efficiency curves up to 440 W; and (**b**) power efficiency curves for different switching frequencies.

4. Conclusions

This paper has proposed a high-efficiency zeta inverter using active clamp and synchronous rectification techniques for single-phase grid-connected applications. The operation principle of the proposed inverter has been described. The active clamp and synchronous techniques adopted in the proposed inverter have been explained. The non-complementary switching scheme has been applied to the active clamp circuit. It effectively reduces the switching losses and circulating current in the primary circuit. The synchronous rectification with SiC MOSFETs alleviates switching losses in the secondary circuit. The control strategy for the CCM operation of the proposed inverter has been presented. A 220-W prototype system has been designed and tested to evaluate the performance of the proposed inverter. Experimental results have shown that the proposed inverter has improved the power efficiency by reducing switching losses, compared to the previous inverter.

Author Contributions: W.-Y.C. managed the project and wrote the manuscript. M.-K.Y. performed the experiments, analyzed the data, and edited the manuscript.

References

1. Dong, D.; Agamy, M.S.; Harfman-Todorovic, M.; Liu, X.; Garces, L.; Zhou, R.; Cioffi, P. A PV residential microinverter with grid-support function: Design, implementation, and field testing. *IEEE Trans. Ind. Appl.* **2018**, *54*, 469–481. [CrossRef]
2. Hu, H.; Harb, S.; Kutkut, N.H.; Shen, Z.H.; Batarseh, I. A single-stage microinverter without using electrolytic capacitors. *IEEE Trans. Power Electron.* **2013**, *28*, 2677–2687. [CrossRef]
3. Feng, X.; Wang, F.; Wu, C.; Luo, J.; Zhang, L. Modeling and comparisons of aggregated flyback microinverters in aspect of harmonic resonances with the grid. *IEEE Trans. Ind. Electron.* **2019**, *66*, 276–285. [CrossRef]

4. Lodh, T.; Pragallapati, N.; Agarwal, V. Novel control scheme for an interleaved flyback converter based solar PV microinverter to achieve high efficiency. *IEEE Trans. Ind. Appl.* **2018**, *54*, 3473–3482. [CrossRef]
5. Han, B.; Lee, J.S.; Kim, M.S. Repetitive controller with phase-lead compensation for Cuk CCM inverter. *IEEE Trans. Ind. Electron.* **2018**, *65*, 2356–2367. [CrossRef]
6. Han, B.; Lai, J.-S.; Kim, M.S. Dynamic modeling and controller design of dual-mode Cuk inverter in grid-connected PV/TE applications. *IEEE Trans. Power Electron.* **2018**, *33*, 8887–8904. [CrossRef]
7. Levin, G.H.; Flores, J.V.; Salton, A.; dos Reis, F.S. Internal model control of the zeta converter for the grid connection of photovoltaic panels. In Proceedings of the Brazilian Power Electronics Conference, Gramado, Brazil, 27–31 October 2013; pp. 666–671.
8. Viero, R.C.; Lopez, H.F.M.; Zollmann, C.A.; dos Reis, F.S. Dynamic modeling of a sinusoidal inverter based on zeta converter working in DCM for PV arrays. In Proceedings of the Annual Conference on IEEE Industrial Electronics Society (IECON), Glendale, AZ, USA, 7–10 November 2010; pp. 439–444.
9. Surapaneni, R.K.; Rathore, A.K. A single-stage CCM zeta microinverter for solar photovoltaic AC module. *IEEE J. Emerg. Sel. Top. Power Electron.* **2015**, *3*, 892–900. [CrossRef]
10. Murthy-Bellur, D.; Kazimierczuk, M.K. Isolated two-transistor zeta converter with reduced transistor voltage stress. *IEEE Trans. Circuits Syst. II Express Briefs* **2011**, *58*, 41–45. [CrossRef]
11. Brown, C.D.; Sarlioglu, B. Reducing switching losses in BLDC motor drives by reducing body diode conduction of MOSFETs. *IEEE Trans. Ind. Appl.* **2015**, *51*, 1864–1871. [CrossRef]
12. Chen, M.; Sun, J. Reduced-order averaged modeling of active-clamp converters. *IEEE Trans. Power Electron.* **2006**, *21*, 487–494. [CrossRef]
13. Zhang, J.; Huang, X.; Wu, X.; Qian, Z. A high efficiency flyback converter with new active clamp technique. *IEEE Trans. Power Electron.* **2010**, *25*, 1775–1785. [CrossRef]
14. Yin, S.; Liu, Y.; Liu, Y.; Tseng, K.J.; Pou, J.; Simanjorang, R. Comparison of SiC voltage source inverters using synchronous rectification and freewheeling diode. *IEEE Trans. Ind. Electron.* **2018**, *65*, 1051–1061. [CrossRef]
15. Mohammadi, M.; Ordonez, M. Synchronous rectification of LLC resonant converters using homopolarity cycle modulation. *IEEE Trans. Ind. Electron.* **2019**, *66*, 1781–1790. [CrossRef]
16. Jordán, J.; Esteve, V.; Sanchis-Kilders, E.; Dede, E.J.; Maset, E.; Ejea, J.B.; Ferreres, A. A comparative performance study of a 1200 V Si and SiC MOSFET intrinsic diode on an induction heating inverter. *IEEE Trans. Power Electron.* **2014**, *29*, 2550–2562. [CrossRef]
17. Han, Y.; Luo, M.; Zhao, X.; Guerrero, J.M.; Xu, L. Comparative performance evaluation of orthogonal-signal-generators-based single-phase PLL algorithms—A survey. *IEEE Trans. Power Electron.* **2016**, *31*, 3932–3944. [CrossRef]
18. Zmood, D.N.; Holmes, D.G. Stationary frame current regulation of PWM inverters with zero steady-state error. *IEEE Trans. Power Electron.* **2003**, *18*, 814–822. [CrossRef]
19. Pereira, L.F.A.; Flores, J.V.; Bonan, G.; Coutinho, D.F.; da Silva, J.M.G. Multiple resonant controllers for uninterruptible power supplies-a systematic robust control design approach. *IEEE Trans. Ind. Electron.* **2014**, *61*, 1528–1538. [CrossRef]
20. Zhang, W.; Huang, X.; Liu, Z.; Lee, F.C.; She, S.; Du, W.; Li, Q. A new package of high-voltage cascade gallium nitride device for megahertz operation. *IEEE Trans. Power Electron.* **2016**, *31*, 1344–1353. [CrossRef]
21. Zhao, S.; Li, Q.; Lee, F.C.; Li, B. High-frequency transformer design for modular power conversion from medium-voltage AC to 400 VDC. *IEEE Trans. Power Electron.* **2018**, *33*, 7545–7557. [CrossRef]

10

Flatness-Based Control for the Maximum Power Point Tracking in a Photovoltaic System

Leopoldo Gil-Antonio [1,2], Belem Saldivar [1,3,*], Otniel Portillo-Rodríguez [1,*], Juan Carlos Ávila-Vilchis [1], Pánfilo Raymundo Martínez-Rodríguez [4] and Rigoberto Martínez-Méndez [1]

1 Faculty of Engineering, Autonomous University of the State of Mexico, Instituto Literario No. 100 Oriente, Toluca 50130, Estado de México, Mexico; leopoldo.gil@tesjo.edu.mx (L.G.-A.); jcavilav@uaemex.mx (J.C.Á.-V.); rigo@ieee.org (R.M.-M.)
2 Tecnológico de Estudios Superiores de Jocotitlán Carretera Toluca-Atlacomulco km. 44.8, Jocotitlán 50700, Estado de México, Mexico
3 Cátedras CONACYT, Av. Insurgentes Sur 1582, Col. Crédito Constructor, Alcaldía Benito Juárez, Ciudad de México 03940, Mexico
4 School of Sciencies, UASLP, San Luis Potosi 78290, SLP, Mexico; pamartinez@ieee.org
* Correspondence: mbsaldivarma@conacyt.mx (B.S.); oportillor@uaemex.mx (O.P.-R.)

Abstract: Solar energy harvesting using Photovoltaic (PV) systems is one of the most popular sources of renewable energy, however the main drawback of PV systems is their low conversion efficiency. An optimal system operation requires an efficient tracking of the Maximum Power Point (MPP), which represents the maximum energy that can be extracted from the PV panel. This paper presents a novel control approach for the Maximum Power Point Tracking (MPPT) based on the differential flatness property of the Boost converter, which is one of the most used converters in PV systems. The underlying idea of the proposed control approach is to use the classical flatness-based trajectory tracking control where a reference voltage will be defined in terms of the maximum power provided by the PV panel. The effectiveness of the proposed controller is assessed through numerical simulations and experimental tests. The results show that the controller based on differential flatness is capable of converging in less than 0.15 s and, compared with other MPPT techniques, such as Incremental Conductance and Perturb and Observe, it improves the response against sudden changes in load or weather conditions, reducing the ringing in the output of the system. Based on the results, it can be inferred that the new flatness-based controller represents an alternative to improve the MPPT in PV systems, especially when they are subject to sudden load or weather changes.

Keywords: MPPT; differential flatness; nonlinear control

---

1. Introduction

For many years fossil fuels have been the primary source of energy in the world, however these resources are finite. The threat of climate change due to global warming (caused in part by the burning of fossil fuels) has prompted the search for renewable energy sources such as the sun and the wind. Solar energy is a sustainable, environmentally friendly, and cost-efficient source of energy available around the world, and for this reason solar technologies using photovoltaic (PV) systems have penetrated the electric power production market, with the additional advantages of working quietly and with low maintenance cost. Nonetheless, as it has been remarked by [1], the increment in the use of PV devices in power systems has generated new challenges, such as those related with control strategies looking to provide good, even optimal, operating conditions of the PV systems. Moreover, the control strategies need to cope with the fact that the performance of PV systems depends on solar irradiance, ambient temperature, and load impedance [2]. As in [1], the approach proposed in this paper is implemented in the open programming environment MATLAB®/Simulink®; nevertheless, the focused problem is the application of a novel control strategy (flatness-based) for the MPPT in PV systems. As mentioned in [3], it is evident that PV components will be essential elements in power electrical systems, where their inclusion considers the MPPT control problem as a strategy that is part of one of the two most important approaches highlighted in literature that corresponds to "the installation and operation of an energy storage system (ESS), while keeping the PVs in the maximum power point tracking (MPPT)..." [3]. To the best knowledge of the authors, no flatness-based control strategies have been reported to cope with the MPPT problem in PV systems. This paper proposes the use of a control strategy based on the differential flatness properties of the PV system model to tackle the MPPT task.

The main drawback of PV systems is their low conversion efficiency. For an optimal operation, it is required for the system to operate at the Maximum Power Point (MPP), which represents the maximum energy that can be extracted from the PV panel. To reach this condition, it is necessary to use Maximum Power Point Tracking (MPPT) systems. A major challenge in PV systems is to handle their non-linear characteristics of current-voltage I-V relation that generate a unique MPP in the power-voltage relation P-V [4]. The MPPT process becomes complicated due to the fact that the P-V relation varies with weather conditions. MPPT methods not only allow an increase in the power delivered by PV systems to the load, but they also give rise to a longer operating life of the system [5].

MPPT algorithms are designed so that the PV system adapts to weather changes in such a way that optimal power is delivered. Typically, these algorithms are integrated into electronic power converter systems where their duty cycle is controlled to deliver the maximum available power to the load [6,7]. Several methods to solve the problem of MPPT have been reported in the literature. Among them are the Perturb and Observe (P&O) method, the Incremental Conductance (IncCond) algorithm, fuzzy logic-based methods, neural networks techniques, and the sliding mode control. The P&O and IncCond are the most popular algorithms to track the MPP [8–13].

The P&O method is inexpensive and relatively simple; its operation is based on periodic measures of the voltage and current of the PV system to calculate the MPP. However, its main disadvantage is that it delivers an oscillatory power around the MPP. Besides, since it is unable to detect if the power variation is caused by weather effects or by inherent perturbations of the algorithm, the P&O method may fail in the presence of abrupt changes of temperature and solar irradiance [14].

The IncCond method is based on the fact that the slope of the P-V curve of a PV system is zero when the MPP is reached, positive to the left of the MPP and negative to its right. Based on that, the method calculates the MPP by comparing the instantaneous conductance with the incremental conductance to modify the required reference voltage. The main disadvantage of the IncCond method is that the response of the system to reach the MPP under certain conditions can be slow. However, the IncCond technique exhibits less oscillatory behavior around the MPP compared to the P&O method [15].

Fuzzy logic-based methods have demonstrated fast convergence and high performance under varying weather conditions [16–23]. These methods do not require a mathematical model and are able to handle the system nonlinearities. However, the main disadvantage of these controllers is that their effectiveness depends on the error calculation and on the definitions of their base rules for the fuzzy inference mechanism. These definitions must be done by a human expert and the effectiveness of the control relies on them.

Neural network-based algorithms require proper training strategies to effectively track the MPP [24–26]. It is worth mentioning that neural networks must be retrained to be applied to different PV systems since they may have different specifications. Besides this, since the system parameters can be modified over time, it is necessary to periodically train a neural network to ensure its effectiveness. Furthermore, according to [27,28], the implementation of fuzzy logic and neural network algorithms may be complex.

The sliding mode control technique sends an on-off signal to control the operation of the power converter to reach the MPP of the PV system. The commutation function is computed based on the fact that the derivative of the power with respect to the voltage is positive to the left of the MPP, negative to its right, and zero at the MPP. Conventional sliding mode control techniques have limitations such as variable operating frequency and the presence of nonzero steady state error [29].

The use of flatness-based control strategies in the area of power electrical systems has been explored in the last decade, where some works have reported the use of flatness-based controllers for power converters. For instance, authors in [30] use a recursive approach were differential flatness properties are used to synthesize a controller that guarantees closed-loop asymptotic stability of a power converter connected to a DC motor. Two flatness-based controllers are proposed in [31] for a Boost inverter (two DC-DC Boost converters connected in differential mode to a grid) in order to control, individually, each of the two output voltages of the Boost converters. A flatness-based control approach is provided in [32] to achieve the maximum power that is captured by a wind generator driving a permanent magnet synchronous generator whose mathematical model is flat and is connected to a battery bank using an AC-DC converter. A differential flatness-based controller is proposed in [33] to improve the stability of the converters used in a distributed PV-energy storage DC generation system under the presence of parameter disturbances, while an extremum seeking algorithm guarantees the tracking of the MPP under irradiance disturbances. Authors in [30–33] validate their flatness-based control strategies through numerical simulations. Nevertheless, as it has been remarked previously, differential flatness-based methods have not been investigated to deal with the MPPT problem in PV systems.

The main contribution of this paper concerns the use of a novel control strategy to perform the tracking task of the MPP in PV systems. A new methodology that exploits the differential flatness property of the Boost converter allows deriving a MPPT controller for a stand-alone PV system constituted by a PV panel and a Boost power converter with resistive load. The strategy is based on the trajectory tracking flatness-based control approach where a reference voltage is defined in terms of the maximum power provided by the PV panel. Like the mentioned works that have used a flatness approach [30–33], the performance of the control law reported in this paper is validated through numerical simulations. Nevertheless, this paper also reports experimental tests confirming the controller performance.

The proposed approach is shown to be capable of operating under changing conditions of solar irradiance, temperature, and load, even when the variations are abrupt. This is an important feature that an efficient MPPT must have since the presence of clouds or trees, for instance, may degrade the performances of certain controllers. The robustness against load variations (normally related in practice to energy consumption demands) is an important aspect disregarded in most of the existing control techniques. This paper presents also a comparative analysis that shows the benefits of the proposed controller compared to the two most popular methods for the MPPT: P&O and IncCond. The flatness-based controller gives rise to an improved system response regarding the power behavior: before reaching the MPP, less oscillations are shown with the proposed control in contrast to the IncCond, and, once reaching the MPP, unlike the P&O method, no oscillations are exhibited.

This paper is organized as follows: Section 2 presents the mathematical model of the PV system that is used in this study. Section 3 details the design of the differential flatness-based controller to track the MPP. Section 4 presents the numerical simulations to evaluate the effectiveness of the controller under variations of solar irradiance, temperature and load. Besides, a comparative analysis shows the benefits of the proposed controller compared to the P&O and IncCond methods. In Section 5, the performance of the proposed control approach is validated through experimental tests conducted in a prototype of the system. Finally, Section 6 gives some concluding remarks.

2. Photovoltaic System Model

The block diagram of the PV system under consideration is shown in Figure 1. The energy source of the PV system is constituted by a photovoltaic module, the Boost converter acts as interface between the solar panel and the load, and the MPPT controller allows reaching the available maximum power. Notice that the current I_{PV} and the voltage V_{PV} provided by the PV module are used as inputs to the Boost converter and to the MPPT controller which is also fed with the output voltage v of the converter. The average control signal u_{av} generated by the MPPT controller is sent to the converter.

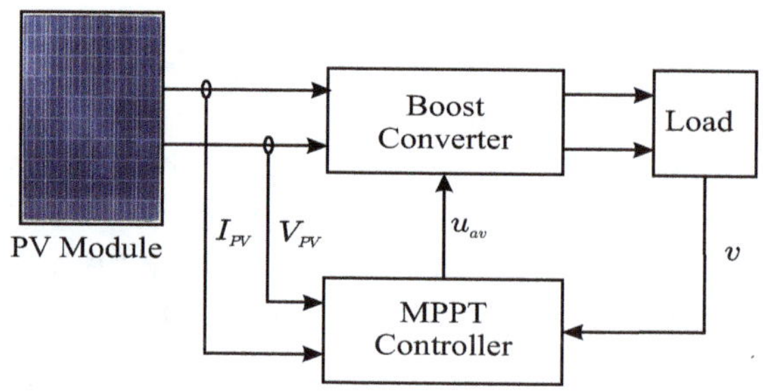

Figure 1. Photovoltaic (PV) system under a Maximum Power Point Tracking (MPPT) controller.

2.1. PV Module Model

A PV module, also known as solar panel, consists of a set of solar cells that are connected in series or in parallel depending on the application for which the PV module is built. If the solar cells are connected in series, the voltage of the PV module is increased while the current keeps constant. Conversely, if a parallel connection is considered, the PV module current is increased while the voltage remains constant.

The equivalent circuit that represents a solar cell is constituted by an energy source that represents the energy generated by the incidence of solar irradiance G which delivers the photogenerated current I_{ph}. This circuit also considers a diode connected in parallel with the energy source, a series resistance R_s that represents the internal resistance to the current flow of the cell and a shunt resistance R_{sh} representing the leakage current (see Figure 2).

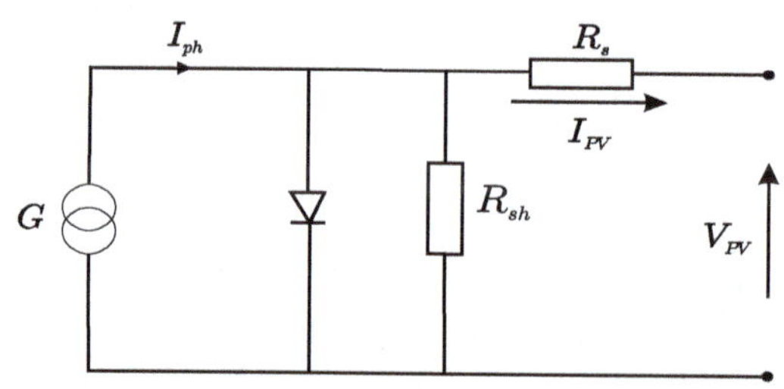

Figure 2. Equivalent circuit of a PV cell.

The equivalent circuit of a PV cell is mathematically described by Equations (1)–(3), where it is assumed that $R_s << R_{sh}$ [34].

$$I_{PV} = I_{ph} - I_o \left[\exp\left(\frac{q\left(V_{PV} + R_s I_{PV}\right)}{KT_c A} \right) - 1 \right] \quad (1)$$

In Equation (1), I_{PV} and V_{PV} denote the output current and voltage of the PV cell, respectively, I_{ph} is the current photogenerated by the solar cell, I_o is the reverse saturation current of the diode, q is the electron charge (1.602×10^{-19} C), R_s is the internal resistance to the current flow of the cell, K is the Boltzmann constant (1.381×10^{-23} J/K), T_c is the temperature of the cell, and A is the diode ideality factor.

The photogenerated current I_{ph} is determined by:

$$I_{ph} = \frac{S}{S_{ref}}\left[I_{SC} + \mu_{SC}\left(T_c - T_{ref}\right)\right] \tag{2}$$

where S is the solar irradiance, S_{ref} is the solar irradiance under reference conditions (1000 W/m^2), I_{SC} is the short-circuit current, μ_{SC} is the coefficient of short-circuit current, and T_{ref} is the reference temperature of the photovoltaic cell [35]. The reverse saturation current I_o is determined by:

$$I_o = I_{o,ref}\left(\frac{T_c}{T_{ref}}\right)^3 \exp\left[\frac{qE_G\left(\frac{1}{T_{ref}} - \frac{1}{T_c}\right)}{AK}\right] \tag{3}$$

where E_G is the semiconductor energy band used in the solar cell and $I_{o,ref}$ represents the reference reverse saturation current given by:

$$I_{o,ref} = \frac{I_{SC}}{\exp\left(\frac{V_{OC}}{N_s K T_c A}\right) - 1} \tag{4}$$

where N_s is the number of cells connected in series and V_{OC} is the open circuit voltage [35]. In view of Equation (1), the current provided by a PV module is given by:

$$I_{PV} = N_p I_{ph} - N_p I_o\left[\exp\left(\frac{q\left(\frac{V_{PV}}{N_s} + \frac{R_s I_{PV}}{N_p}\right)}{K T_c A}\right) - 1\right] \tag{5}$$

where N_p is the number of cells connected in parallel [36]. The characteristics of the PV module Renesola Virtus II JC250M-24/Bb, used in this research work, are plotted in Figures 3a,b and 4. Figure 3a shows the *I-V* curves for different values of solar irradiance considering a constant temperature of 25 °C, Figure 3b shows the *P-V* curves for different values of solar irradiance considering a constant temperature of 25 °C, and Figure 4 shows the *I-V* curves considering different temperature values and a solar irradiance of 1000 W/m^2. Notice that the voltage is most likely to be influenced by temperature variations, and the current by solar irradiance changes. As illustrated in Figures 3a,b and 4, high solar irradiance values increase the current, and high temperatures reduce the output voltage of a PV panel. Then, the power conversion capability can be affected under conditions of high temperature and low solar irradiance.

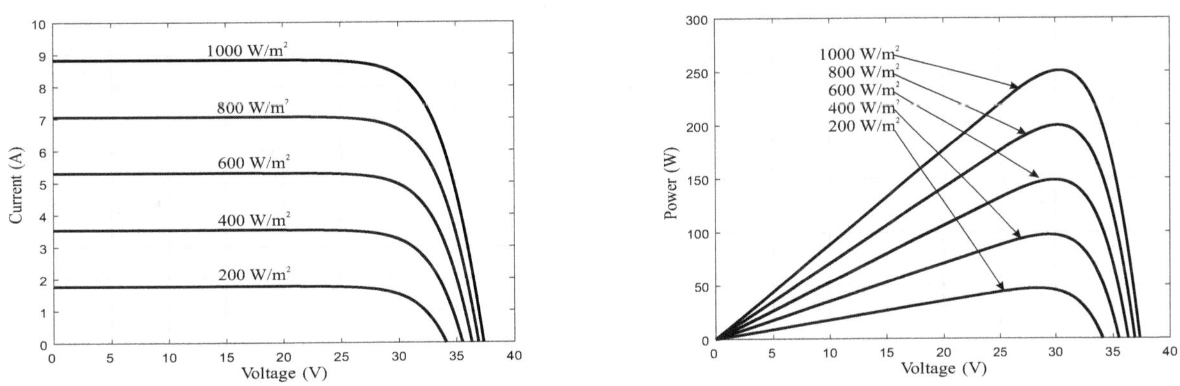

(**a**) *I-V* curves. (**b**) *P-V* curves.

Figure 3. Curves of the PV panel Virtus II JC250M-24/Bb under different irradiance levels and a constant temperature of 25 °C.

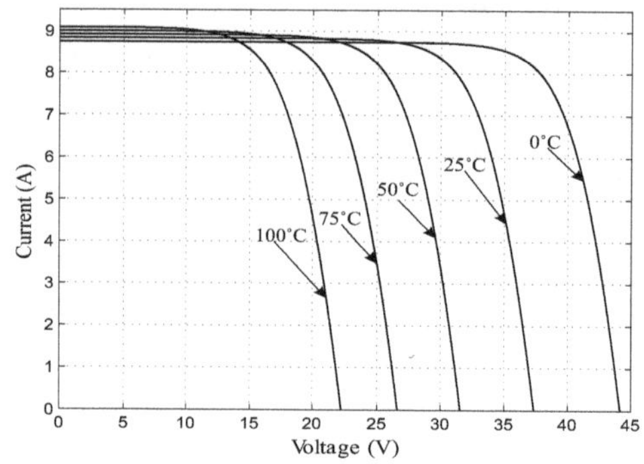

Figure 4. *I-V* curves of the PV panel Virtus II JC250M-24/Bb under different temperature levels and a constant irradiance of 1000 W/m^2.

2.2. Boost Converter Model

A Boost converter is a DC-DC power converter that increases the voltage from its input source to its output. Figure 5 shows the electronic diagram of a Boost power converter in which the input source is the voltage supplied by a photovoltaic panel. This circuit is constituted by a capacitance C, a load resistance R, and an inductance L. The current across the inductor is denoted by i and the output at the capacitor terminals by v. Note that i is assumed to be equal to the PV current I_{PV}.

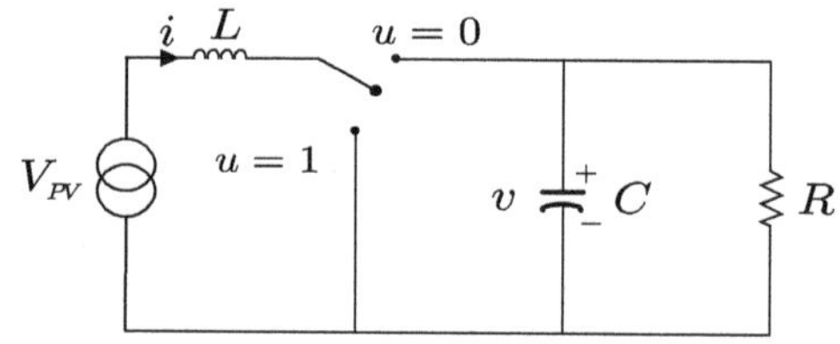

Figure 5. Block diagram of a Boost converter.

Note that the Boost converter includes a switching function u that commutes between two operation modes. When $u = 1$ there is no connection between the voltage source V_{PV} and the system load R. In this case, the system dynamics is described by Equations (6).

$$\begin{aligned} \frac{dv}{dt} &= -\frac{v}{RC} \\ \frac{di}{dt} &= \frac{V_{PV}}{L} \end{aligned} \tag{6}$$

When $u = 0$, there is energy transmission between the power supply and the load, then the system dynamics is described by Equations (7).

$$\begin{aligned} \frac{dv}{dt} &= \frac{i}{C} - \frac{v}{RC} \\ \frac{di}{dt} &= -\frac{v}{L} + \frac{V_{PV}}{L} \end{aligned} \tag{7}$$

Equations (6) and (7) can be unified into the bilinear system defined by Equations (8),

$$\begin{aligned} \frac{dv}{dt} &= \frac{i}{C}(1-u) - \frac{v}{RC} \\ \frac{di}{dt} &= -\frac{v}{L}(1-u) + \frac{V_{PV}}{L} \end{aligned} \tag{8}$$

where i is the inductor current, v is the output voltage, and V_{PV} is the power supply voltage. The control input u represents the switch position function which takes a binary value $u \in \{0,1\}$.

In order to design a control law for the Boost converter, an *average model* must be defined. The *average model* of the Boost converter is represented by the same Equations (8), but the control variable u is redefined as a sufficiently smooth function u_{av} taking values in the real interval [0, 1]. Then, the model given in Equations (8) is referred to as *average model* replacing the switched control input $u \in \{0, 1\}$ by the average control input $u_{av} \in [0, 1]$.

The steady state value of the current $\bar{i}$, for constant values of the input, can be obtained in a parameterized form in terms of the corresponding output voltage equilibrium value $\bar{v}$ as follows:

$$\bar{i} = \frac{\bar{v}^2}{RV_{PV}} \tag{9}$$

3. Flatness-Based Control Design

In nonlinear systems theory, the flatness property is due to the ability of dynamic systems to support an accurate linearization through endogenous feedback. A system that satisfies the flatness property is named a differentially flat system. The main attribute of flat systems is that state and input variables can be expressed directly (without the integration of a differential equation) in terms of a particular set of variables called flat outputs (or linearizing outputs) and a finite number of their derivatives [37]. This section presents the main contribution of this paper which is the development of a controller based on the differential flatness property of the Boost converter for the MPPT of a PV system.

The flatness property of the Boost converter has been established in [38]. A flat output F can be determined by the total energy stored in the system, that is

$$F = \frac{1}{2}\left(Li^2 + Cv^2\right) \tag{10}$$

Using Equations (8), the differential parametrization of the states and the input of the Boost converter are obtained as follows:

$$i = -\frac{RCV_{PV}}{2L} + \beta \tag{11}$$

$$v = \sqrt{\frac{2}{C}F - \frac{L}{C}\left(-\frac{RCV_{PV}}{2L} + \beta\right)} \tag{12}$$

$$u_{av} = 1 - \frac{i\dot{V}_{PV} + \frac{1}{L}V_{PV}^2 + \frac{2}{R^2C}v^2 - \ddot{F}}{\left(\frac{1}{L}V_{PV} + \frac{2}{RC}i\right)v} \tag{13}$$

$$\beta = \frac{1}{2}\sqrt{\left(\frac{RCV_{PV}}{L}\right)^2 + \frac{4}{L}\left(RC\dot{F} + 2F\right)} \tag{14}$$

The derivation of Equations (11)–(14) is presented in Appendix A. Note that the model defined by Equations (8) is represented by the differential parameterization given by Equations (11)–(14) stated in terms of the flat output F and its time derivatives $\dot{F}$ and $\ddot{F}$. In Equation (13), $\dot{V}_{PV}$ is the time derivative of the output voltage of the PV cell.

The underlying idea of the proposed control approach for the MPPT is to use the classical flatness-based trajectory tracking control but instead of predefine a reference path, a reference voltage v^* will be obtained from the maximum power provided by the PV panel. From this reference voltage, using Equation (10), a reference flat output F^* is constructed. The controller is designed to track the reference signal F^* with a guaranteed convergence of the error to zero. As shown below, this ensures an efficient tracking of the MPP.

The control design is explained as follows. To obtain the reference voltage corresponding to the maximum power provided by the PV panel, the power delivered by the Boost converter P_C must match the power supplied by the PV panel P_{PV}, i.e,

$$P_C = P_{PV} \tag{15}$$

Note that, for a pure resistive load, P_C can be written in terms of the output voltage v and the load resistance R, and P_{PV} can be expressed in terms of I_{PV} and V_{PV}, as shown in Equation (16).

$$P_C = \frac{v^2}{R}, \qquad P_{PV} = V_{PV} I_{PV} \tag{16}$$

Then, in view of Equations (15) and (16), the reference voltage v^* that the controller must take into account to guarantee the tracking of the MPP is given by,

$$v^* = \sqrt{V_{PV} I_{PV} R} \tag{17}$$

In order to steer the converter output voltage v to the reference voltage v^* defined in Equation (17), the control must force the flat output F to follow the reference flat output F^* calculated as:

$$F^* = \frac{1}{2}\left(L\left(\frac{v^{*2}}{RV_{PV}}\right)^2 + Cv^{*2}\right) \tag{18}$$

In order to establish the closed-loop dynamics, in Equation (13), the highest order derivative of the flat output is replaced by an auxiliary input $\mu_{aux} = \ddot{F}$. Regarding the tracking problem under consideration, the error is defined as $e = F - F^*$. To guarantee stable closed-loop error dynamics, the auxiliary input μ_{aux} can be defined as:

$$\mu_{aux} = \ddot{F}^* - \beta_1(\dot{F} - \dot{F}^*) - \beta_0(F - F^*) \tag{19}$$

with $\beta_0 > 0$ and $\beta_1 > 0$ are constant gains to be determined. Equation (19) allows one to define the error dynamics as:

$$\ddot{e} + \beta_1\dot{e} + \beta_0 e = 0 \tag{20}$$

which is associated with the characteristic polynomial,

$$p(s) = s^2 + \beta_1 s + \beta_0 \tag{21}$$

For the sake of convenience, the characteristic polynomial can be written in terms of the damping factor ζ and the natural frequency ω_n as follows,

$$p(s) = s^2 + 2\zeta\omega_n s + \omega_n^2 \tag{22}$$

where $\omega_n > 0$ and $\zeta > 0$. Notice that β_1 and β_0 are determined by

$$\beta_1 = 2\zeta\omega_n, \qquad \beta_0 = \omega_n^2 \tag{23}$$

The error will asymptotically converge to zero if the gains of the feedback tracking controller are chosen such that the roots of the characteristic polynomial defined in Equation (21) lie in the left half complex plane. The constants β_1 and β_0 must be chosen accordingly. In view of the parametrization of the control input given in Equation (13) and the auxiliary input dynamics defined in Equation (19), the flatness-based controller that ensures the tracking of the MPP with an asymptotic convergence of the error to zero is given by

$$u_{av} = 1 - \frac{i^* \dot{V}_{PV} + \frac{1}{L} V_{PV}^2 + \frac{2}{R^2 C}(v^*)^2 - \mu_{aux}}{\left(\frac{1}{L} V_{PV} + \frac{2}{RC} i^*\right) v^*} \tag{24}$$

where i^* is obtained from Equation (11) where β is given in terms of F^* and $\dot{F}^*$. As it was previously described, the switching function of the Boost converter is able to take binary values, then, the average control law defined in Equation (24) cannot be directly implemented. As usual practice in this scenario, a Pulse Width Modulation (PWM) has been used for which the average control u_{av} acts as a modulation signal. Figure 6 presents a schematic diagram illustrating the control system implementation.

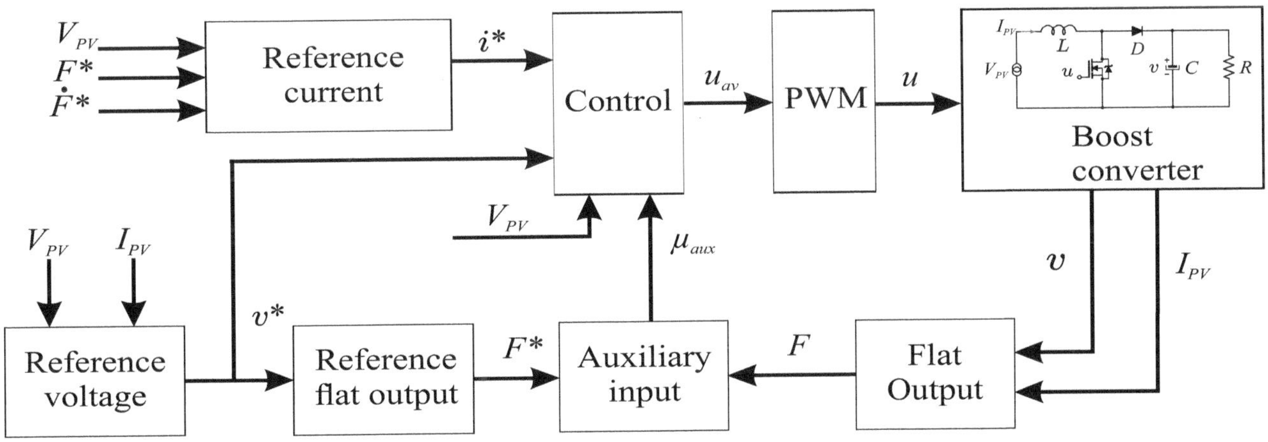

Figure 6. MPPT system block diagram.

4. Simulation Results

In this section the performance of the proposed control approach is evaluated through numerical simulations. The PV system consisting of a photovoltaic module, a Boost converter, a load resistance, and the flatness-based controller is implemented in the MATLAB®/Simulink®environment (R2015a, The MathWorks, Inc, Natick, MA, USA). Table 1 shows the characteristics of the Virtus II photovoltaic module (JC250M-24/Bb, ReneSola, Shanghai, China) used for this experimental setting.

Table 1. Virtus II Electrical characteristics.

Parameters	Value
Maximum Power (P_{MAX})	250 W
Current at Maximum Power (I_{MP})	8.31 A
Voltage at Maximum Power (V_{MP})	30.1 V
Short-Circuit Current (I_{SC})	8.83 A
Open-Circuit Voltage (V_{OC})	37.4 V

Table 2 presents the parameters of the Boost power converter and the proposed values of the natural frequency and the damping factor for the flatness-based controller. Notice that with the proposed values of damping ζ and natural frequency ω_n, the controller gains are $\beta_0 = 90{,}000$ and $\beta_1 = 60$. Then, the roots of the characteristic polynomial given in Equation (21) are $s_1 = -30 + 2.985\text{j}$ and $s_2 = -30 - 2.985\text{j}$. Since the roots lie on the left side of the complex plane, the stability of the error dynamics is ensured, i.e., the error $e = F - F^*$ asymptotically converges to zero as $t \to \infty$.

Table 2. Boost converter parameters.

Parameters	Value
Capacitance C	470 µF
Resistance R	12 Ω
PWM operation frequency	45 kHz
Inductance L	10 mH
Natural frequency ω_n	300
Damping factor ζ	0.1

Through simulations, the robustness of the differential flatness-based controller under variations of the solar irradiance, temperature and load is evaluated. To develop such evaluation, four study cases were considered:

Case 1. At 1.1 s, the solar irradiance is increased from 500 W/m^2 to 1000 W/m^2 within a time interval of 0.1 s. A constant temperature of 25 °C is considered (see Figure 7c).

Case 2. At 0.8 s, the temperature is increased from 15 °C to 25 °C within a time interval of 0.1 s. A constant solar irradiance of 800 W/m^2 is considered (see Figure 8c).

Case 3. Considering a constant temperature of 25 °C, at 0.6 s, the solar irradiance is decreased from 1000 W/m^2 to 500 W/m^2 within a time interval of 0.5 s. Then, at 1.2 s, the solar irradiance is increased from 500 W/m^2 to 1000 W/m^2 within a time interval of 0.5 s. In a different time instant (at 2 s), the temperature increases from 25 °C to 40 °C within a time interval of 0.1 s (see Figure 9c).

Case 4. A solar irradiance of 1000 W/m^2 and a temperature of 25 °C are kept constant while the load instantly decreases from 12 Ω to 6 Ω at 0.5 s (see Figure 10c).

In the first three cases, a constant load of 12 Ω is considered.

Note that the proposed time scales are short (tests were performed within time intervals less than 3 s). This aims at demonstrating the strength of the proposed approach for the MPPT, i.e., to evaluate the performance of the flatness-based control law against sudden variations of weather conditions. Clearly, if the controller is efficient in tracking the MPP in spite of the abrupt changes considered in the study cases described above, it will also have a satisfactory performance under realistic operating conditions involving smoother variations of solar irradiance and of temperature, occurring within longer time intervals.

Figures 7–10 illustrate the performance of the proposed control approach under the conditions stated above. Table 3 presents the theoretical maximum power for the different values of temperature and solar irradiance considered in the four study cases. It is important to mention that, theoretically, load variations do not affect the maximum power.

Table 3. Theoretical maximum power.

Solar Radiation (W/m^2)	Temperature (°C)	P_{MAX} (W)
500	25	125
510	23	118
800	25	200
800	15	210
900	45	203
1000	25	250
1000	40	232

The trajectories generated under the conditions described in **Case 1** are depicted in Figure 7. The output power of the Boost converter and the power delivered by the PV panel are shown in Figure 7a. The output voltage of the converter and the current and voltage of the PV panel are presented in Figure 7b. Figure 7c shows the variation of the solar irradiance. Note that the theoretical values of the maximum power for a constant temperature of 25 °C are reached: 125 W for a solar irradiance of 500 W/m^2, and 250 W for 1000 W/m^2. Observe that, as expected, during the time period in which the solar irradiance increase takes place (from 1.1 s to 1.2 s), an increment of the current of the PV panel occurs. However, the voltage V_{PV} is kept at a constant value (except for the variation observed within the time lapse in which the solar irradiance variation occurs).

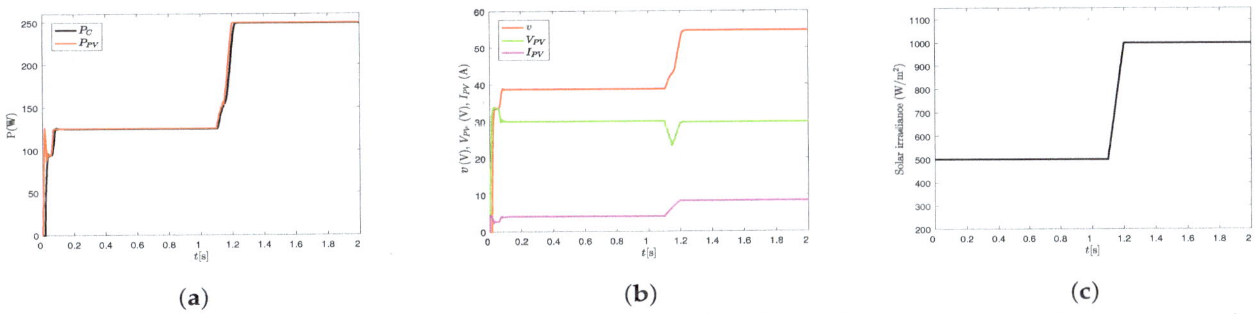

Figure 7. Simulation results under a variation of solar irradiance (under a constant temperature of 25 °C). Case 1. (**a**) Power of the PV panel (P_{PV}) and output power of the converter (P_C); (**b**) Voltage of the converter (v), voltage and current of the PV panel (V_{PV}, I_{PV}); (**c**) Solar irradiance variation.

The system response obtained by considering the conditions stated in **Case 2** is illustrated in Figure 8. Figure 8a depicts the Boost converter output power as well as the power delivered by the PV panel. Figure 8b shows the output voltage of the converter and the current and voltage provided by the PV panel. The variation of temperature is presented in Figure 8c. In this case, an increase of temperature occurs from 0.8 s to 0.9 s. In this period of time, the power values are decreased from 210 W/m^2 to 200 W/m^2. These values are in accordance with the theoretical maximum power that can be achieved (see Table 3). As expected, the current of the PV module remains constant and the value of V_{PV} decreases.

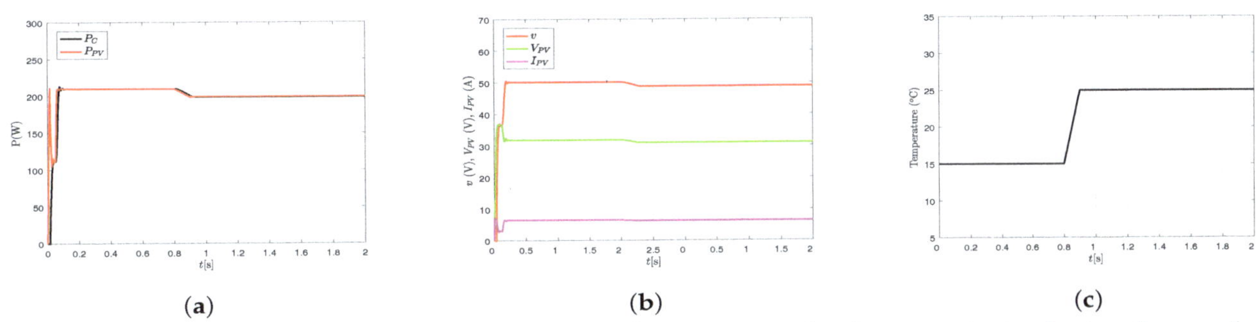

Figure 8. Simulation results under a variation of temperature (under a constant solar irradiance of 800 W/m^2). Case 2. (**a**) Power of the PV panel (P_{PV}) and output power of the converter (P_C); (**b**) Voltage of the converter (v), voltage and current of the PV panel (V_{PV}, I_{PV}); (**c**) Temperature variation.

The controller performance under the conditions described in **Case 3** is illustrated in Figure 9. Figure 9a presents the output power of the Boost converter and the power delivered by the PV panel. Figure 9b shows the trajectories of the converter output voltage as well as the voltage and current values delivered by the PV panel. Figure 9c illustrates the solar irradiance and temperature variations. It is obvious that when a decrease of solar irradiance occurs, the power provided by the PV panel P_{PV} and the current I_{PV} also decrease meanwhile the voltage V_{PV} does not vary significantly. Conversely, when solar irradiance increases, the power provided by the PV panel P_{PV} and the current I_{PV} are also incremented meanwhile the voltage V_{PV} undergoes variations but after a short period of time it reaches the previous value again. The temperature increase leads to a slight reduction of the voltage V_{PV} and a decrease of the power P_{PV} meanwhile the current I_{PV} remains almost unchanged. Notice that in all cases the theoretical maximum value of the power provided by the PV panel is achieved (see Table 3).

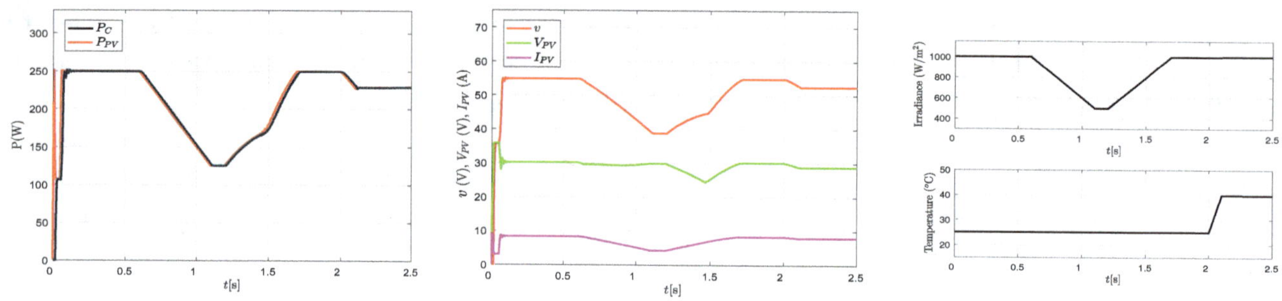

(a) (b) (c)

Figure 9. Simulation results under solar irradiance and temperature variations. Case 3. (**a**) Power of the PV panel (P_{PV}) and output power of the converter (P_C); (**b**) Voltage of the converter (v), voltage and current of the PV panel (V_{PV}, I_{PV}); (**c**) Solar irradiance and temperature variations.

The controller performance under the conditions described in **Case 4** is illustrated in Figure 10. Figure 10a shows the output power of the Boost converter and the power delivered by the PV panel. Figure 10b shows the trajectories of the converter output voltage as well as the voltage and current values delivered by the PV panel. Figure 10c illustrates the load variation. Note that the load is diminished at 0.5 s leading to minor variations of the signals coming from the PV panel P_{PV}, I_{PV}, and V_{PV}. However, the output voltage of the Boost converter v is significantly decreased. It is noteworthy that variations of the load do not prevent the proposed algorithm from successfully tracking the maximum power.

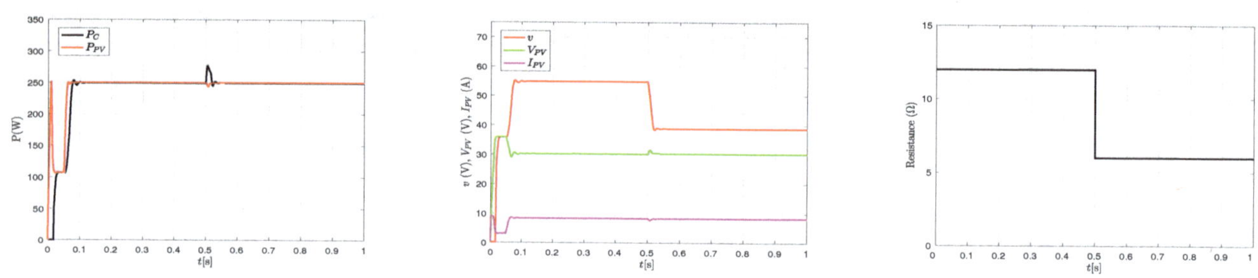

(a) (b) (c)

Figure 10. Simulation results under a load variation. Case 4. (**a**) Power of the PV panel (P_{PV}) and output power of the converter (P_C); (**b**) Voltage of the converter (v), voltage and current of the PV panel (V_{PV}, I_{PV}); (**c**) Load variation.

Simulation results highlight the performance of the proposed control approach. **Case 1** is defined to demonstrate the robustness of the control law against solar irradiance abrupt changes; **Case 2** aims to show the controller performance under a sudden variation of temperature; **Case 3** allows evaluating the performance of the controller under abrupt variations of solar irradiance and temperature; finally, **Case 4** considers the transit from a lightly loaded system to a heavily loaded one.

Simulation results show that the proposed control scheme is robust against variations of external conditions and that the corresponding values of maximum power are reached in all cases. From Figures 7–10, it can be noticed that the control action takes about 0.1 s to be reflected in the system trajectories and that the system reaches the steady state in a very short period of time regardless the variation of weather conditions and load. This stabilization time can be considered acceptable since most of the PV applications do not require a faster response.

As can be seen, the sudden variations of temperature, solar irradiance and load do not degrade the performance of the flatness-based controller. It can be concluded that the flatness-based controller is effective in maintaining the output power of the PV panel at the corresponding optimum values. Furthermore, it is robust against variations of external conditions and operates under low or high temperatures.

Comparative Analysis

In order to determine the benefits that the proposed control scheme offers in contrast to the most popular MPPT techniques: Perturb and Observe (P&O) and Incremental Conductance (IncCond), numerical simulations were performed. Under the conditions stated in **Case 3**, the conventional P&O method (see for instance [12]) and the IncCond with integral regulator (IncCond+IR) algorithm proposed in [39] were evaluated. Figure 11 shows the performance of these three techniques; $P_{P\&O}$ corresponds to the power obtained with the P&O method, P_{FBC} is the power obtained with the flatness-based control technique proposed in this paper, and $P_{IncCond+IR}$ is the power obtained with the IncCond+IR algorithm.

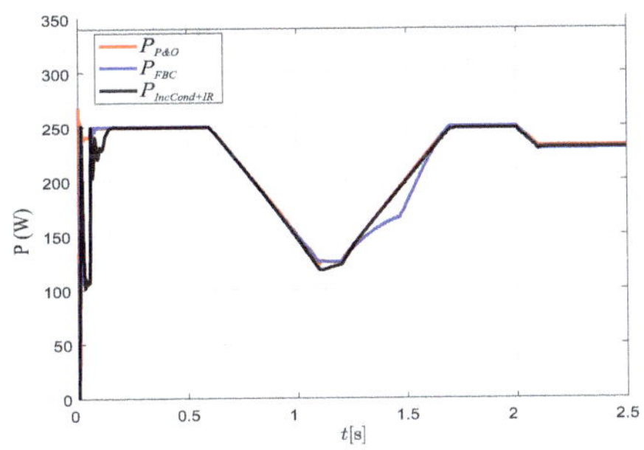

Figure 11. Comparison of three techniques for the MPPT. Simulation results under conditions stated in **Case 3**. Red line: power obtained with the Perturb and Observe (P&O) method. Blue line: power obtained with the flatness-based controller. Black line: power obtained with the IncCond+IR algorithm.

Note that within the time interval where the solar irradiance increase occurs (between 1.2 s and 1.7 s), the response produced with the flatness-based approach differs from the ones produced with the other techniques. However, the theoretical power is reached when the solar irradiance is set at 1000 W/m^2.

At first glance, it might seem that the three methods have almost the same performance, however, a zoom of Figure 11 shows that, although the time to reach the MPP is almost equal (about 0.13 s), the response obtained with the IncCond+IR method exhibits oscillations of significant magnitude before reaching the MPP (see Figure 12a), and that, as expected, the response produced with the P&O technique presents small amplitude oscillations around the MPP (see Figure 12b).

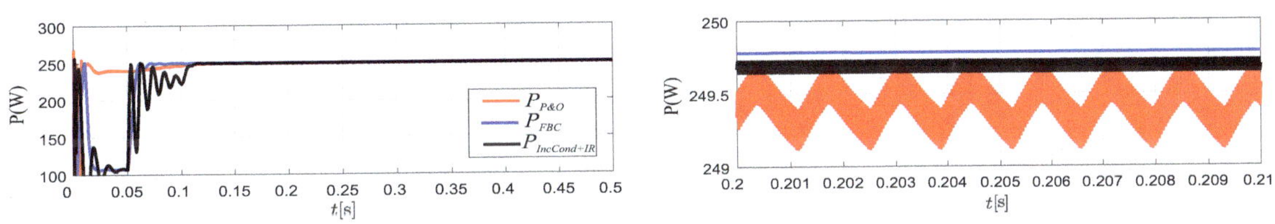

(a) (b)

Figure 12. Zoom of Figure 11. (**a**) Zoom of Figure 11 between 0 and 0.5 s; (**b**) Zoom of Figure 11 between 0.2 and 0.21 s.

In view of the simulation results, it can be concluded that one important advantage of the proposed approach is that less oscillations are produced before reaching the MPP compared to the IncCond+IR method. Furthermore, the flatness-based method does not give rise to oscillations around the MPP, like the ones observed for the P&O technique.

Another disadvantage of the P&O technique is that its accuracy depends on the perturbation size: if it is small, the method takes more time to reach the MPP, if it is big, the amplitude of the oscillations around the MPP increases [26]. This does not occur with the proposed flatness-based method.

5. Experimental Results

Experimental results obtained with a PV panel Virtus II model JC250M-24/Bb show the performance of the flatness-based controller under realistic scenarios. For the implementation of this MPPT controller, a Boost power converter is designed to operate in continuous mode at a switching frequency of 45 kHz; the parameter values used in the experimental set up are shown in Table 2.

The MPPT flatness-based control algorithm was implemented with the Simulink Real-Time® software (R2015a, The MathWorks, Inc, Natick, MA, USA) using a Humusoft® MF 624 multifunction I/O card (MF624, Humusoft Ltd., Prague, Czechia) that is installed on a computer with an Intel® Core™ processor (i7-2600, 3.40 GHz, Intel, Santa Clara, CA, USA) and 8GB RAM memory (DDR3 1333 MHz, Hynix, Incheon, Korea). The complete scheme operates at a fixed sampling frequency of 10 kHz. A PWM generator set to a frequency of 45 kHz is employed. Four analog inputs are used to sense the current I_{PV} and voltage V_{PV} of the PV panel, the output voltage of the converter v, and the current through the load resistor i_R, which allows monitoring possible variations of the load. The control system can be described as follows. The input signals I_{PV} and V_{PV} are acquired by sensors S22P (current) and AMC1100 (voltage) respectively. The average control signal u_{av} adjusts the duty cycle of the PWM generator which activates the MOSFET 19NF20 through the optocoupler circuit IX3180. The auxiliary input μ_{aux} defined in Equation (19) is generated with the use of the reference signal F^* (derived from the output voltage reference) and the flat output F. The Vantage Pro2^{TM} weather station is used to measure solar radiation; it is remotely connected with an indicator console. The experimental configuration diagram, the nominal values of the components, and the information of electronic devices and sensors are shown in Figure 13.

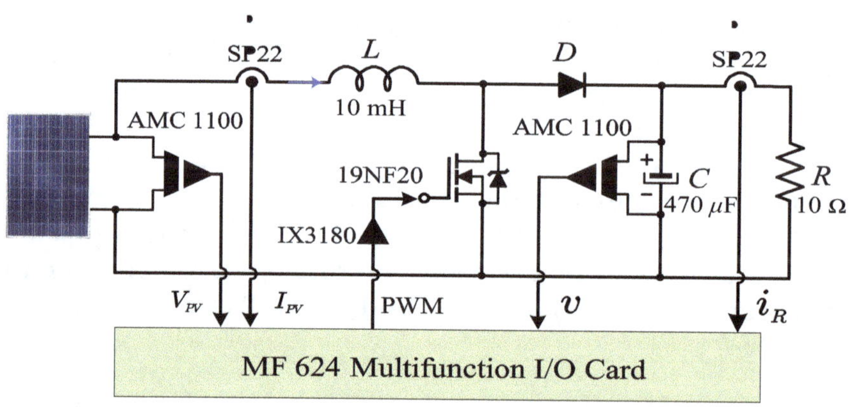

Figure 13. General scheme of a experimental setup of MPPT.

It is important to mention that the true MPP is unknown during a real-time test under natural sunlight. This is because the temperature cannot be accurately measured since PV panels have a multicell distribution and an insulated laminated structure [40].

During the development of experiments, temperature differences of more than 5 °C were found in different sections of the PV panel. Higher temperatures were identified near to the edges of the PV panel due to the heating generated by its laminated structure. To cope with this, approximate values of the panel temperature were calculated from the average of the measured temperatures. Based on these values, the corresponding values of the theoretical maximum power were obtained.

The first experimental test was carried out when a solar irradiance of 510 W/m^2 was measured and an average temperature of 35 °C was calculated. Figure 14 shows the closed-loop response of the system under these conditions. Figure 14a corresponds to the experimental result and Figure 14b to the simulation response.

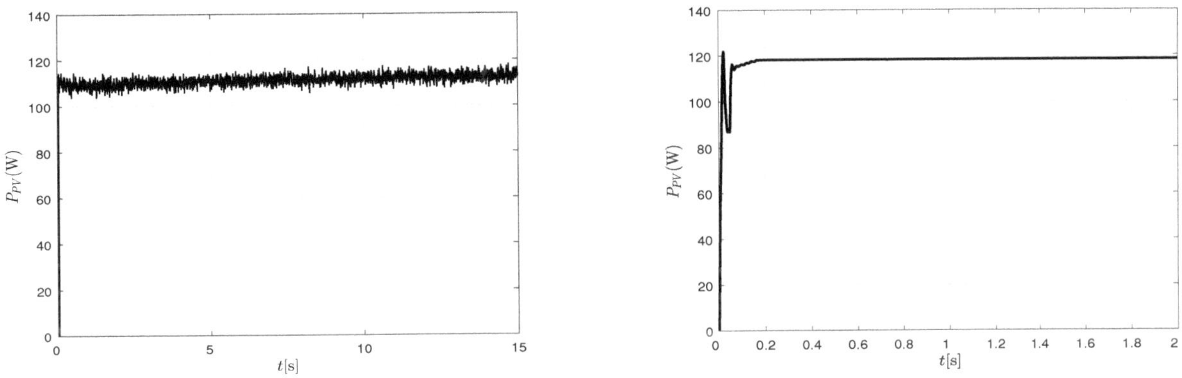

(**a**) Experimental result. (**b**) Simulation result.

Figure 14. System response under a solar irradiance of 510 W/m^2 and an average temperature of 35 °C. Power of the PV panel (P_{PV}).

The second test was conducted under a solar irradiance of 900 W/m^2 and an average temperature of 45 °C. The system response is shown in Figure 15. Figure 15a corresponds to the experimental result and Figure 15b to the simulation response.

Notice that, in simulation, the theoretical maximum powers are reached (118 W for the first case, 203 W for the second one, see Table 3), meanwhile the powers obtained with the experimental tests are below these values (around 110 W for the first case, around 190 W for the second one). As explained before, these differences between the power values are due to the fact that it is impossible to estimate precisely the true MPP value based on the weather conditions because of the complexity of module lamination and measurement accuracy [40]. Besides, there are power losses in the connection cable between the photovoltaic module and the converter that should be considered. In [41], it is shown that the theoretical MPP values can be either higher or lower than the practical measured tracking values. Despite this, from Figures 14a and 15a it can be inferred that the flatness-based controller allows delivering all the available power to the load.

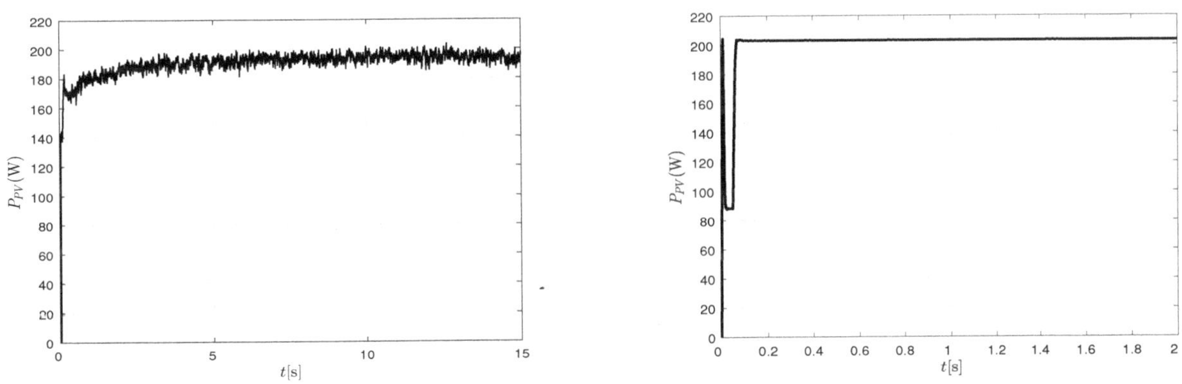

(**a**) Experimental result. (**b**) Simulation result.

Figure 15. System response under a solar irradiance of 900 W/m^2 and an average temperature of 45 °C. Power of the PV panel (P_{PV}).

6. Conclusions

In this paper a novel control strategy for the MPPT was presented. The differential flatness-based property of the Boost converter is exploited to design a controller that allows an effective tracking of the MPP. Numerical simulations were performed in MATLAB®/Simulink® to assess the effectiveness of the flatness-based control law against sudden variations of weather conditions and load considered in four study cases. For all of them, the theoretical maximum power was reached; it can be concluded that the sudden variations of temperature, solar irradiance, and load do not degrade the performance of the proposed MPPT technique. Furthermore, it was shown that the flatness-based technique has some advantages compared to the P&O method and to the IncCond algorithm related to the PV system response. Simulation results show that the oscillatory behavior around the MPP is avoided with the proposed approach.

Experimental tests were conducted in a prototype of the system consisting of a PV panel Virtus II JC250M-24/Bb. It was noticed that, during the real-time tests, the temperature is not uniform on the PV panel surface. Differences of more than 5 °C were found in different sections of the PV panel. As a result, the theoretical MPP is unknown. Despite this, from the experimental results, it can be inferred that the flatness-based controller allows delivering all the available power from the PV panel to the load. In fact, the power experimentally obtained is close to the theoretical maximum power calculated from average values of the panel temperature.

Author Contributions: Investigation, L.G.-A., B.S., O.P.-R.; methodology, B.S., O.P.-R.; resources, O.P.-R., P.R.M.-R.; supervision, B.S.; validation, J.C.Á.-V., R.M.-M., P.R.M.-R.; writing—original draft, L.G.-A.; writing—review and editing B.S., R.M.-M., J.C.Á.-V.

Appendix A

This section presents the parametrization of the states and control input of the Boost converter model derived from the flat output. Consider the flat output given by

$$F = \frac{1}{2}\left(Li^2 + Cv^2\right) \tag{A1}$$

Its time derivative is:

$$\dot{F} = Li\frac{di}{dt} + Cv\frac{dv}{dt} \tag{A2}$$

substituting the system dynamics defined by Equations (8) yields,

$$\begin{aligned} \dot{F} &= i[-v\mathbf{u} + V_{PV}] + v[i\mathbf{u} - \frac{v}{R}] \\ \dot{F} &= iV_{PV} - \frac{v^2}{R} \end{aligned} \tag{A3}$$

where:

$$\mathbf{u} = 1 - u_{av} \tag{A4}$$

The parametrization of i is obtained by substituting v^2 (that can be derived from Equation (A3)) into Equation (A1),

$$i = -\frac{RCV_{PV}}{2L} + \beta \tag{A5}$$

where:

$$\beta = \frac{1}{2}\sqrt{\left(\frac{RCV_{PV}}{L}\right)^2 + \frac{4}{L}\left(RC\dot{F} + 2F\right)} \tag{A6}$$

From Equation (A1), one obtains:

$$v = \sqrt{\frac{2F - Li^2}{C}} \tag{A7}$$

substituting Equation (A5) into Equation (A7) yields the parametrization of v, which is obtained as follows:

$$v = \sqrt{\frac{2}{C}F - \frac{L}{C}\left(-\frac{RCV_{PV}}{2L} + \beta\right)^2} \tag{A8}$$

The second derivative of F is:

$$\ddot{F} = i\dot{V}_{PV} + V_{PV}\frac{di}{dt} - \frac{2v}{R}\frac{dv}{dt} \tag{A9}$$

Substituting the system dynamics defined by Equations (8) into Equation (A9) yields the parametrization of the average control as follows:

$$u_{av} = 1 - \frac{i\dot{V}_{PV} + \frac{1}{L}V_{PV}^2 + \frac{2}{R^2C}v^2 - \ddot{F}}{\left(\frac{1}{L}V_{PV} + \frac{2}{RC}i\right)v} \tag{A10}$$

References

1. Bueno, P.G.; Hernandez, J.C.; Sanchez-Sutil, F. Stability assessment for transmission systems with large utility-scale photovoltaic units. *IET Renew. Power Gener.* **2016**, *10*, 584–597. [CrossRef]
2. Chu, C.C.; Chen, C.L. Robust maximum power point tracking method for photovoltaic cells: A sliding mode control approach. *Sol. Energy* **2009**, *83*, 1370–1378. [CrossRef]
3. Hernandez, J.C.; Bueno, P.G.; Sanchez-Sutil, F. Enhanced utility-scale photovoltaic units with frequency support functions and dynamic grid support for transmission systems. *IET Renew. Power Gener.* **2017**, *11*, 361–372. [CrossRef]
4. Rezk, H.; Eltamaly, A.M. A comprehensive comparison of different MPPT techniques for photovoltaic systems. *Sol. Energy* **2015**, *112*, 1–11. [CrossRef]
5. Bahgat, A.B.G.; Helwa, N.H.; Ahmad, G.; El Shenawy, E.T. Maximum power point tracking controller for PV systems using neural networks. *Renew. Energy* **2005**, *30*, 1257–1268.
6. Brunton, S.L.; Rowley, C.W.; Kulkarni, S.R.; Clarkson, C. Maximum power point tracking for photovoltaic optimization using ripple-based extremum seeking control. *IEEE Trans. Power Electron.* **2010**, *25*, 2531–2540. [CrossRef]
7. Mastromauro, R.A.; Liserre, M.; Kerekes, T.; Dell'Aquila, A. A single-phase voltage-controlled grid-connected photovoltaic system with power quality conditioner functionality. *IEEE Trans. Ind. Electron.* **2009**, *56*, 4436–4444. [CrossRef]
8. Babaa, S.E.; Armstrong, M.; Pickert, V. Overview of Maximum Power Point Tracking Control Methods for PV Systems. *J. Power Energy Eng.* **2014**, *2*, 59–72. [CrossRef]
9. Hohm, D.P.; Ropp, M.E. Comparative Study of Maximum Power Point Tracking Algorithms. *Prog. Photovolt. Res. Appl.* **2003**, *11*, 47–62. [CrossRef]
10. Zainudin, H.N.; Mekhilef, S. Comparison Study of Maximum Power Point Tracker Techniques for PV Systems. In Proceedings of the 14th International Middle East Power Systems Conference, Cairo, Egypt, 19–21 December 2010; pp. 750–755.
11. Kollimalla, S.K.; Mishra, M.K. Variable Perturbation Size Adaptive P&O MPPT Algorithm for Sudden Changes in Irradiance. *IEEE Trans. Sustain. Energy* **2014**, *5*, 718–728.
12. Elgendy, M.A.; Zahawi, B.; Atkinson, D.J. Operating Characteristics of the P&O Algorithm at High Perturbation Frequencies for Standalone PV Systems. *IEEE Trans. Energy Convers.* **2015**, *30*, 189–198.
13. Zakzouk, N.E.; Elsaharty, M.A.; Abdelsalam, A.K.; Helal, A.A.; Williams, B.W. Improved Performance Low-Cost Incremental Conductance PV MPPT Technique. *IET Renew. Power Gener.* **2016**, *10*, 561–574. [CrossRef]

14. Jain, S.; Agarwal, V. A new algorithm for rapid tracking of approximate maximum power point in photovoltaic systems. *IEEE Power Electron. Lett.* **2004**, *2*, 16–19. [CrossRef]
15. Salas, V.; Olías, E.; Barrado, A.; Lázaro, A. Review of the maximum power point tracking algorithms for stand-alone photovoltaic systems. *Sol. Energy Mater. Sol. Cells* **2006**, *90*, 1555–1578. [CrossRef]
16. Ngan, M.S.; Tan, C.W. A Study of Maximum Power Point Tracking Algorithms for Stand-alone Photovoltaic Systems. In Proceedings of the 2011 IEEE Applied Power Electronics Colloquium (IAPEC), Johor Bahru, Malaysia, 18–19 April 2011.
17. Rakhshan, M.; Vafamand, N.; Khooban, M.H. Maximum Power Point Tracking Control of Photovoltaic Systems: A Polynomial Fuzzy Model-Based Approach. *IEEE J. Emerg. Sel. Top. Power Electron.* **2018**, *6*, 292–299. [CrossRef]
18. Tang, S.; Sun, Y.; Chen, Y.; Zhao, Y.; Yang, Y.; Szeto, W. An Enhanced MPPT Method Combining Fractional-Order and Fuzzy Logic Control. *IEEE J. Photovolt.* **2017**, *7*, 640–650. [CrossRef]
19. Rezaei, A.; Gholamian, S.A. Optimization of New Fuzzy Logic Controller by Genetic Algorithm for Maximum Power Point Tracking in Photovoltaic System. *ISESCO J. Sci. Technol.* **2013**, *9*, 9–16.
20. Alajmi, B.N.; Ahmed, K.H.; Finney, S.J.; Williams B.W. Fuzzy-Logic-Control Approach of a Modified Hill-Climbing Method for Maximum Power Point in Microgrid Standalone Photovoltaic System. *IEEE Trans. Power Electron.* **2011**, *26*, 1022–1030. [CrossRef]
21. Dounis, A.I.; Kofinas, P.; Alafodimos, C.; Tseles, D. Adaptive fuzzy gain scheduling PID controller for maximum power point tracking of photovoltaic system. *Renew. Energy* **2013**, *60*, 202–214. [CrossRef]
22. Alajmi, B.N.; Ahmed, K.H.; Finney, S.J.; Williams, B.W. A Maximum Power Point Tracking Technique for Partially Shaded Photovoltaic Systems in Microgrids. *IEEE Trans. Ind. Electron.* **2013**, *60*, 1596–1606. [CrossRef]
23. Ramaprabha, R.; Mathur, B.L. Intelligent Controller based Maximum Power Point Tracking for Solar PV System. *Int. J. Comput. Appl.* **2011**, *12*, 37–41. [CrossRef]
24. Younis, M.A.; Khatib, T.; Najeeb, M.; Ariffin, A.M. An Improved Maximum Power Point Tracking Controller for PV Systems Using Artificial Neural Network. *Prz. Elektrotech. Electr. Rev.* **2012**, *88*, 116–121.
25. Syafaruddin; Karatepe, E.; Hiyama, T. Artificial neural network-polar coordinated fuzzy controller based maximum power point tracking control under partially shaded conditions. *IET Renew. Power Gener.* **2009**, *3*, 239–253. [CrossRef]
26. Esram, T.; Chapman, P.L. Comparison of photovoltaic array maximum power point tracking techniques. *IEEE Trans. Energy Convers.* **2007**, *22*, 439–449. [CrossRef]
27. Veerachary, M.; Senjyu, T.; Uezato, K. Neural-network-based maximum-power-point tracking of coupled-inductor interleaved-boostconverter-supplied PV system using fuzzy controller. *IEEE Trans. Ind. Electron.* **2003**, *50*, 749–758. [CrossRef]
28. Subudhi, B.; Pradhan, R. A Comparative Study on Maximum Power Point Tracking Techniques for Photovoltaic Power Systems. *IEEE Trans. Sustain. Energy* **2013**, *4*, 89–98. [CrossRef]
29. Salmi, T.; Bouzguenda, M.; Gastli, A.; Masmoudi, A. MATLAB/Simulink Based Modelling of Solar Photovoltaic Cell. *Int. J. Renew. Energy Res.* **2012**, *2*, 2013–2018.
30. Rigatos, G.; Siano, P.; Ademi, S.; Wira, P. Flatness-Based Control of DC–DC Converters Implemented in Successive Loops. *Electric Power Compon. Syst.* **2018**, *46*, 673–687. [CrossRef]
31. Renaudineau, H.; Lopez, D.; Flores-Bahamonde, F.; Kuoro, S. Flatness-based control of a boost inverter for PV microinverter application. In Proceedings of the IEEE 8th International Symposium on Power Electronics for Distributed Generation Systems, Florianopolis, Brazil, 17–20 April 2017; pp. 1–6.
32. Alhamed Aldwaihi, H.; Delaleau, E. Maximum Power Point Tracker of a Wind Generator Based on the Flatness-Based Control. In Proceedings of the IEEE Energy Conversion Congress and Exposition, Phoenix, AZ, USA, 17–22 September 2011; pp. 883–888.
33. Xue, H.; Li, H.; Wang, Y. Nonlinear Differential Flatness Control of a Distributed PV-ES DC Generation System. *WSEAS Trans. Syst. Control* **2016**, *40*, 53–61.
34. Rekioua, D.; Matagne, E. *Optimization of Photovoltaic Power Systems—Modelization, Simulation and Control*; Springer: London, UK; Dordrecht, The Netherlands; Heidelberg, Germany; New York, NY, USA, 2012.
35. Kalogirou S.A. *Solar Energy Engineering*; Elsevier Inc.: San Diego, CA, USA, 2009.
36. Granda-Gutiérrez, E.E.; Orta, O.A.; Díaz-Guillén, J.C.; Jimenez, M.A.; Osorio, M.; Gonzleáz, M.A. Modelado y simulación de celdas y paneles solares. *Congreso Internacional de Ingeniería Electrónica ELECTRO* **2013**, *35*, 17–22.

37. Fliess, M.; Lévine, J.L.; Martin, P.; Rouchon, P. Flatness and defect of non-linear systems: Introductory theory and examples. *Int. J. Control* **1995**, *61*, 1327–1361. [CrossRef]
38. Sira-Ramírez, H.; Agrawal, S.K. *Differentially Flat Systems*; Marcel Dekker: New York, NY, USA, 2004.
39. Nguyen, D.T.; Goro, F. PV-Active Power Filter Combination Supplies Power to Nonlinear Load and Compensates Utility Current. *IEEE Power Energy Technol. Syst. J.* **2016**, *2*, 32–42. [CrossRef]
40. Weidong, X.; Zeineldin, H.H.; Zhang, P. Statistic and Parallel Testing Procedure for Evaluating Maximum Power Point Tracking Algorithms of Photovoltaic Power Systems. *IEEE J. Photovolt.* **2013**, *3*, 1062–1069.
41. Fu-Sheng, P.; Ru-Min, C.; Shin, H.K.; Tai-Sheng, L. Performance Evaluation of Parabolic Prediction to Maximum Power Point Tracking for PV Array. *IEEE Trans. Sustain. Energy* **2011**, *2*, 60–68.

11

6.78-MHz, 50-W Wireless Power Supply Over a 60-cm Distance using a GaN-Based Full-Bridge Inverter

Seung-Hwan Lee [1], Kyung-Pyo Yi [2] and Myung-Yong Kim [2,*]

1 School of Electrical and Computer Engineering, University of Seoul, Seoul 02504, Korea; seunghlee16@uos.ac.kr
2 Smart Electrical & Signaling Division, Korea Railroad Research Institute, Uiwang 16105, Korea; kpyi82@krri.re.kr
* Correspondence: mykim@krri.re.kr

Abstract: An inductive wireless power transfer system is proposed as a power supply for an on-line monitoring system for an overhead catenary. Because of the high voltage (25 kV_{rms}) applied to the catenary, galvanic isolation was required to supply power to the attached monitoring system. The proposed wireless power system was able to transmit 50 W over a distance of 60 cm at 6.78 MHz. Design methodologies for the transmitter and the receiver coils, 6.78-MHz GaN-based full-bridge inverter, and rectifier are proposed in this paper. Pareto optimality, a multi-objective optimization technique, was used to determine optimal solutions in terms of efficiency and copper usage. A 100-W, 6.78-MHz full-bridge inverter was developed using 100 V, 35 A, E-HEMT GaN MOSFETs. Because of the high operating frequency, two factors were considered in the design of the full-bridge inverter, (1) close placement of the gate driver and the switch to minimize parasitic inductance and the resulting fluctuation of the drive signal and (2) effective heat dissipation from the switches and gate drivers for a high power rating. In addition, a full-wave rectifier was built using Schottky barrier diodes with a reverse recovery time of a few tens of nano-seconds. The developed wireless power system was experimentally evaluated. The measured coil-to-coil efficiency was 77%, and the measured efficiencies of the inverter and the rectifier were 92% and 93%, respectively. The overall system efficiency was 57% for a transfer of 47 W. Finally, the dependences of the efficiency on the distance, operating frequency, and load were evaluated.

Keywords: wireless power transfer; inductive power transfer; Pareto optimality; coil design; magnetics design; GaN-based inverter and converter

---

1. Introduction

Applications of inductive wireless power transfer (WPT) technology have transitioned from W level biomedical devices to MW level transportation systems [1–8]. Among the various applications, application that the WPT system is useful as a power supply: an on-line monitoring system for a high-voltage catenary (see Figure 1a) [9]. Because of the high-voltage (25 kV_{rms}) of the catenary, the monitoring system cannot obtain power from a low-voltage grid (110 V_{rms} or 220 V_{rms}) without having a sufficient insulation or isolation gap. According to [10], the minimum isolation distance between a 25-kV_{rms} overhead wire and ground is 25 cm (30-cm isolation distance is recommended). This is the reason that the lengths of the insulators used for 25-kV_{rms} catenaries are longer than 30 cm (60 cm or longer insulators are common). Therefore, any system that supplies a low voltage power to the 25-kV_{rms} catenary needs to transmit power over the lengths of the insulators.

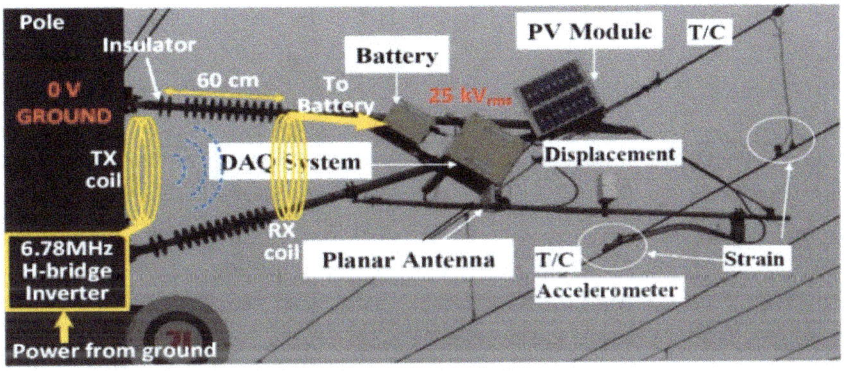

(a)

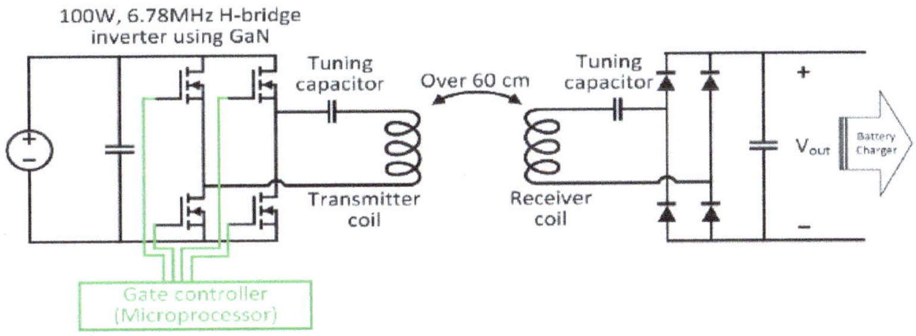

(b)

Figure 1. Configurations of the target catenary monitoring system and the WPT system: (**a**) Target system configuration; (**b**) Block diagram of the wireless power transfer system.

Previously, a 12-V, 12-Ah Pb-battery and a photovoltaic (PV) battery charger were used as an energy source. The PV charger was attached to an arm of a support pole (see Figure 1a). It solved the high voltage isolation problem. However, the size and weight of the PV module required to supply sufficient power (larger than 50 W) to the monitoring system were too large and heavy to be installed on the arm.

Instead of the PV module, an inductive WPT system was proposed that is able to wirelessly transmit energy over the insulator in [11,12]. According to [13], the inductive WPT system was more efficient compared to other technologies like a capacitive WPT system or a far-field power radiating system. The distance between the coils of the target WPT system was over 60 cm and output power was 50 W. The operating frequency of the system was 6.78 MHz (the highest frequency allowed for a WPT system [14]) to achieve efficient power transmission for this large air-gap. Because of the high operating frequency and large output power, the development of low-loss transmitter and receiver coils and a high efficiency 6.78-MHz inverter are required to construct an effective WPT system. There have been many previous studies that worked on such large air-gap, MHz frequencies WPT systems.

Sample et al. proposed a 7.65-MHz, 70-cm distance WPT system in 2011 [15]. The authors used two circular coils with diameters of 59 cm as a transmitter and a receiver. The measured coil-to-coil efficiency was 70% at 30 W, 70-cm power transmission. The main focus of the paper was to determine the optimal operating frequencies depending on the change of the distance between the coils. Duong and Lee published a paper demonstrating 30-mW transmission over a distance of 60 cm with a peak efficiency of 46% [16]. In this paper, the main focus of the paper was a mathematical analysis of the coil-to-coil system using an equivalent circuit modeling technique. Lee et al. proposed a coil design methodology for a large air-gap system in [17]. The design methodology identified a feasible design space as its first step. Then the coil-to-coil efficiency, copper usage, control stability, and other performance metrics were calculated in the entire design space, and a weighted sum of the metrics was calculated to determine an optimal design for the transmitter and the receiver coils. Based on the proposed methodology, a 95% coil-to-coil efficiency was achieved at a distance of 30 cm and 3.74-MHz operation. However, these aforementioned papers did not consider the feasibility and efficiency of MHz operation inverters, even though it is challenging to build an efficient high-power converter at these high frequencies.

The literature [18–22] focused on the development of high efficiency MHz switching class-E amplifiers for loosely-coupled WPT systems. Class-E amplifiers are very common in radio-frequency (RF) application since the amplifier requires only one low-side switch for power conversion. However, the class-E inverter requires a bulky DC inductor, and the optimal circuit parameters (e.g., shunt capacitances) are highly dependent on the operating conditions (load, duty cycle, and operating frequency, etc.) [23]. Therefore, the class-E inverter requires fine tuning of the circuit parameters depending on the load and operation conditions.

Instead of the class-E inverter, recent rapid evolution of large bandgap semiconductor switches (Silicon-carbide (SiC) and Gallium-nitride (GaN) Metal-Oxide-Semiconductor Field-Effect Transistor (MOSFET)) and their gate drivers have enabled significant advancement of radio-frequency half- and full-bridge switch-mode inverters. Multiple papers reported on GaN-based full-bridge inverters for WPT systems in 2017 [24–26]. Zhao et al. proposed a multi-frequency pulse-width-modulation (MFPWM) scheme which was implemented using a GaN-based full-bridge inverter [24]. Using MFPWM, an inverter can simultaneously transmit power to two receivers that operate at 100 kHz and at 6.78 MHz. Since the focus of the research was the transmission of power to multiple receivers, rather than improvement of the inverter efficiency using GaN switches, the efficiency of the developed inverter was not sufficiently high to be used for a system operating at a few tens of Watts. Bonache-Samaniego et al. proposed a self-oscillating resonant converter that operates at 6.78 MHz [25]. The gate signals of the developed full-bridge inverter's GaN switches were generated using the sensed current of the resonant tank. The proposed system was useful in the construction of a low cost and simple inverter. However, it was not suitable for the control of the operating frequency, duty, or power level. Xue and Zhang demonstrated a 50-W, 6.78-MHz, 90% efficiency GaN-based inverter for a WPT system. The main focus of the paper was the design of a filter network to achieve a wider zero-voltage-switching (ZVS) range and higher efficiency at light loads. The research used a 13.56-MHz oscillator and a D-flip-flop to generate a 6.78-MHz gate drive signal. The output power was controlled using a phase-shift control of the legs of the inverter based on a phase delay element.

In this paper, a novel 50-W, 60-cm distance wireless power system operating at a frequency of 6.78 MHz is proposed. The design and fabrication methodologies for the transmitter and the receiver coils, a 100-W, 6.78-MHz H-bridge inverter using GaN switches, and a 50-W, 6.78-MHz full-wave rectifier design are investigated. The power level and efficiency of the proposed system were experimentally evaluated. In the first part of the paper, design methodology for the transmitter and receiver coil is proposed using a Pareto optimality, a multi-objective optimization technique. The designed coils are fabricated on a printed-circuit-board (PCB) for ease of mass-production. The dominant sources of losses of the PCB coils are identified and the methods for fabrication of low-loss coils at 6.78-MHz frequencies are presented. The details of a 100-W, 6.78-MHz operating E-HEMT GaN-based H-bridge inverter is explained in the following section. The H-bridge inverter is developed to supply power to the transmitter coil. The gate signals of the switches are controlled using a commercial 32-bit digital-signal-processor, and the details of the PCB layout techniques for low parasitic inductance and an effective heat dissipation are discussed. Finally, the performance evaluation of the developed system is presented using experimental results. Efficiencies and losses of each component of the system are measured and compared. Also, the dependence of the DC-to-DC efficiency on the distance, load, and operating frequency is evaluated with the test-bed.

2. Target System Configuration

A photo of the target catenary monitoring system is shown in Figure 1a. The catenary monitoring system has multiple sensors that were installed on the catenary, a data acquisition (DAQ) system that gathers information from the sensors and wirelessly transmits the information to a data center, and a battery pack as an energy source. A WPT system that can transmit 50 W over a 60-cm distance (the length of the insulator) is proposed. A schematic of the proposed WPT system is shown in Figure 1b. The proposed WPT system collects power from a 220-V_{rms} single-phase grid and converts it to DC with a diode rectifier. Then, it is converted to a 6.78-MHz AC voltage and supplied to the transmitter (TX) coil using a GaN-based H-bridge inverter. The induced voltage of a receiver coil that is weakly coupled to the transmitter is rectified and supplied to the battery using a full-wave diode rectifier. The design details of the transmitter and the receiver coils, 6.78-MHz resonant inverter, and receiver side rectifier are described in following sections.

3. Development of Transmitter and Receiver coils

3.1. Transmitter and Receiver coil Design

There were three main constraints with regard to the design of the transmitter (TX) and the receiver (RX) coils; (1) the maximum sizes of the TX and RX coils were limited to 40 cm × 40 cm, (2) the minimum distance between the TX and RX coil was 60 cm. (3) the required output power was 50 W, and load resistance was 6 Ω. For full utilization of the given space, square coils were assumed for the TX and RX coils as shown in Figure 2. The design parameters to be determined for a square coil were the inner diameter (d_{in}), turn spacing (s), width of the turn (w), and the number of turns of the transmitter and the receiver coils (n_{tx} and n_{rx}). The design spans for the parameters are summarized in Table 1.

Table 1. Design spans of transmitter and receiver coils.

Parameters	Design Limits	Parameters	Design Limits
Inner diameter, d_{in}	10–30 cm	Trace width, w	1–2.5 cm
Turn spacing, s	0.5–2 cm	# of turns, n	1–8 turns

Pareto optimality [27], a multi-objective optimization theory, was applied in the coil design process. The maximization of the power transfer efficiency (η) and minimization of the total usage of the copper (m) were two objectives during optimization. The power transfer efficiency and total mass of the copper depending on the coil shape change were calculated using (1) [28] and (2).

$$\eta = \frac{1}{1 + \frac{R_2}{R_L} + \left(\frac{R_L + R_2}{\omega_0 M}\right)^2 \frac{R_1}{R_L}} \tag{1}$$

$$mass\ (m) = \rho * l * w * t \tag{2}$$

where, ω_0 is the operating frequency of the system, M is the mutual inductance between the transmitter and the receiver, R_1 and R_2 are the equivalent series resistances (ESRs) of the transmitter and the receiver coils, respectively, R_L is the load resistance of the system, ρ is the density of the copper, l, w and t are the total length, width, and the thickness of the coil trace, respectively. The mutual inductance and the ESRs of the TX and RX coils were calculated for each combination of the inner diameter, turn-spacing, width of the turn, and the number of the turns. The mutual inductance was calculated using a numerical integration of the flux linkage between the coils. According to [29], the z-directional magnetic field intensity at point P (H_z) generated by a square transmitter coil follows (3).

$$H_z = \frac{I}{4\pi}\sum_{i=1}^{n_{tx}} \oint \frac{\overrightarrow{dl}_i \times \overrightarrow{r}_i}{r_i^3},\ \lambda = \sum_{k=1}^{n_{rx}} \oint \mu_0 H_z\, dA_k,\ M = \frac{\lambda}{I} \tag{3}$$

Then, the total flux linkage (λ) at the receiver coil and resulting mutual inductance (M) can be calculated. H_z is the z-directional magnetic field intensity at point P, I is the current flowing in the transmitter coil, n_{tx} is the number of turns of the transmitter coil, $\overrightarrow{dl}_i$ is the infinitesimal displacement vector of i-th turn of the transmitter and $\vec{r}_i$ is a direction vector from $d\vec{l}_i$ to the point (see Figure 3), dA_k is the infinitesimal area element in k-th turn of the receiver coil, μ_0 is the permeability of air, λ is the total flux linkage at receiver coil, and M is the resulting mutual inductance. The ESRs of the transmitter and the receiver coils were calculated using (4).

$$R = \frac{l}{\sigma \delta w} \tag{4}$$

where l is the total length of the coil, w is the width of the trace of the coil, σ is the conductivity of copper, and δ (= 25.4 μm) is the skin-depth of the copper at 6.78 MHz. Figure 4 shows a $\eta - m$ distribution in the entire coil design space.

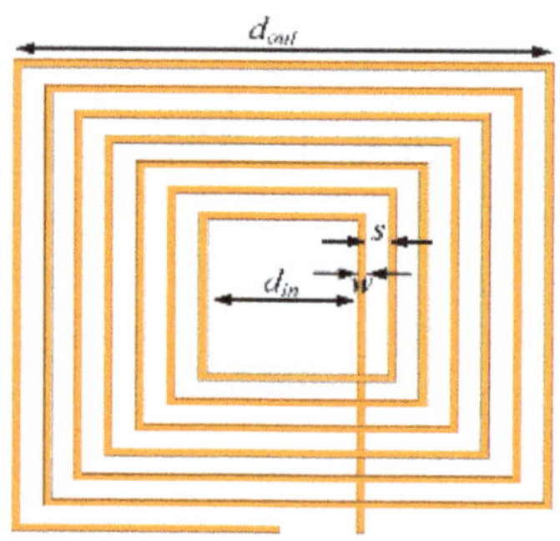

Figure 2. Geometric design parameters of the square coils.

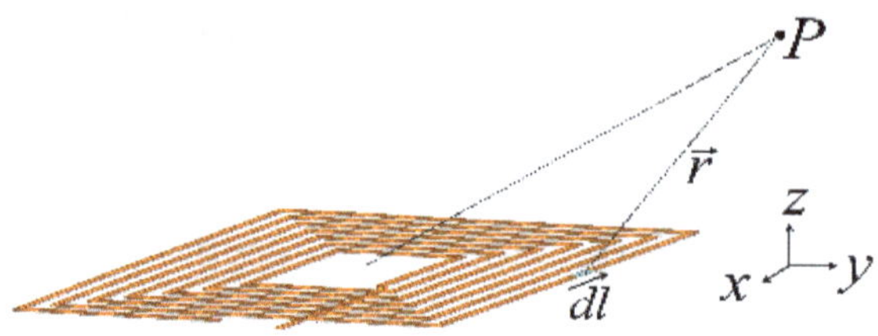

Figure 3. H-field calculation diagram.

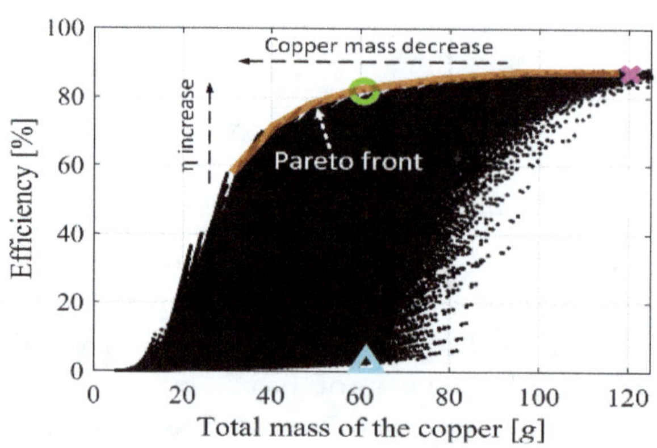

Figure 4. Pareto front of the WPT system.

From the figure, it should be noted that the WPT coils with the same total mass of copper have very different efficiencies depending on their design. For example, the first WPT system (identified by the cyan triangular point in Figure 4 had an 8-turn transmitter coil with a 10 cm inner diameter, 40 cm outer diameter, 6 mm of turn spacing, and 13 mm turn width. The receiver of the first WPT system was a 1-turn coil with a 10 cm inner diameter, 12.6 cm outer diameter, 6 mm of turn spacing, and 13 mm turn width. The second WPT system (identified by green circular point in Figure 4 had a 3-turn transmitter coil with a 30 cm inner diameter, 40 cm outer diameter, 5 mm of turn spacing and 13 mm turn width. Its receiver coil was a 3-turn with a 30 cm inner diameter, 40 cm outer diameter,

5 mm of turn spacing, and 13 mm turn width. The total mass of both systems was 60 g; however, the power transfer efficiency of the first system was 2.4% at 60 cm of air-gap while the efficiency of the second system was 82%.

Since the goal function was to maximize the efficiency while minimizing copper usage, non-dominated solutions (Pareto front) were located at the upper left corner of the plot. The brown line in the figure represents the Pareto front. Among the non-dominated solutions, one of the solutions with 87% efficiency and 110 g of copper usage was selected for the TX and RX coil design. This solution is denoted as a magenta X in the plot. It is notable that the decision was made for high efficiency rather than low copper mass. The selected design parameters for the TX and RX coils are summarized in Table 2.

Table 2. Selected design of the transmitter and receiver coils.

Parameters	Transmitter Coil	Receiver Coil
Inner diameter d_{in}	18 cm	12 cm
Outer diameter d_{out}	39.8 cm	39.6 cm
Turn spacing, s	0.7 cm	0.7 cm
Trace width, w	2.2 cm	2.2 cm
Trace thickness	71 μm	71 μm
Number of turns, n	4	5
Substrate thickness	1.6 mm	1.6 mm

3.2. Finite Element Analysis of Transmitter and Receiver

The selected design was evaluated by finite element analysis (FEA) results using the full-wave simulation software ANSYS HFSS (19.0, ANSYS, Canonsburg, PA, USA) and ANSYS DESIGNER (19.0, ANSYS, Canonsburg, PA, USA). Figure 5a shows a simulation model of the WPT system and Figure 5b shows the resulting coil-to-coil efficiency vs. operating frequency for a 6 Ω load. The theoretical efficiency was calculated using (1). It should be noted that the theoretical and the simulated results are in good agreement with each other.

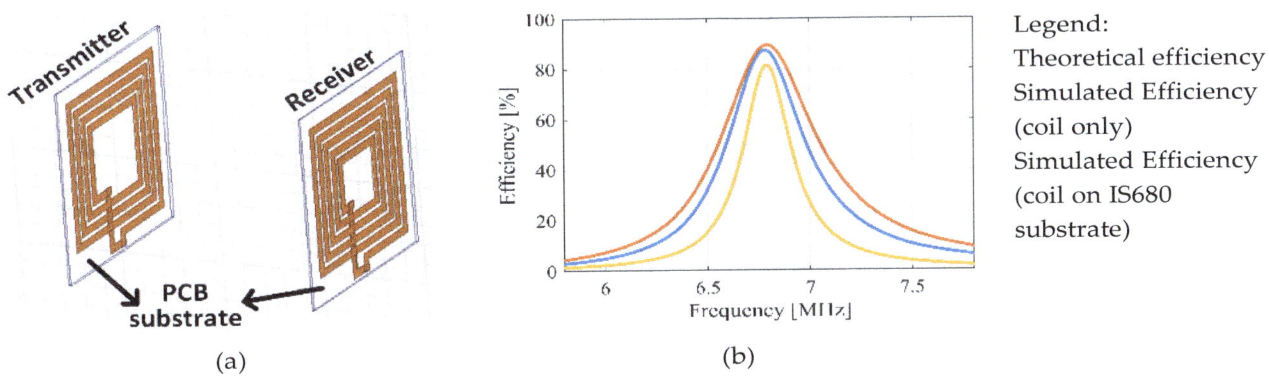

Figure 5. Finite element simulation model and simulated efficiencies: (**a**) FEA model; (**b**) Simulated efficiency.

3.3. High Q coil Fabrication

Because of the large diameter and a small number of turns of the TX and RX coils, the coils were fabricated using a PCB for ease of mass-production. Fabrication of a high-quality factor transmitter and the receiver coils on the PCB is discussed in a previous paper [12]. A summary of the previous paper is provided in this chapter. At high frequencies, especially in the MHz frequency ranges, the reduction of three primary types of losses was important to achieve high quality factor (Q-factor) coils on the PCB. These include skin-effect loss, proximity-effect loss, and dielectric loss. Because of the target WPT system's high operating frequency, skin- and proximity-effect losses were critical in the process of analyzing the Ohmic loss of the copper trace. In addition, the dielectric loss of the PCB substrate was important in determining the entire loss of the PCB coil because it can dominate other losses when the coil is not designed properly. According to the full-wave simulation results in [12], the dielectric loss was approximately 10 times greater than the Ohmic loss when the PCB substrate was a very common material, FR-4. A low dissipation factor PCB material (IS680 [30]) was very effective in reducing dielectric loss (see Table 3). Figure 6a,b show the distributions of the dielectric losses at the transmitter and the receiver coils when they transfer 50 W to a load.

Table 3. Comparison of material properties of FR-4 and IS680.

Parameters	FR-4	IS680
Relative permittivity	4.8	3.2
Loss tangent	0.017	0.0032
Dielectric strength	20 kV/mm	30 kV/mm
Thermal conductivity	0.29 W/m/K	0.32 W/m/K

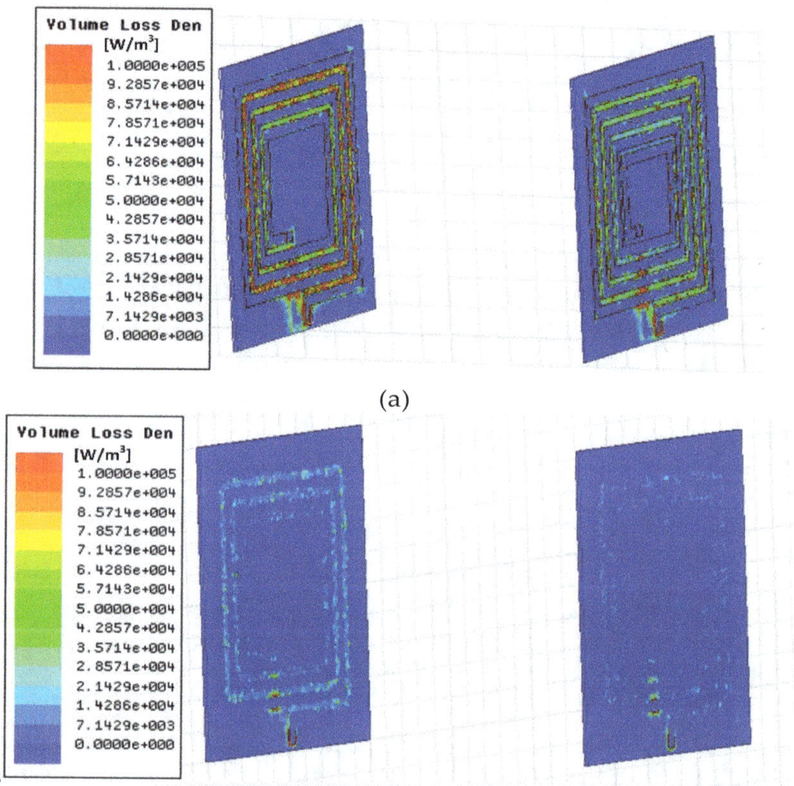

(b)

Figure 6. Configuration of the WPT system: (**a**) Dielectric loss density-FR4; (**b**) Dielectric loss density-IS680.

It should be noted that the transmitter and receiver coils using FR-4 substrates have much higher dielectric losses compared to the losses associated with the coils using the IS680 substrates. For further reduction of the dielectric loss, compensation capacitors were distributed in the middle of the coil windings as shown in Figure 7. This was helpful in reducing the electric field intensities between the adjacent turns and additional improvements of the Q-factors of the coils [12]. Figure 7 shows photos of the printed transmitter and receiver coils on IS680 substrates with distributed capacitors.

Circuit parameters of the transmitter and the receiver coils were measured using an impedance analyzer and were compared with theoretical and simulation results. Table 4 shows a summary of the parameters. Theoretical self-inductances of the coils were calculated using Wheeler's formula (5) in [31] while mutual inductance and resistances were calculated using (3) and (4).

$$L_S = K_1 \mu_0 \frac{n^2 d_{avg}}{1 + K_2 \rho} \quad (5)$$

wher $K_1 = 2.34$, $K_2 = 2.75$, $\rho = (d_{out} - d_{in})/(d_{out} + d_{in})$ is the fill ratio, $d_{avg} = (d_{out} + d_{in})/2$ is the average diameter, d_{out} is an outer diameter, and n is the number of turns. The measured self- and mutual-inductances and the ESRs of the coils matched the simulation results very well. It was notable that the measured Q-factors of the transmitter and the receiver coils were greater than 300, which is remarkably high compared to conventional coils on the PCBs.

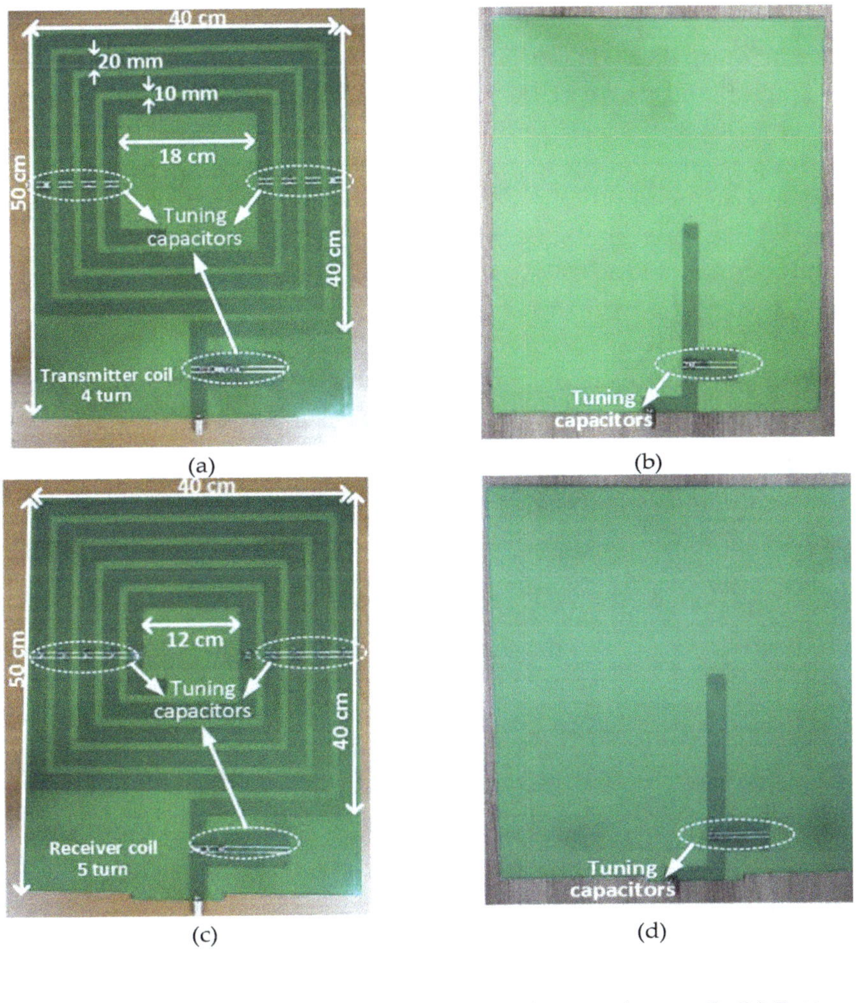

Figure 7. Photos of transmitter and receiver coils: (**a**) Top of transmitter coil; (**b**) Bottom of transmitter coil; (**c**) Top of receiver coil; (**d**) Bottom of receiver coil.

Table 4. Circuit parameters of the designed coils at 6.78 MHz.

Circuit Parameters	Theoretical		FEA Results		Measurement	
	TX	RX	TX	RX	TX	RX
Self-inductance [μH]	6.67	7.15	6.5	7.53	6.24	7.2
AC resistance [Ω]	0.31	0.38	0.48	0.70	0.58	0.81
Q-factor @ 6.78 MHz	916	801	576	458	458	379
Mutual-inductance [μH]	0.11		0.11		0.09	
Coupling coefficient	0.016		0.016		0.013	

3.4. Coil-to-Coil Power Transfer Performance

From the full-wave simulation results, an efficiency in excess of 81% was expected for a 50-W power transfer. To experimentally evaluate the fabricated coils, they were tuned using ceramic capacitors. The resulting tuned frequencies of the coils were measured using a network analyzer, E5061B (Keysight). Figure 8 shows the measured magnitude of the scattering parameters (S-parameters) of the transmitter and the receiver coils when they were 60 cm away from each other. S11 (blue line) and S22 (green line) had their minimum values at the tuned frequency, 6.87 MHz. S21 (red line), the transmission coefficient, was also maximized at the tuned frequency. This tuned frequency was 1.3% higher than the target frequency of 6.78 MHz, but this tuning frequency error was ignored in the experiment because the error has a negligible impact on the peak power transfer efficiency as shown in Figures 9 and 10. Moreover, a GaN-based H-bridge inverter that will be explained in a following section could adjust its output frequency in 0.01-MHz steps.

The coil-to-coil efficiency was measured using two different tests. The first measurement was performed using the E5061B and its real-time WPT coil analysis software that can simulate coil-to-coil efficiency in real-time with arbitrary load settings based on the measured S-parameters of the resonant coils [32]. Using the S-parameters of the coils, the input and output power, efficiency, and the input impedance of the coil-to-coil system were simulated in real-time. The simulation results are shown in Figure 9. The blue and red lines show the input and the output power of the coils when the input voltage of the transmitter coil is set to 20 V_{rms} and a virtual load of 6 Ω is connected to the receiver coil. As shown in the figure, the peak power transfer efficiency was approximately 77% when the operating frequency was 6.87 MHz with input and output powers of 84 W and 65 W, respectively. This result was a little lower than the FEA results in Figure 5b. This is because the measured ESRs of the coils were a little higher than the simulated results of the FEA as shown in Figure 7.

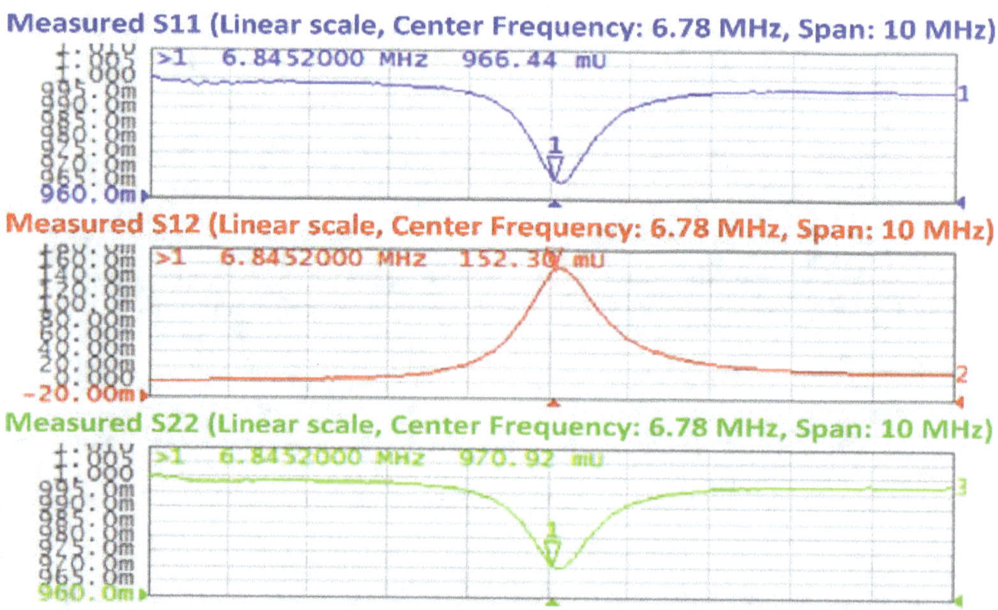

Figure 8. Measured S-parameters of the transmitter and receiver coils.

The coil-to-coil efficiency was evaluated experimentally as well. Using a variable frequency H-bridge inverter, the power transfer efficiency between the transmitter and the receiver coils was measured at various frequencies. In regard to the measurement of the coil-to-coil efficiency, an oscilloscope (HDO8108A, TeledyneLecroy, Chestnut Ridge, NY, USA) with a 12-bit ADC resolution, 10 GS/s sample rate, 1 GHz bandwidth, a differential voltage probe (HVD3605A, TeledyneLecroy, Chestnut Ridge, NY, USA) and a current probe (CP031, TeledyneLecroy, Chestnut Ridge, NY, USA) with a measurement bandwidth of 100 MHz (14 times higher than 6.87 MHz) were used. The measured efficiency is plotted in Figure 10. A peak power transfer efficiency occurred when the operating frequency was 6.86 MHz, which was very close to the result in Figure 9. The peak value was 77%, which was very similar to the real-time simulated results in Figure 9. These results demonstrate the validity of the coil design and fabrication methodologies described in Sections 3.1 and 3.3.

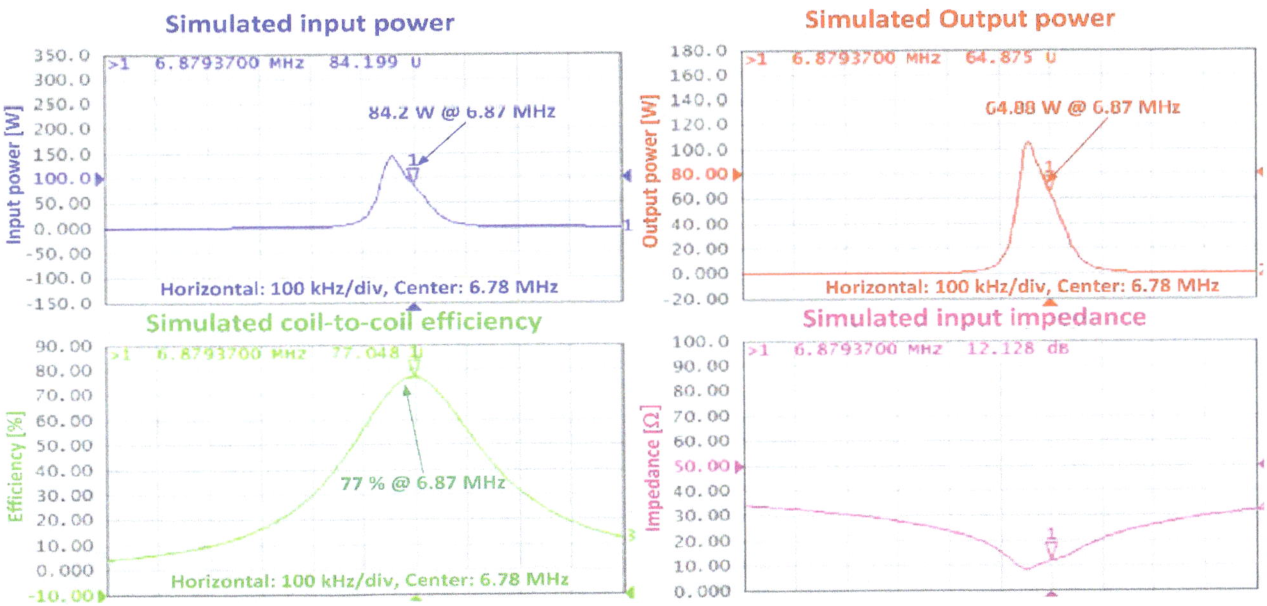

Figure 9. Real-time simulation results using Keysight's WPT analysis software (with 20 V_{rms} input voltage and 6 Ω load resistance setting).

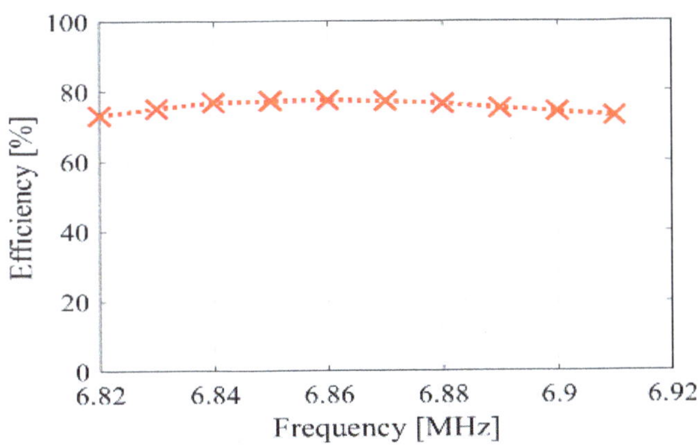

Measurement conditions:

Distance: 60 cm
Input voltage: 20 V_{rms}
Load: 6 Ω

Figure 10. Measured coil-to-coil efficiency with 6 Ω load resistor.

4. 6.78-MHz, 100-W, 92% Efficiency H-Bridge Inverter

In order to supply 50 W using the 77% efficient transmitter and receiver coils, the development of a high efficiency, 6.78-MHz, 100-W, H-bridge inverter was investigated. Four GS61004B, 100-V, 35-A, enhancement mode HEMT (High-electron-mobility transistor) GaN switches (GaN Systems, Ottawa, ON, Canada), and two TI's LM5113 half-bridge drivers (TI, Dallas, TX, USA) were used. A TI's 32-bit microcontroller, TMS320F28377S (TI, Dallas, TX, USA), was used as a gate signal generator and a power controller. The details of the developed inverter are presented in the following sections.

4.1. PCB Layout

After multiple failures in an attempt to develop a high-frequency inverter, the authors discovered that the PCB layout of the driver and GaN switches were the most important factors in determining its performance. Two features are of notable importance in the PCB design. The first is a parasitic inductance between the gate driver and the GaN switch. Due to the high operating frequency (6.78 MHz), parasitic inductances between the gate drivers and the switches form resonance circuits with the input capacitances of the GaN switches. This resulted in unwanted ringing of the gate drive signal. This ringing caused undesirable on/off and failures of the GaN switches.

The second aspect to be considered is a heat dissipation of the GaN switches to achieve high power generation (100 W) compared to the size of the switch package. The GS61004B switch was designed to sink heat to the bottom side of its package, especially through its source pad. Per its datasheet, the required PCB pattern under its source and drain pin is over 25 × 25 mm^2 for optimum heat dissipation. However, this large pad requirement leads to a large gap between the gate driver and the GaN switch and resulted in a large parasitic inductance between the driver and switch. Figure 11a shows the initial PCB layout of the inverter that focused on the heat dissipation of the four switches. In this initial design, the PCB pattern under the source pin of the switches was 25 mm × 25 mm. To avoid additional thermal burden on the PCB, the gate drivers (LM5113) were positioned 4 cm away from the GaN switches. However, this 4-cm separation caused significant ringing of the gate drive signal and the operation of the GaN switches was not stable even at 0.5 MHz.

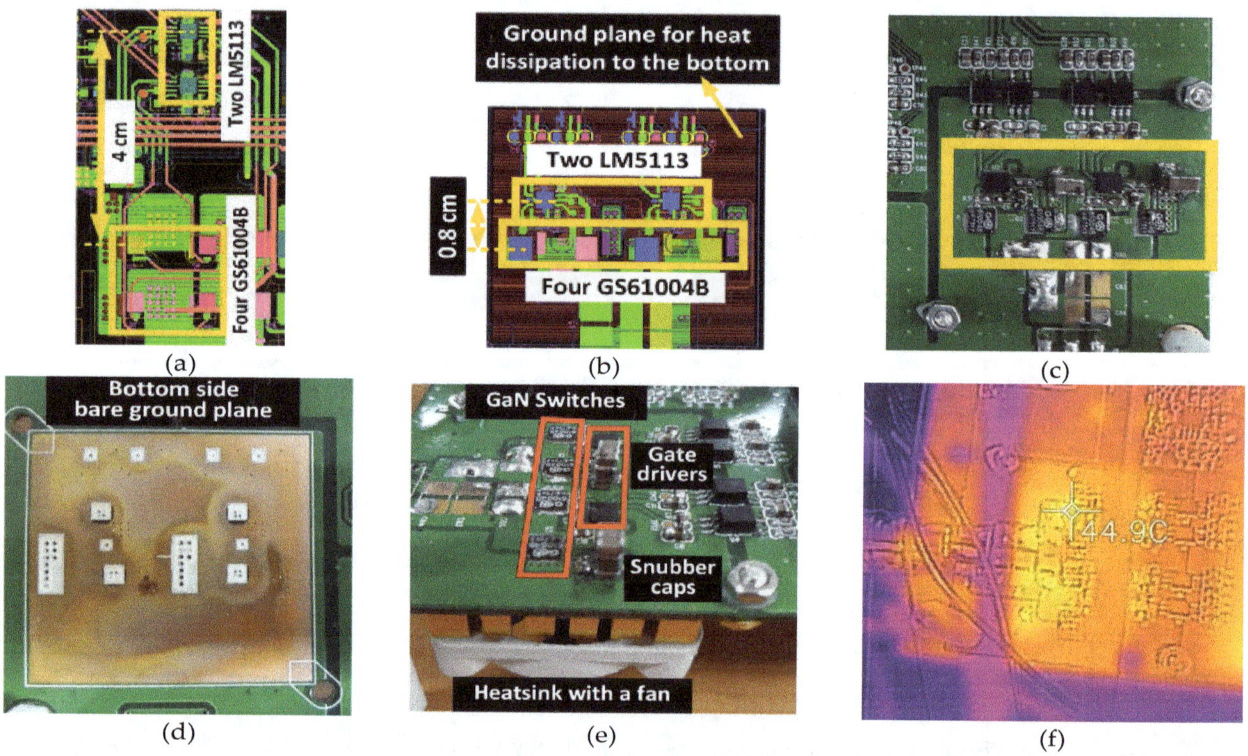

Figure 11. Configuration of the WPT system: (**a**) Initial design; (**b**) Revised design; (**c**) Fabricated inverter; (**d**) Bottom side ground plane; (**e**) Assembled heatsink and a fan; (**f**) Thermal image of the inverter at 81-W output.

Figure 11b shows a revised version of the inverter. The gate drivers were placed 0.8 cm above the switches and only 5 mm × 5 mm of the PCB pattern was allocated to the source pin of the high-side GaN switch. With this compact layout of the driver and switches, the gate signal ringing problem was eliminated. However, the revised version did not allocate sufficient pad area for heat dissipation of the switches. Instead of a large pad, a heat sink was installed at the bottom layer of the PCB in this revised version. The ground plane at the bottom layer of the PCB was uncovered as shown in Figure 11c. The generated heat from the switches and gate drivers was transferred to this bottom plane and cooled using a heat sink that was assembled with a fan (see Figure 11d). A thermal image of the fabricated inverter is shown in Figure 11e. The maximum temperature rise of the GaN switch's package was 15 °C compared to its initial temperature when the inverter output power was 81 W. Figure 12 shows a measured input and output voltages and currents of the inverter with 25 V_{rms} input voltage and 6 Ω load. The measured input power was 88 W and output power was 81 W. Therefore, the efficiency of the resonant inverter was 92%.

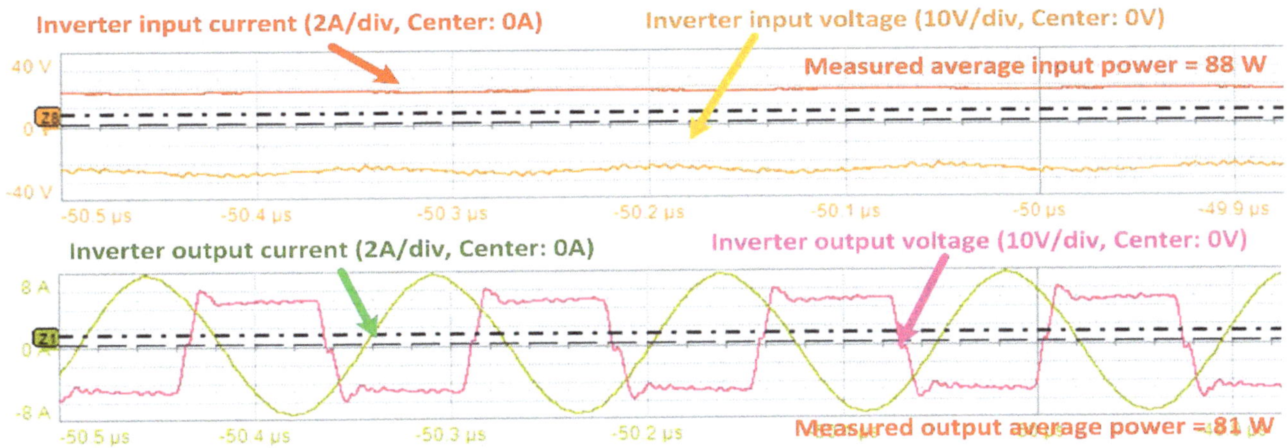

Figure 12. Input and output power of the H-bridge inverter. Measurement conditions: distance: 60 cm, input voltage: 25 V_{rms}, load: 6 Ω.

4.2. Switch Control

The gate drive signals of the inverter were digitally controlled using a TI's 200-MHz, 32-bit microcontroller unit (MCU), TMS320F28377S. The resolution of a PWM (Pulse width modulation) module of the MCU was 5 ns when the system clock was 200 MHz. The PWM module was able to generate pulses with periods of 145 ns (6.67 MHz) or 150 ns (6.89 MHz) with a single count change. PWM signals in the frequency range of 6.67 MHz to 6.89 MHz could not be generated. However, the target WPT system had very high Q coils and the transmitted power and efficiency were very sensitive to the operating frequency as shown in Figure 9. Therefore, the regular PWM module was not suitable for 6.78-MHz operation. Instead of the regular PWM module, a high-resolution pulse width modulation (HRPWM) module capable of enhancing the resolution of the PWM to 180 ps was used for fine adjustment of the PWM frequency. Using the HRPWM, it was possible to change the operating frequency in 0.01-MHz steps. Phase-shift control and frequency control of the inverter switches were easily implemented using the MCU. Figure 13 shows a photo of the entire assembled inverter board.

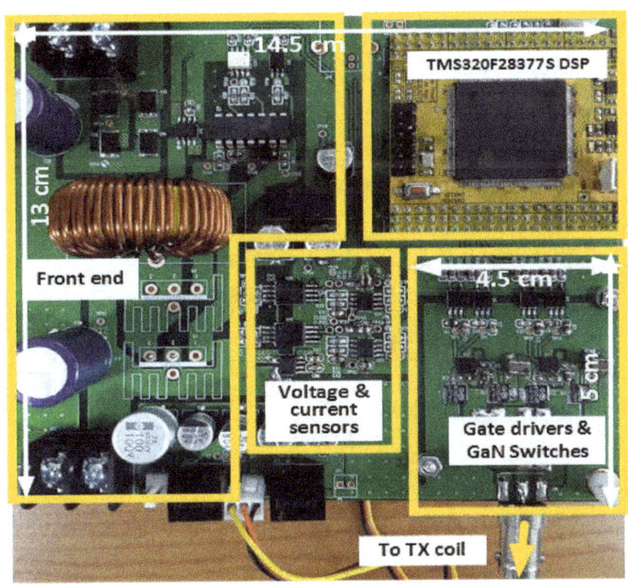

Figure 13. A photo of the assembled H-bridge inverter.

5. Receiver Side Full-Wave Rectifier Design

The series-tuned receiver coil was coupled to a full-bridge rectifier. The rectifier diodes were 100-V, 8-A, Schottky barrier diodes (PMEG100V080ELPD from NXP, Eindhoven, Netherlands) with a 10-ns reverse recovery time when the forward current is 0.5 A. To minimize unwanted parasitic inductances and capacitances that can alter the tuning frequency of the RX coil, four diodes were placed very close to the BNC terminals. Two low-side diodes were connected on top of the PCB and high-side diodes were connected at the bottom of the PCB as shown in Figure 14a. Multiple ceramic capacitors and two polymer aluminum capacitors were used as a filter. The diodes were designed for heat dissipation through its cathode pad. Therefore, a heat sink was installed at the rectifier's positive DC plane as shown in Figure 14b.

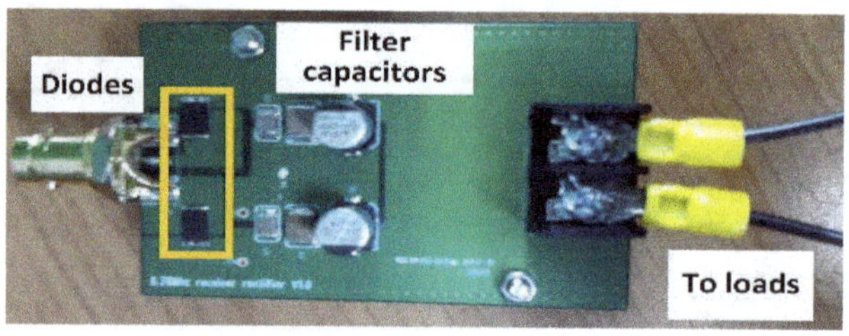

(a)

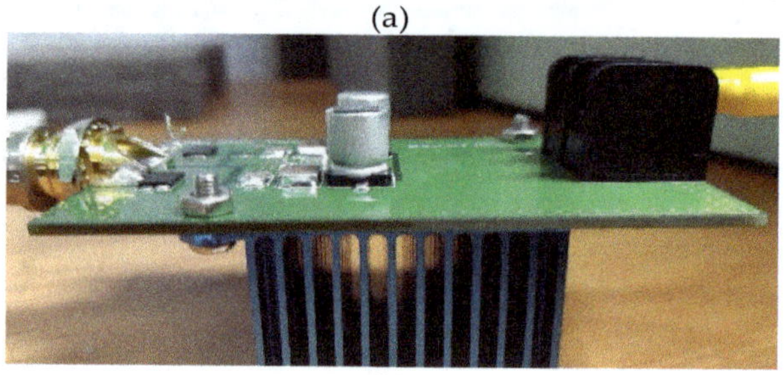

(b)

Figure 14. Photos of full-wave rectifier: (**a**) Top-view (**b**) Side-view.

6. Experimental Evaluation

6.1. Experimental Setup

Figure 15 shows a configuration of the test-bed. The H-bridge inverter was connected to the transmitter resonant tank through a coaxial cable (16 AWG, American wire guage) with a BNC termination. Also, the receiver coil was connected to the diode rectifier with the same coaxial cable and termination. A Chroma's DC electronic load, 63102 (Chroma, Foothill Ranch, CA, USA) was used with a constant resistor mode setting.

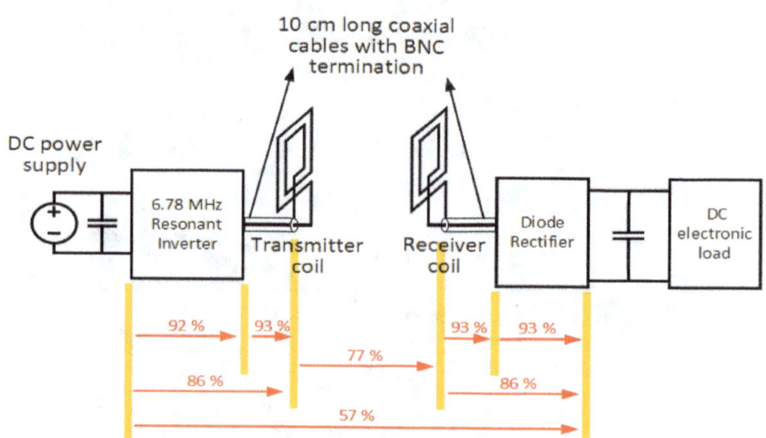

Figure 15. Configuration of the test-bed.

Figure 16 shows a photo of the entire test-bed. Instead of the DC electronic load, two 20-W, 12-V DC light bulbs were connected to the rectifier's output terminal to demonstrate wireless energy transmission over 60 cm. Using this test-bed, the power transfer efficiency at various distances, load power, and operating frequency was measured.

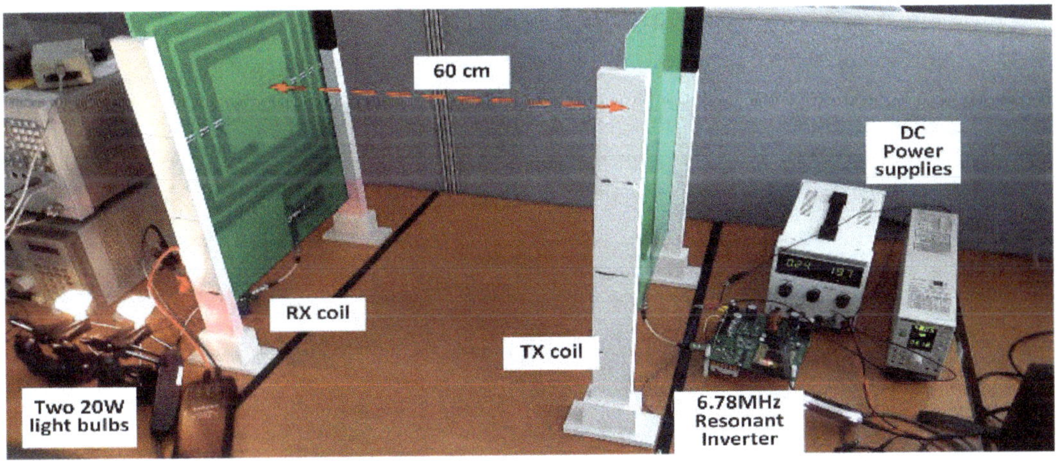

Figure 16. A photo of the experimental setup.

6.2. Efficiency Analysis of the Developed System

Figure 17 shows the measured input DC voltage and current of the H-bridge inverter and output DC voltage and current of the rectifier when the load resistance was 6 Ω. Because the efficiency was maximized when the operating frequency was 6.86 MHz, the switching frequency of the inverter was 6.86 MHz. The measured output power was 47.1 W while the measured input power was 82.3 W. Therefore, the measured DC-to-DC power transfer efficiency was 57%. This result was relatively low compared to the measured efficiency between the coils which was 77% at its rated operation.

For a loss analysis of the developed system, the efficiency of every component of the system was measured and denoted in Figure 15. It was found that additional losses were coming from the 16 AWG coaxial cables. This is a notable point since a coaxial cable of only 10 cm in length had a comparable power transfer efficiency to the inverter and the rectifier. Although the cross-sectional area of the coaxial cable (1.3 mm^2) was sufficient to carry a 5-A_{rms} transmitter current at low frequencies, it was not appropriate to carry the same current at 6.78 MHz.

Figure 18 shows the estimated loss percentages of the developed system. The conduction losses of the H-bridge inverter and the RX side rectifier were calculated using $R_{DS(on)}$ (37 mΩ) and the forward voltage drop of the rectifying diodes based on their datasheet. The switching loss of the inverter was calculated by subtracting the conduction loss from the total loss associated with the inverter. The losses of the transmitter and receiver coils were calculated using their measured resistance and measured root-mean-square (RMS) of the currents of the coils. It should be noted that the losses of the transmitter and receiver coils overwhelmed other losses. The switching loss of the inverter was second largest loss in the developed system. Therefore, the coil design and fabrication method were key in improving the efficiency of this loosely-coupled WPT system. To evaluate the effectiveness of the developed system on large distance power transmission, efficiency values were measured at various distances of the transmitter and the receiver coils. The distance was increased from 60 cm to 80 cm in 5-cm steps.

Figure 19 shows the measured efficiencies when the load was 6 Ω, the input power was 80 W, and the operating frequency was 6.86 MHz. When the distance was 60 cm, the efficiency was 57%. However, this value monotonically decreased to 29% when the distance was 80 cm. It is notable that the total efficiency is reduced by approximately 7% for every 5-cm increase in the distance of the coils. This shows that the efficiency of the proposed system is sensitive to the distance of the coils. For this outdoor large air-gap WPT system, appropriate fixtures and installation guidelines are required to prevent a significant degradation of its performance due to wind and other severe weather conditions. The impact of the operating frequency on the efficiency of the test-bed was also evaluated as shown in Figure 20. The measured overall efficiency was also maximized when the operating frequency was 6.86 MHz. The efficiency decreased as the operating frequency moved away from the tuned frequency. The efficiency was decreased to 38% when the operating frequency was 6.94 MHz. This is 0.08 MHz higher than the tuned frequency. This result demonstrates that fine control of the inverter operating frequency is a critical factor of power transfer efficiency.

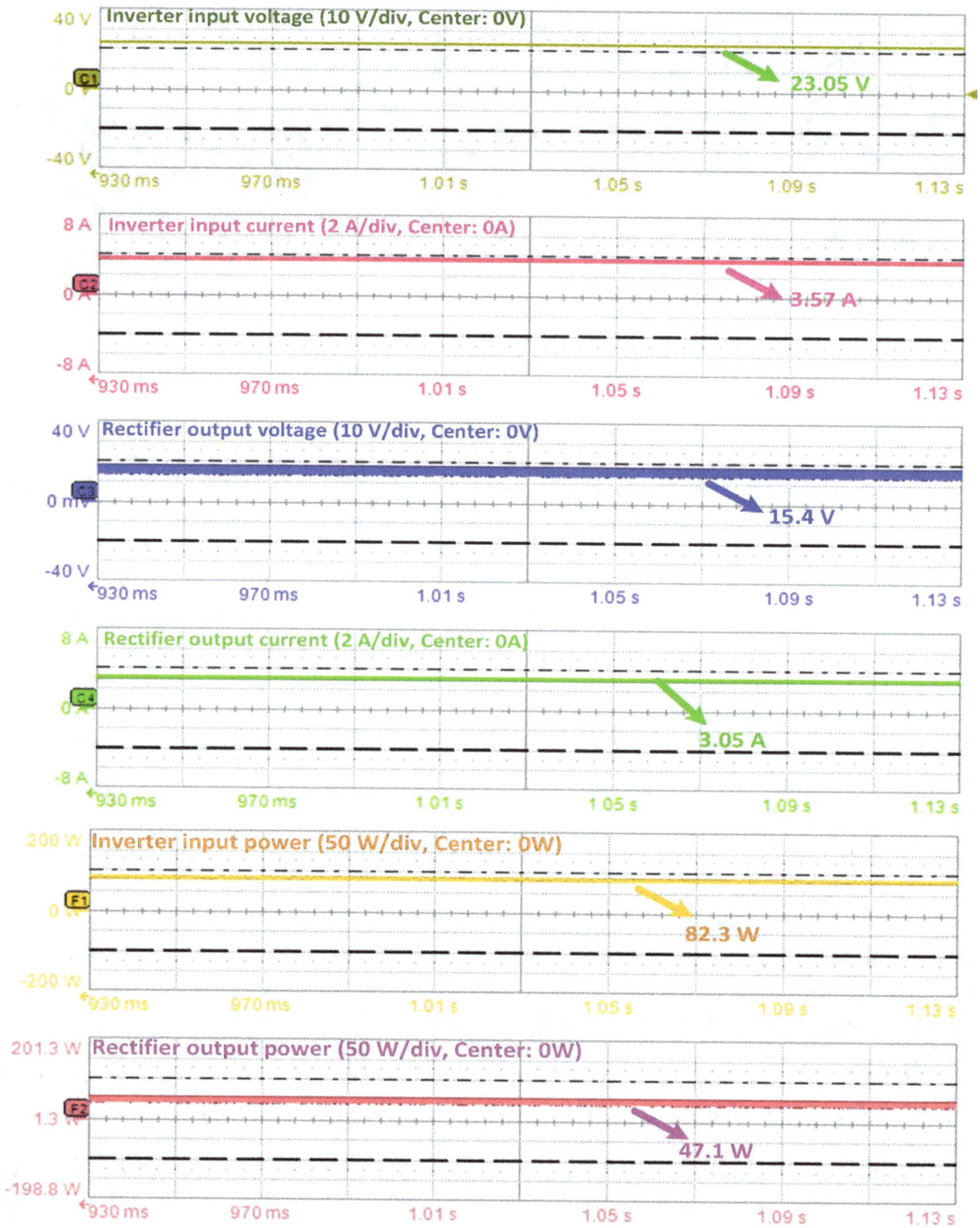

Figure 17. Measured input and output DC voltages and currents at the rated output power. Measurement conditions: distance: 60 cm, operating frequency = 6.86 MHz, load: 6 Ω.

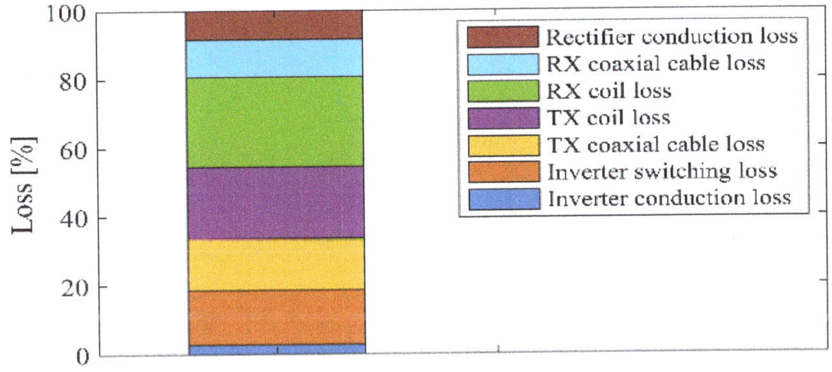

Figure 18. Estimated percentage losses of the developed system.

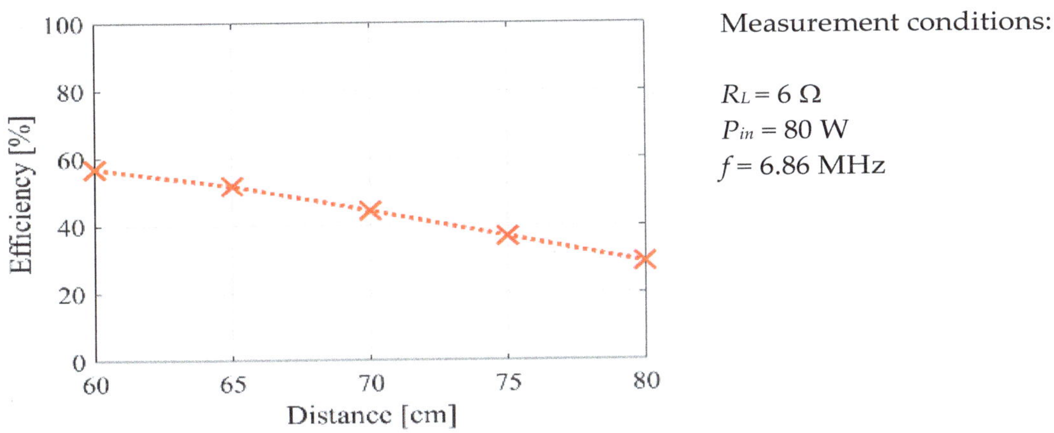

Figure 19. Efficiency dependent on distance.

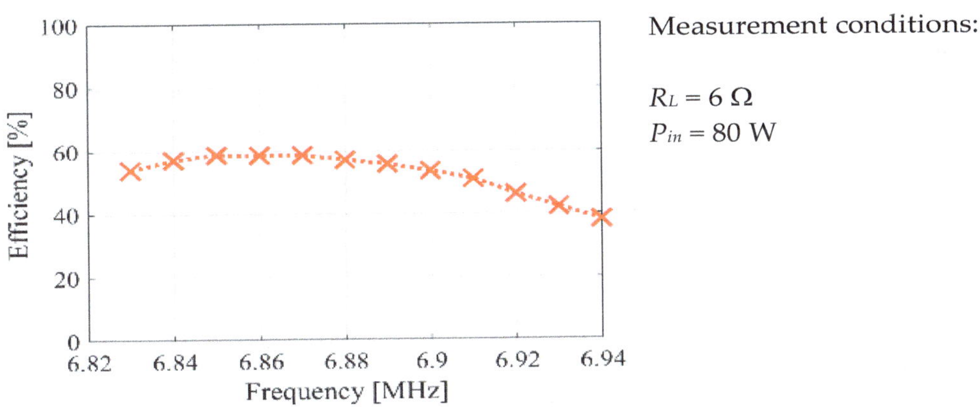

Figure 20. Efficiency dependence on frequency.

Finally, the impact of load resistance on the efficiency of the system was evaluated. Figure 21 shows the result of an efficiency vs. load resistance plot. When the load resistance has a rated value $R_L = 6\ \Omega$, the power transfer efficiency is maximized to 57%. As the load resistance increased or decreased, the power transfer efficiency decreased, and the efficiency is only 35% when the load resistance is 30 Ω. This plot shows that there is an optimal load resistance for a maximum power transmission.

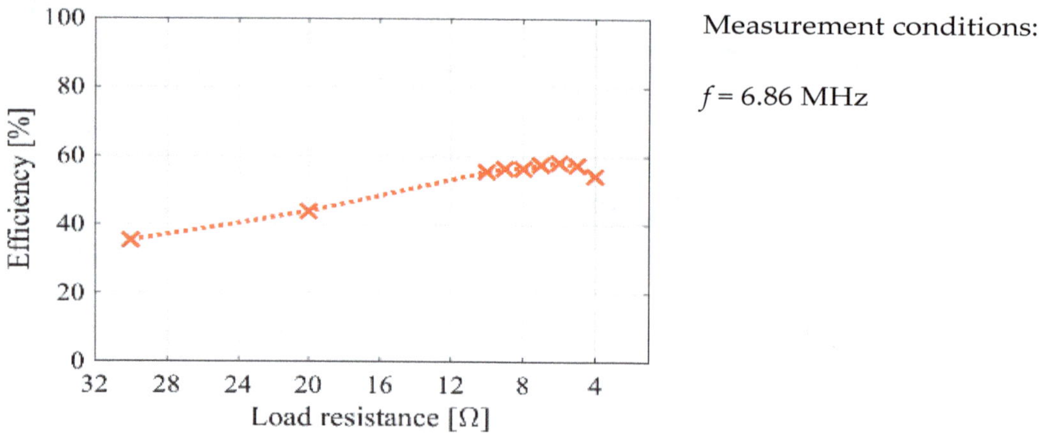

Figure 21. Efficiency dependence on load resistance.

7. Conclusions

In this paper, a new 6.78-MHz, 50-W, over 60 cm wireless power transfer system is proposed. The design and fabrication methodologies for the coils, a GaN-based 6.78-MHz H-bridge inverter, and a rectifier were explained. Using a multi-objective optimization method, Pareto optimality, the transmitter and the receiver coil were determined to a 4-turn and a 5-turn, 40 cm × 40 cm square coils. The selected transmitter and receiver coils were fabricated on a PCB. It was shown that the dielectric loss of the PCB substrate can dominate Ohmic loss of copper traces at MHz frequencies. A PCB substrate with very low dissipation factor was used to reduce the dielectric loss of the coils. The fabricated coils had over 350 Q-factors at 6.78 MHz. The peak coil-to-coil efficiency for the WPT system was 77% at 60 cm air-gap.

In addition, a 6.78-MHz, 100-W, 92% efficiency, GaN-based H-bridge inverter was developed. Two important points were considered in the development of the inverter: a low parasitic inductance of the gate drive circuit and a good heat dissipation of the GaN switch at its rated output power. Parasitic inductances between the gate drivers and GaN switches caused significant ringing of the on/off signals and poor operation performances even at sub-MHz frequencies. A close positioning of the gate driver to the switch was the key for 6.78-MHz operation. However, the congregating of the gate drivers and switches was inferior in cooling and heat dissipation of the inverter. To resolve the thermal issues, the ground plane at the bottom of the PCB was uncovered, and a heatsink with a fan was attached to the PCB. The fabricated 6.78-MHz resonant inverter had a 92% efficiency and the peak temperature was lower than 45 °C at its rated output. Using a high-resolution PWM module of the TI's MCU, the switching signals could be changed in 0.01-MHz steps. This was very helpful in fine tuning of the high Q-factor coils. Using a full-wave rectifier that was built using four Schottky barrier diodes, the induced voltage and current of the receiver coil were converted to DC values.

The measured DC-to-DC efficiency of the developed WPT system was 57% when the output power was 47 W. The efficiencies of individual components of the system were measured and lossy parts were identified. The impacts of the coil distance, operating frequency, and load resistance on the power transfer efficiency were also evaluated. As the distance increased from 60 cm to 80 cm, the efficiency decreased from 57% to 29% (about a half of the initial value). Therefore, the distance was important in the efficiency of the WPT system. Also, the efficiency decreased from 57% to 38% when the operating frequency was increased from 6.86 MHz to 6.94 MHz (a change of only 0.08 MHz). Finally, the dependence of efficiency on the load resistance was tested. It was shown that there was an optimal load value that maximizes the efficiency, and this value was determined to be 6 Ω, which corresponds to the target-rated load resistance of the developed system.

Author Contributions: Conceptualization, S.-H.L. and M.-Y.K.; methodology, S.-H.L. and M.-Y.K.; software analysis, K.-P.Y.; validation, S.-H.L., K.-P.Y., and M.-Y.K..; formal analysis, S.-H.L.; investigation, M.-Y.K., writing—original draft preparation, S.-H.L.; writing—review and editing, K.-P.Y. and M.-Y.K.; supervision, M.-Y.K..; project administration, S.-H.L. and K.-P.Y.; funding acquisition, S.-H.L. and M.-Y.K..

References

1. RamRakhyani, A.K.; Mirabbasi, S.; Chiao, M. Design and Optimization of Resonance-Based Efficient Wireless Power Delivery Systems for Biomedical Implants. *IEEE Trans. Biomed. Circuits Syst.* **2011**, *5*, 48–63. [CrossRef] [PubMed]
2. Salim, A.; Baldi, A.; Ziaie, B. Inductive link modeling and design guidelines for optimum power transfer in implantable wireless microsystems. In Proceedings of the 25th Annual International Conference of the IEEE Engineering in Medicine and Biology Society (IEEE Cat. No. 03CH37439), Cancun, Mexico, 17–21 September 2003; Volume 4, pp. 3368–3371.
3. Lenaerts, B.; Puers, R. An inductive power link for a wireless endoscope. *Biosens. Bioelectron.* **2007**, *22*, 1390–1395. [CrossRef] [PubMed]
4. Ko, Y.D.; Jang, Y.J. The optimal system design of the online electric vehicle utilizing wireless power transmission technology. *IEEE Trans. Intell. Transp. Syst.* **2013**, *14*, 1255–1265. [CrossRef]
5. Kim, J.H.; Lee, B.S.; Lee, J.H.; Lee, S.H.; Park, C.B.; Jung, S.M.; Lee, S.G.; Yi, K.P.; Baek, J. Development of 1-MW Inductive Power Transfer System for a High-Speed Train. *IEEE Trans. Ind. Electron.* **2015**, *62*, 6242–6250. [CrossRef]
6. Guidi, G.; Suul, J.A.; Jenset, F.; Sorfonn, I. Wireless Charging for Ships: High-Power Inductive Charging for Battery Electric and Plug-In Hybrid Vessels. *IEEE Electrif. Mag.* **2017**, *5*, 22–32. [CrossRef]
7. Lee, S.H.; Kim, J.H.; Lee, J.H. Development of a 60 kHz, 180 kW, over 85% efficiency inductive power transfer system for a tram. *Energies* **2016**, *9*, 1075. [CrossRef]
8. Wang, Z.; Wei, X.; Dai, H. Design and Control of a 3 kW Wireless Power Transfer System for Electric Vehicles. *Energies* **2015**, *9*, 10. [CrossRef]
9. Park, Y.; Cho, Y.H.; Lee, K.; Jung, H.; Kim, H.; Kwon, S.; Park, H. Development of an FPGA-based Online Condition Monitoring System for Railway Catenary Application. In Proceedings of the World Congress on Railway Research (WCRR), Seoul, Korea, 18–22 May 2008; p. 143.
10. Korea Rail Network Authority. *Railway Design Guidelines and Handbooks: Guidace on Overhead Line Clearances (KR E-03140)*; Korea Rail Network Authority: Daejeon, Korea, 2012.
11. Lee, S.-H.; Lee, J.-H.; Yi, K.-P. A New Design Methodology for a 1 Meter Distance, 6.78 MHz Wireless Power Supply System for Telemetries. In Proceedings of the IEEE Energy Conversion Congress and Exposition, ECCE 2016, Milwaukee, WI, USA, 18–22 September 2016.
12. Lee, S.-H.; Yi, K.-P. Development of 50 W High Quality Factor Printed Circuit Board Coils for a 6.78 MHz, 60 cm Air-gap Wireless Power Transfer System. *J. Korean Soc. Railw.* **2016**, *19*, 468–479. [CrossRef]
13. Lee, S.-H. Design Methodologies for Low Flux Density, High Efficiency, kW Level Wireless Power Transfer Systems with Large Air Gaps. Ph.D. Dissertation, University of Wisconsin-Madison, Madison, WI, USA, 2013.
14. AirFuelAllience. *AirFuel Resonant Wireless Power Transfer (WPT) System Baseline System Specifications*; AirFuelAllience: Beaverton, OR, USA, 2017.
15. Sample, A.P.; Meyer, D.A.; Smith, J.R.; Member, S.; Meyer, D.A.; Smith, J.R. Analysis, Experimental Results, and Range Adaptation of Magnetically Coupled Resonators for Wireless Power Transfer. *IEEE Trans. Ind. Electron.* **2011**, *58*, 544–554. [CrossRef]
16. Duong, T.P.; Lee, J.-W. Experimental Results of High-Efficiency Resonant Coupling Wireless Power Transfer Using a Variable Coupling Method. *IEEE Microw. Wirel. Compon. Lett.* **2011**, *21*, 442–444. [CrossRef]
17. Lee, S.-H.; Lorenz, R.D. A Design Methodology for Multi-kW, Large Airgap, MHz Frequency, Wireless Power Transfer Systems. In Proceedings of the IEEE Energy Conversion Congress and Exposition: Energy Conversion Innovation for a Clean Energy Future (ECCE 2011), Phoenix, AZ, USA, 17–22 September 2011.
18. Calder, R.J.; Lee, S.H.; Lorenz, R.D. Efficient, MHz Frequency, Resonant Converter for Sub-Meter (30 cm) Distance Wireless Power Transfer. In Proceedings of the 2013 IEEE Energy Conversion Congress and Exposition, ECCE 2013, Denver, CO, USA, 15–19 September 2013; pp. 1917–1924.
19. Akuzawa, Y.; Tsuji, K.; Matsumori, H.; Ito, Y.; Ezoe, T.; Sakai, K. A 95% Efficient Inverter with 300-W Power Output for 6.78-MHz Magnetic Resonant Wireless Power Transfer System. In Proceedings of the 2015 IEEE MTT-S International Microwave Symposium, Phoenix, AZ, USA, 17–22 May 2015; pp. 60–62.
20. Sakai, K.; Ito, Y.; Ezoe, T.; Akuzawa, Y. A 99%-Efficiency GaN Converter for 6.78 MHz Magnetic Resonant Wireless Power Transfer System. *J. Eng.* **2014**, *2014*, 598–600.
21. Yang, J.-R.; Kim, J.; Park, Y.-J. Class E Power Amplifiers using High-Q Inductors for Loosely Coupled Wireless Power Transfer System. *J. Electr. Eng. Technol.* **2014**, *9*, 569–575. [CrossRef]

22. Narayanamoorthi, R.; Juliet, A.V.; Chokkalingam, B.; Padmanaban, S.; Leonowicz, Z.M. Class E Power Amplifier Design and Optimization for the Capacitive Coupled Wireless Power Transfer System in Biomedical Implants. *Energies* **2017**, *10*, 1409.
23. Grebennikov, A. Load Network Design Techniques for Class E RF and Microwave Amplifier. *High Freq. Electron.* **2004**, *3*, 18–32.
24. Zhao, C.; Costinett, D. GaN-Based Dual-Mode Wireless Power Transfer Using Multifrequency Programmed. *IEEE Trans. Ind. Electron.* **2017**, *64*, 9165–9176. [CrossRef]
25. Bonache-Samaniego, R.; Olalla, C.; Martinez-Salamero, L.; Maksimovic, D. 6.78 MHz Self-Oscillating Parallel Resonant Converter Based on GaN Technology. In Proceedings of the 2017 IEEE Applied Power Electronics Conference and Exposition (APEC), Tampa, FL, USA, 26–30 March 2017; pp. 1594–1599.
26. Xue, L.; Zhang, J. Single-stage 6.78 MHz Power-Amplifier Design using High-Voltage GaN Power ICs for Wireless Charging Applications. In Proceedings of the 2017 IEEE Applied Power Electronics Conference and Exposition (APEC), Tampa, FL, USA, 26–30 March 2017; pp. 3743–3750.
27. Ngatchou, P.; Zarei, A.; El-Sharkawi, A.; El-Sharkawi, M.A. Pareto Multi Objective Optimization. In Proceedings of the 13th International Conference on Intelligent Systems Application to Power Systems, Arlington, VA, USA, 6–10 November 2005; pp. 84–91.
28. Lee, S.H.; Lee, B.S.; Lee, J.H. A New Design Methodology for A 300-Kw, Low Flux Density, Large Air Gap, Online Wireless Power Transfer System. *IEEE Trans. Ind. Appl.* **2016**, *52*, 4234–4242. [CrossRef]
29. Fotopoulou, K.; Flynn, B.W. Wireless Power Transfer in Loosely Coupled Links: Coil Misalignment Model. *IEEE Trans. Magn.* **2011**, *47*, 416–430. [CrossRef]
30. Isola IS680. Available online: http://www.isola-group.com/products/all-printed-circuit-materials/is680/ (accessed on 3 August 2018).
31. Mohan, S.S.; Hershenson, M.D.M.; Boyd, S.P.; Lee, T.H. Simple Accurate Expressions for Planar Spiral Inductances. *IEEE J. Solid-State Circuits* **1999**, *34*, 1419–1420. [CrossRef]
32. Keysight E5061B-006 Wireless Power Transfer Analysis. Available online: https://www.keysight.com/en/pd-2571977-pn-E5061B/wireless-power-transfer-analysis?cc=US&lc=eng (accessed on 19 August 2018).

12

Modeling of Magnetic Elements Including Losses—Application to Variable Inductor †

Sarah Saeed *, **Ramy Georgious** and **Jorge Garcia**

LEMUR Research Group, Department of Electrical, Electronics, Computers and Systems Engineering, University of Oviedo, 33204 Gijon, Spain; georgiousramy@uniovi.es (R.G.); garciajorge@uniovi.es (J.G.)

* Correspondence: saeedsarah@uniovi.es

† This paper is an extended version of our paper published in 2018 IEEE Applied Power Electronics Conference and Exposition (APEC), San Antonio, TX, USA, 4–8 March 2018, pp. 1750–1755.

Abstract: This paper proposes and develops a circuit-based model aiming to simulate variable magnetic power elements in power electronic converters. The derived model represents the magnetic element by a reluctance-based equivalent circuit. The model takes into consideration device core losses, with the main emphasis given to hysteresis losses, which are modeled using the Jiles-Atherton model. The core loss model is further validated on different ferromagnetic materials to prove its range of applicability. The winding losses of the magnetic device are also taken into consideration, which are obtained using Dowell empirical formulas. In addition, the frequency dependence of the device losses is also considered. The proposed modeling procedure has been applied to study and characterize a double E-core variable power inductor structure in a 1 kW SiC full bridge DC-DC converter. The procedure has been verified by comparing the simulation results to the experimental measurements, confirming the validity and accuracy of the full circuit-based model.

Keywords: magnetics modeling; variable inductor; hysteresis; eddy currents; saturable core

---

1. Introduction

Understanding the behavior of a magnetic device in a Power Electronic Converter (PEC) is essential to optimize the design and to foster the performance of the whole system. Variable magnetic elements allow for additional degrees of freedom in the design and control of PECs. This is particularly useful in resonant converters where the usual frequency control has some drawbacks due to Electro-Magnetic Interference (EMI) issues, synchronization, variable sampling time, etc., especially for a large range of variation. If variable magnetics are used, the same control margins can be obtained at a constant switching frequency, therefore allowing for an optimization of the EMI filters and sampling procedures. In other applications, such as the Dual-Active-Bridge (DAB) converter, in addition to adding a new degree of freedom to the control, the inclusion of variable magnetics can increase operation parameters, such as the soft switching margins [1–3].

The recent growing applications of variable magnetic elements have implied the need for developing accurate models to define the magnetic device behavior. The magnetic core as well as the device windings must be characterized to achieve an accurate device model.

Some models define the magnetic core material in terms of a relationship between magnetic flux density and field intensity referred to as a hysteresis curve. In [4], an initial survey has been conducted classifying the existing magnetic material models, according to different frequencies, bias conditions, and temperatures of interest. It aims to provide comparable information for models and their availability in some circuit simulators. More recently, a literature review on the fundamentals, modeling, and design of magnetic regulators has been comprehensively presented in [5]. After a careful review of these and other references, the modeling methods are confined to analytical and numerical methods. Specific to variable magnetic devices, modeling strategies are confined to three directions: Finite Elements Analysis (FEA), gyrator-capacitor model [6], and reluctance equivalent circuit [7]. The FEA model is based on the numerical method, while the gyrator-capacitor model and the reluctance model are based on the analytical method.

As the complexity of magnetic devices increases, the analytical method becomes too complicated to predict the behavior of the device in a simple and practical manner. Therefore, incorporating those concepts in a computer-based simulation provides a good compromise between convenience, accuracy, and numerical efficiency. On the other hand, developing such simulations enables real-time applications on the modeling at the converter controller level, which can provide an on-line calculation of model parameters for real-time control [8]. Consequently, many efforts have been directed towards computer-based simulations, especially time-domain models [9,10]. Although the analytical methods of calculation are generally known and can be implemented in simulation models in a straight-forward manner, the selection of the suitable methods is critical from the circuit simulation perspective. The computation methods are expected to achieve good convergence with an acceptable compromise of accuracy to the time of processing the simulation results.

Henceforth, the aim of this work is to develop a circuit-based time-domain model of the variable magnetic element. The proposed model includes the device losses, mainly core and winding losses. Also, this circuital model is able to work in different platforms, for e.g., LTSpice® (Linear Technology Corporation, Milpitas, CA, USA) [11], MATLAB-Simulink® (MathWorks, Natick, MA, USA) [12], and PSIM® (Powersim Inc., Rockville, MD, USA) [13], with equally valid accuracy. Thereby, the whole electromagnetic system design and simulation can be carried out using only one simulator environment. This provides an acceptable accuracy in compromise with the complication and time required for FEA models. Consequently, it allows the investigation of the overall PEC performance incorporating the variable magnetic device. Section 2 presents an overview of the variable magnetic device structure that will be used, together with the models of interest in the literature. In Section 3, the magnetic core losses are studied and the method used to model those losses is presented. This section also provides an idea on the implementation of the model equations, the validation of the model against experimental measurement, and the approach to estimate the model parameters as a function of the operation frequency. Later, in Section 4, the model of the winding losses is presented, and validated against experimental results. After that, Section 5 explains the use of the loss models to implement the full device model. In Section 6, the proposed simulation model is validated in comparison to the previous models that does not include losses, and experimental results are provided. Finally, Section 7 summarizes the conclusions of the work.

2. Modeling of Variable Magnetic Elements

From the study of the state of the art, the double E-core structure, depicted in Figure 1, is selected to be the most appropriate for the implementation of the variable inductor in this study and the most comprehended in literature [14]. The basic principle of operation of a double E-core variable inductor is described in this section.

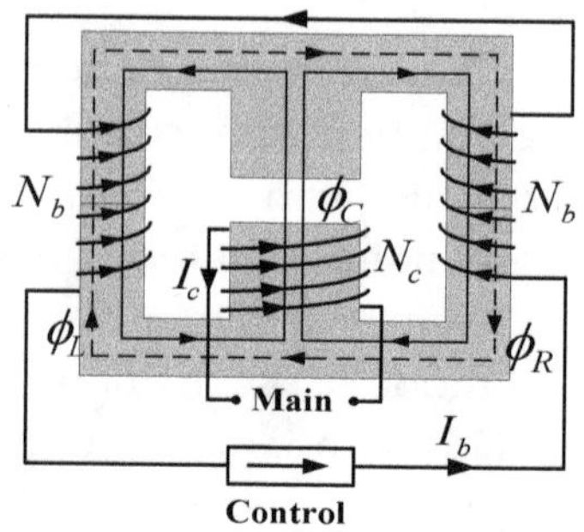

Figure 1. Variable inductor based on a double E-core structure.

Due to the current (I_c) flowing through the main winding (N_c) of the inductor, as clarified in Figure 1, an AC flux (ϕ_C) circulates through the center arm of the typical E-core structure of the magnetic core and splits to the outer arms. Applying a relatively small DC current (I_b) to the bias control windings (N_b), a DC flux (ϕ_R or ϕ_L) is produced, which circulates mainly through the outer (ungapped) closed path of the core [14]. This DC flux can bias the operation of the magnetic material towards the nonlinear region on the $B(H)$ curve, thus causing the inductance seen from the main winding terminals to vary as a function of the DC bias current, Equation (1).

$$L = f(I_b), \tag{1}$$

In order to model the variable magnetic device, the operation regions of the magnetic material on the $B(H)$ curve must be considered. These are divided into three areas: the linear region, the saturation knee, and the non-linear region [9].

In addition, to attain accurate results in the modeling procedure, the different involved device losses are included. There are two main types of losses associated to the magnetic device; core losses and winding losses. For MnZn-ferrite materials, the core losses are composed of three fractions. A fraction of the losses is related to the crystal composition, which is called hysteresis losses. Another fraction is related with the structural form of the core, which is called eddy current losses. As the frequency increases, the eddy currents dominate, and for a frequency range of approximately 1 MHz, those losses can contribute with more than 50% of the total power losses. There is also a third fraction known as resonance losses [15]. This fraction is related to the ferrimagnetic resonance of the material and is pertinent to both the crystal and microstructure.

On the other hand, the winding losses depend on the frequency of operation of the magnetic device. At DC operation, the winding losses are due to the resistance of the copper wire. However, as the frequency of operation increases, two effects take place, which are the skin and the proximity effects. Those effects cause the wire resistance to vary, and this modified resistance is referred to as the AC (or high-frequency) winding resistance.

Although ready-made modules are available in many circuit simulation environments [12,13], it is not considered comprehensive to study the magnetic device. The SPICE-based reluctance equivalent circuit provided by authors in [9] is, thus, found to be of more interest. Figure 2 shows the reluctance equivalent circuit for the double E-core variable inductor.

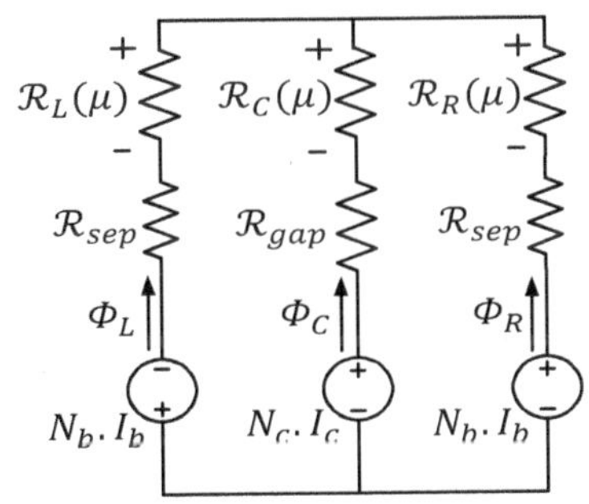

Figure 2. Reluctance equivalent circuit of the double E-core variable inductor shown in Figure 1.

The voltage sources in the circuit represent the magnetomotive forces due to the bias control windings ($N_b \cdot I_b$), as well as the main winding ($N_c \cdot I_c$). $\Re_{gap}$ is a constant reluctance that represents the air gap, while $\Re_{sep}$ is a constant reluctance that represents the air separation between the two E-cores due to non-idealities in the manufacturing process. The components $\Re_L$, $\Re_C$, and $\Re_R$ are the reluctances of the magnetic paths of the left, center, and right arms, respectively. These reluctance values are represented as a function of the permeability of the magnetic material (μ), and it must be noticed that, unlike the usual case, in variable magnetic elements, these permeability values can vary depending on the operation point of the magnetic core material on the characteristic $B(H)$ curve. Thus, for calculating these reluctances, the referred model uses Brauer's equation [16], which defines the $B(H)$ characteristic curve of the magnetic material neglecting the hysteresis effect, as stated by Equation (2).

$$H(B) = (k_1 e^{k_2 B^2} + k_3)B\,, \tag{2}$$

where k_1, k_2, and k_3 are constants that depend on the considered magnetic material. For the present work, the afore-mentioned SPICE-based model has been replicated, specifically, in the MATLAB-Simulink® platform. This has been carried out in order to take advantage of the feature of this environment to integrate MATLAB script with existing Simulink® library tools, aiming to combine complex, accurate simulations with digital processing of information. This allows including the device design calculations into the overall model of the system in one integrated environment.

Two key limitations are found in this model, which restrict its applicability range and accuracy. Firstly, the hysteresis effect is not taken into consideration. Secondly, the model input quantities are dependent on the output ones, which introduces difficulty in the implementation of the computations. In particular, the latter issue implies the necessity for implementing system calculation delays, with an adequate initialization of parameters, especially when including the device in a switching converter simulation. This, furthermore, complicates the simulation in different test platforms. In Simulink®, for instance, the solver applies a numerical method to solve the set of ordinary differential equations that represent the model; therefore, the model causality must be decided [17].

In this paper, the model of the variable inductor has been extended to include device losses. The main loss components that have been taken into consideration are the hysteresis losses of the magnetic core and the winding losses. The following sections will justify the studied loss components and discuss the implementation and validation of the full model [18].

It is worth noting that, throughout the discussion hereafter, the vectorial nature of the magnetic quantities are disregarded to reduce the analysis to a simple unidimensional statement of equations by assuming: (1) specific symmetrical magnetic core geometries, of which geometrical references and paths are well-defined, and (2) the homogeneous nature of typical magnetic core materials together with the uniform distribution of the magnetic core properties. Such assumptions are often used in the analysis and design of magnetic devices for power electronics converters.

3. Model of Core Losses

The core losses in a magnetic core are due to two phenomena: the hysteresis loss, and the eddy currents loss. The hysteresis loss is due to energy required to rotate the magnetic domains when aligning with the applied magnetic field. On the other hand, the eddy current losses are due to currents induced in the magnetic core, which opposes the changing flux in the core. Previous studies in literature [19] have shown that for a ferrite magnetic material operating in a range of frequency up to 100 kHz, the eddy current losses are a very small part of the total core losses. Therefore, for the range of frequencies under study herein, the hysteresis losses will always be dominant. For this reason, the eddy current losses in the magnetic core will be neglected in this study for the sake of simplicity.

There are several methods to calculate the hysteresis losses in a ferro-magnetic material, which were grouped by the authors in [20] to be three main approaches: hysteresis models, empirical equations, and loss separation. This paper undertakes the first approach, specifically the Jiles-Atherton (JA) hysteresis model [21]. The main strengths of the JA model compared to its counterpart approaches are: being the most suitable for development from a circuital simulation perspective, besides having good convergence and acceptable accuracy among a variety of materials and operation conditions [22].

This section is dedicated to explaining in detail the model of core losses. First, the JA equations are defined and implemented using block diagram modeling in Simulink®. Second, a test setup is built for measuring the magnetic flux density and field intensity to validate the model in comparison to experimental measurements. Finally, to include the effect of the switching frequency on the core loss model, expressions are obtained for the JA parameters as a function of the switching frequency based on an empirical approach.

3.1. Model Implementation

The JA model separates the magnetization, M, into reversible, M_{rev}, and irreversible magnetizations, M_{irr}, which correspond to the reversible or irreversible phenomena, which take place within the magnetic material during the magnetization [22]. By computing the latter components, the total magnetization, M, is defined as the summation of both components as explained by Equation (3).

$$M = M_{irr} + M_{rev} \,. \tag{3}$$

The reversible component of magnetization is defined as a fraction, c, of the difference of the anhysteretic magnetization and the irreversible one, as explained by Equation (4).

$$\frac{dM_{rev}}{dH} = c\left(\frac{dM_{an}}{dH} - \frac{dM_{irr}}{dH}\right), \tag{4}$$

where c is referred to as the reversibility coefficient. The irreversible component of magnetization is defined by the differential Equation (5).

$$\frac{dM_{irr}}{dH} = \frac{M_{an} - M_{irr}}{\frac{\delta k}{\mu_o} - \alpha\left(M_{an} - M_{irr}\right)}, \tag{5}$$

where k is the loss coefficient, and α is the interdomain coupling. δ indicates the direction of the magnetizing field H, such that $\delta = 1$ for increasing field, and $\delta = -1$ for decreasing field, as defined by Equation (6).

$$\delta = \begin{cases} 1, & \frac{dH}{dt} > 0. \\ -1, & \frac{dH}{dt} < 0. \end{cases} \tag{6}$$

M_{an} is the anhysteretic magnetization. The anhysteretic magnetization describes the magnetization of an ideal ferromagnet that does not have a loss effect, and thus, its magnetization curve does not present hysteresis. Langevin's function [23] is used within the JA model for defining the anhysteretic magnetization, in which case, an effective field, H_e, replaces the magnetic field, H, as explained by Equation (7) [24].

$$M_{an} = M_s\left(\coth\frac{H_e}{a} - \frac{a}{H_e}\right), \tag{7}$$

where M_s is the saturation magnetization, a is the shape parameter for anhysteretic magnetization, and H_e is defined by Equation (8).

$$H_e = H + \alpha M \,. \tag{8}$$

Consequently, the magnetic flux density can be calculated using Equation (9).

$$B = \mu_0 H_e \,. \tag{9}$$

Table 1 summarizes the definition of the model parameters, M_s, a, c, α and k. The parameters are initially estimated by an iterative procedure to fit the model to the magnetic material $B(H)$ curve data provided by the manufacturer.

Table 1. Jiles-Atherton model parameters.

M_s	Saturation magnetization	A/m
a	Shape parameter for anhysteretic magnetization	A/m
k	Pinning (or loss) coefficient	A/m
c	Reversibility coefficient	–
α	Interdomain coupling	–

Figure 3 shows a block diagram of the detailed implementation of the JA equations [22]. Following this block diagram, the model equations have been implemented in Simulink®. Therefore, for a given core size and magnetic material, the instantaneous magnetic flux density (B) can be estimated for a certain instantaneous magnetic field intensity (H) applied to the magnetic core.

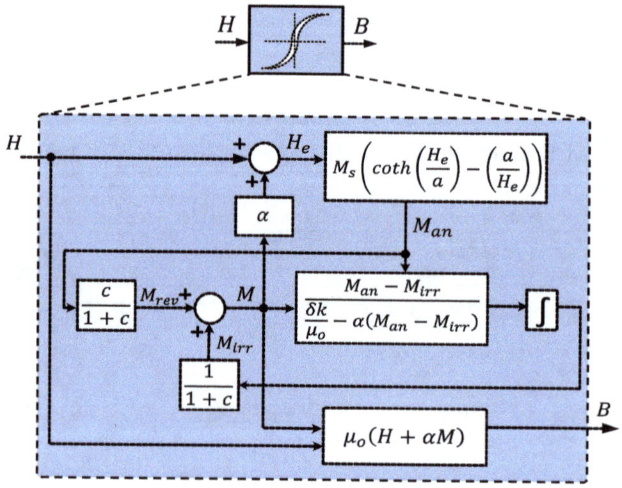

Figure 3. Schematic of the implementation of the Jiles-Atherton (JA) model.

3.2. Test Setup for Measuring Core Losses

In order to validate the implemented JA model and study the effect of frequency on the model parameters, a test setup is developed. The purpose of the experiments is to measure the hysteresis losses of the magnetic core and compare it to the one obtained by the implemented JA model. In literature, mainly two approaches for measuring core losses can be found [25]: electrical methods, and calorimeter-based methods. One of the former approaches has been selected, which is the $B(H)$ curve electrical measurement technique. Specific to the selected technique, the two-winding measurement method [26] is used since it is reported to be accurate for the frequency range under test in this study (<100 kHz) [27].

Figure 4 illustrates a schematic to clarify the test setup used for measuring the core losses, while Figure 5 shows the experimental platform developed.

The power stage used in the tests is a simple half-bridge converter suitable for low power levels. A square-waveform excitation voltage is sought to verify the loss study under non-sinusoidal conditions. The core used for the identification tests is a toroidal core with the design parameters listed in Table 2. This toroidal geometry is used in order to ensure that the model results are not only valid for the selected double E-core structure but also for other geometrical schemes.

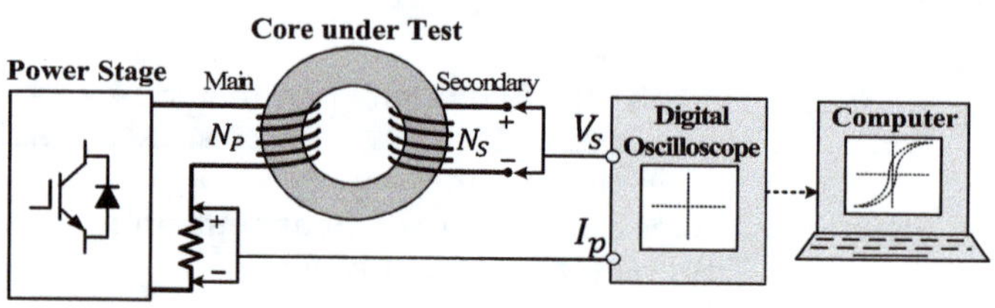

Figure 4. Schematic diagram of the test setup used to measure the $B(H)$ curve.

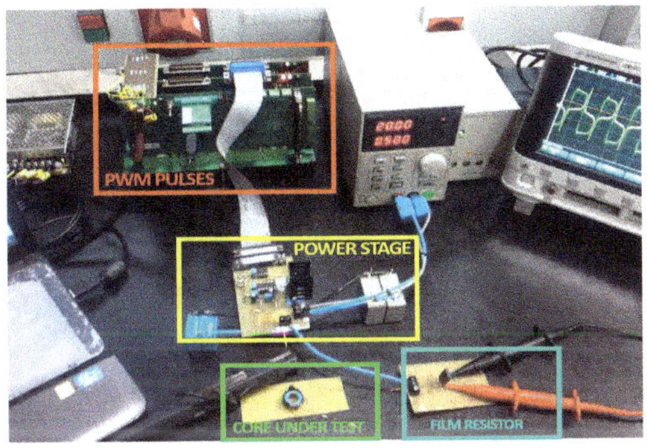

Figure 5. Experimental setup used to measure the $B(H)$ curve.

Table 2. Specifications of the test setup developed to measure the $B(H)$ curves.

Magnetic Core	
Core material	N87
Core type and size	Toroidal core: R16.0 × 9.60 × 6.30
Main winding no. of turns (N_p)	5 turns
Sensing winding no. of turns (N_s)	5 turns
Power Stage Ratings	
Power level	100 W
Input voltage	30 V
Peak current	3 A
Frequency (f)	50 kHz

In addition to the main excitation winding (N_p), a secondary sensing winding (N_s) is added to sense the induced voltage due to the flux in the main one. The advantage of a separate winding is to exclude the voltage drop due to the resistance of the main winding. Therefore, the magnetic flux density can be computed using Equation (10).

$$B = \frac{1}{N_s A} \int_0^T V_s dt\,, \tag{10}$$

where A is the cross-section area of the toroid, and V_s is the open-circuit secondary winding voltage. The integration is performed over one switching period, T. The field intensity can, thus, be computed using Equation (11).

$$H = \frac{N_p I_p}{l}\,, \tag{11}$$

where I_p is the current flowing through the main winding, and l is the effective length of the core. The measured voltage (V_s) and current (I_p) are illustrated in Figure 6a,b, respectively. By applying these measurements to Equations (10) and (11), the B and H quantities can be calculated. In this manner, the measured $B(H)$ curve of the N87 magnetic material is compared to the $B(H)$ curve obtained from the implemented JA model simulation under the same operation conditions. To apply the JA model, the model parameters are estimated by iterative fitting as mentioned previously, the obtained parameters for N87 magnetic material at 50 kHz are stated in Table 3. The resulting comparison is shown in Figure 7. It can be observed that the JA simulation model predicts the $B(H)$ curve of the N87 magnetic material with an error of less than 5%. It can, thus, be concluded that the model is valid and provides acceptable accuracy. Furthermore, the model has been tested under different magnetic materials. The previous measurements were repeated at the exact same operation conditions while using a similar toroidal core from 3C90 magnetic material, the JA parameters of which are stated in Table 3. The B and H quantities were again measured, and the $B(H)$ curves were compared together with the modeled ones, as illustrated in Figure 7. Similar to the previous results, the model shows a clear coincidence with the experimental measurements.

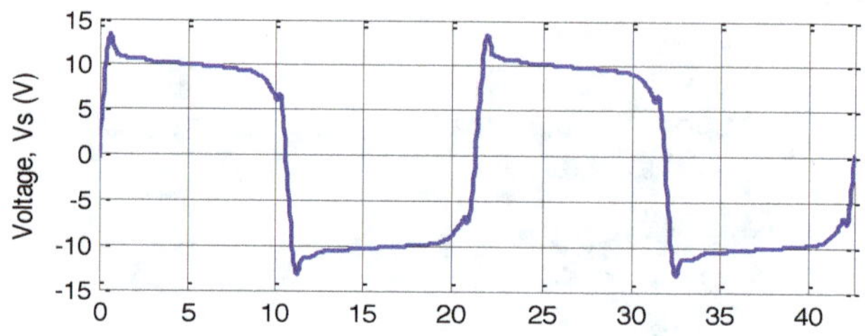

(a)

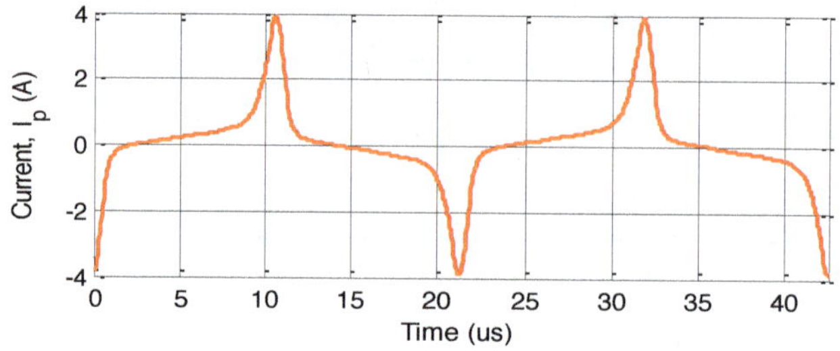

(b)

Figure 6. Experimental measurements. (**a**) Voltage applied to the toroid, and (**b**) current flowing through the main winding.

Table 3. Estimated Jiles-Atherton parameters for N87 and 3C90 magnetic materials at 50 kHz.

JA Parameter	N87	3C90
M_s	$4.0481e^5$	$3.7547e^5$
a	17.7019	19.5349
k	12.5883	12.8057
c	0.3210	0.3210
α	$2.0000e^{-5}$	$2.0000e^{-5}$

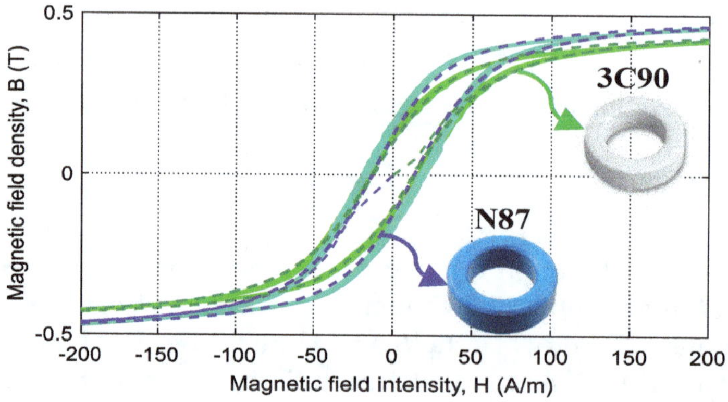

Figure 7. Model (dotted line) versus experimental (solid line) $B(H)$ curves for two different magnetic materials: N87 (blue) and 3C90 (green).

The hysteresis losses are estimated in terms of the area enclosed by the $B(H)$ loop and, therefore, can be expressed as shown by Equation (12).

$$P_{core} = V_e f \oint B(H) dH, \tag{12}$$

where V_e is the core volume, and f is the switching frequency. The integral is performed over a complete cycle of the magnetic field intensity.

3.3. Frequency Dependence of JA Parameters

Since the behavior of the magnetic core material depends on the excitation frequency, different voltage waveforms in the magnetic element imply notable variations in the associated trajectories on the $B(H)$ characteristic of the material. Specifically, it changes the area enclosed in the hysteresis loop, which is directly related to the core losses. This variation of the trajectory in itself implies a variation of the parameters of the JA model parameters. It is, thus, of interest to develop a model that can be used for any frequency without having to readjust the parameters each time different operating conditions are considered.

A practical approach has been taken by conducting a few experiments to characterize the variation of the JA model parameters as a function of the frequency. For the same core size and material, a number of independent tests were carried out, each test for a given frequency of operation. Then, a simple procedure is followed to obtain the parameters, as listed below:

1. The instantaneous waveforms of the voltage (V_s) and the current (I_p), as explained in the previous section, are captured.
2. Fast Fourier Transform (FFT) is applied to these waveforms in order to identify the waveform frequency.
3. For each switching frequency test, the $B(H)$ curve is measured using Equations (10) and (11). Then the measured $B(H)$ curve is fitted to the modeled one by using iterative trial-and-error steps in the JA parameters.
4. For each test, the set of obtained JA model parameters is collected.
5. Using curve fitting techniques (specifically the Curve Fitting Toolbox from Matlab®), an expression is obtained for each of the JA model parameters $M_s(f)$, $a(f)$, and $k(f)$ as a function of switching frequency. The parameters c and α show insignificant variation with frequency.

The obtained expressions for the JA parameters are stated by Equations (13)–(15).

$$M_s(f) = 5.189e^{-8} f^{2.334} + 4e^5 , \tag{13}$$

$$a(f) = 6.004e^{-15} f^{3.002} + 16.935 , \tag{14}$$

$$k(f) = -3.398e^{-7} f^{1.458} + 15 . \tag{15}$$

The expressions are intrinsic to a certain magnetic material. In this case, the procedure has been implemented for the N87 material; however, the same procedure can be followed for extracting a set of expressions for the JA parameters of any other ferrite material.

To test the validity of the obtained expressions for the JA parameters as a function of frequency, several experiments were carried out using the N87 toroidal core described in Table 2. The switching frequency of the converter was changed and the $B(H)$ curves corresponding to different frequency values were measured and compared to the modeled ones, as illustrated in Figure 8. The results of the comparison show that the modeled $B(H)$ curves match the obtained experimental measured curves at a wide range of the operation frequency with acceptable accuracy.

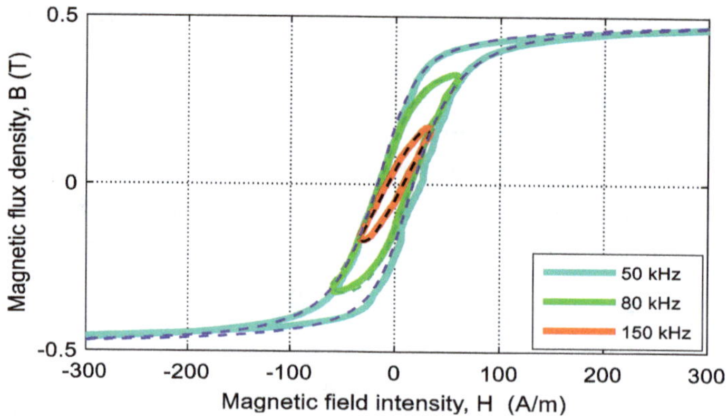

Figure 8. $B(H)$ curves for the prototype under different operation frequencies. Model (dotted line) versus experimental results (solid line).

4. Model of Winding Losses

4.1. Winding Eddy Current Losses

The winding losses are due to the resistance of the copper wire. At DC operation currents or relatively low operation frequencies, this resistance component is constant and calculated using Equation (16).

$$R_{dc} = \rho_{cu} \frac{l_{wire}}{A_{wire}}, \tag{16}$$

where ρ_{cu} is the resistivity of copper material at 20°C, which is equal to 1.68×10^{-8} Ωm, l_{wire} is the total length of the winding wire, and A_{wire} is the cross-section area of the wire.

However, as the switching frequency increases, two effects start to appear, which are the skin effect and proximity effect. These effects induce eddy currents in the winding conductors, altering the resistance of the winding, and significantly contributing to the overall winding losses. It is necessary in this case to calculate the AC resistance of the winding. Dowell provided a method that computes the equivalent winding resistance using a one-dimensional analytical approach [28]. Initially, this method was intended to describe high-frequency loss in foil windings; however, it has been extended to multilayer windings with round conductors by introducing the porosity factor [28,29]. Dowell's method uses a sinusoidal approach, and the calculations are limited to non-gapped cores [30]. On the other hand, the method presents a great advantage of simplicity and a fast computation of the AC winding resistance; thus, it can be easily integrated into the magnetic device model without extra complexity. Dowell estimates the AC resistance (R_{ac}) by scaling the DC winding resistance by a factor, as shown in Equation (17).

$$R_{ac} = R_{dc} \left(M' + \frac{(m^2 - 1)\, D'}{3} \right), \tag{17}$$

where M' and D' are coefficients defined based on the geometrical dimensions of the winding, material characteristics, and frequency of operation, and m is the number of layers. The accuracy of the full winding model will be provided in the following sections to assess the validity of using Dowell's method for the application herein.

4.2. Winding Stray Capacitance

As the operation frequency increases, the parasitic capacitance of the inductor winding becomes more significant, causing the impedance of the inductor to change and introduce the resonant frequencies. In order to attain a full comprehensive model of the device, the stray capacitance of the windings is added to the model. The calculation of the stray capacitance is based on an analytical approach previously presented in literature [31]. This method is valid for multi-layer inductors with ferromagnetic cores, as well as being simple and reliable for simulation purposes. Briefly, the inductor winding is divided into partitions, and the turn-to-turn and turn-to-core capacitances of the winding are predicted as a function of a few geometrical parameters of the device. Accordingly, the overall stray capacitance of the coil (C_s) converges to the expression stated in Equation (18).

$$C_s \cong 1.366 C_{tt}, \tag{18}$$

where C_{tt} is the turn-to-turn capacitance of the coil and is defined by Equation (19).

$$C_{tt} = \varepsilon_0 l_t \left(\frac{\varepsilon_r \theta^*}{ln \frac{D_o}{D_c}} + \cot\left(\frac{\theta^*}{2}\right) - \cot\left(\frac{\pi}{12}\right) \right), \tag{19}$$

where l_t is the turn length, θ^* is the angular coordinate, ε_0 and ε_r are the permittivity of air and relative permittivity of the insulation medium, respectively, and D_o and D_c are the diameters of the wire with and without the insulation coating, respectively, as clarified by Figure 9.

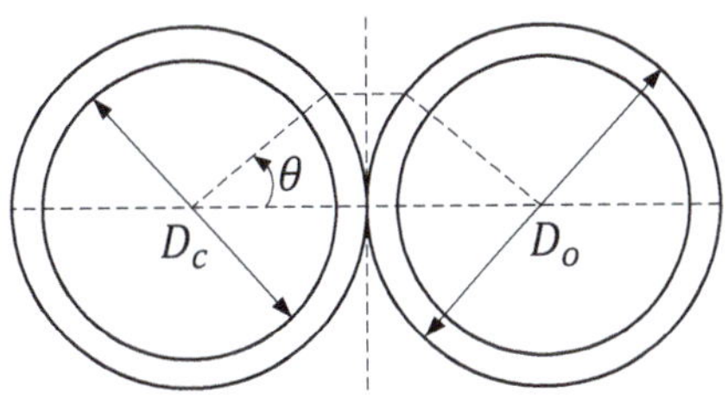

Figure 9. Two adjacent turns of the winding.

4.3. Full Winding Model

The full model of the inductor winding is expressed by the circuit diagram in Figure 10. Therefore, the total impedance of the winding is calculated by Equation (20).

$$Z_T = \frac{(sL + R_{ac} + R_{dc})\frac{1}{sC_s}}{sL + R_{ac} + R_{dc} + \frac{1}{sC_s}}. \tag{20}$$

An inductor prototype was implemented based on the specifications summarized in Table 4. The inductance value is not of specific importance in the design; however, it is interesting to distribute the winding on several layers to emphasize the proximity effect. Figure 11 illustrates the total winding impedance as a function of frequency to compare the developed winding model against the experimental measurements. As it can be observed, the error between the measured and modeled impedances is less than 1%, which represents a quite high accuracy for the study in context.

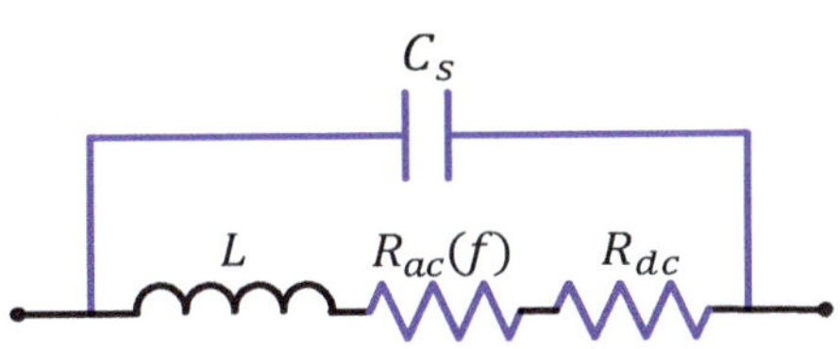

Figure 10. Full winding model.

Table 4. Specifications of the inductor prototype developed to validate the winding model.

Magnetic Core	N87
Core shape and size	ETD49
Main winding no. of turns	80 turns
No. of winding layers, m	4 layers
Diameter of wire, D_{wire}	1 mm

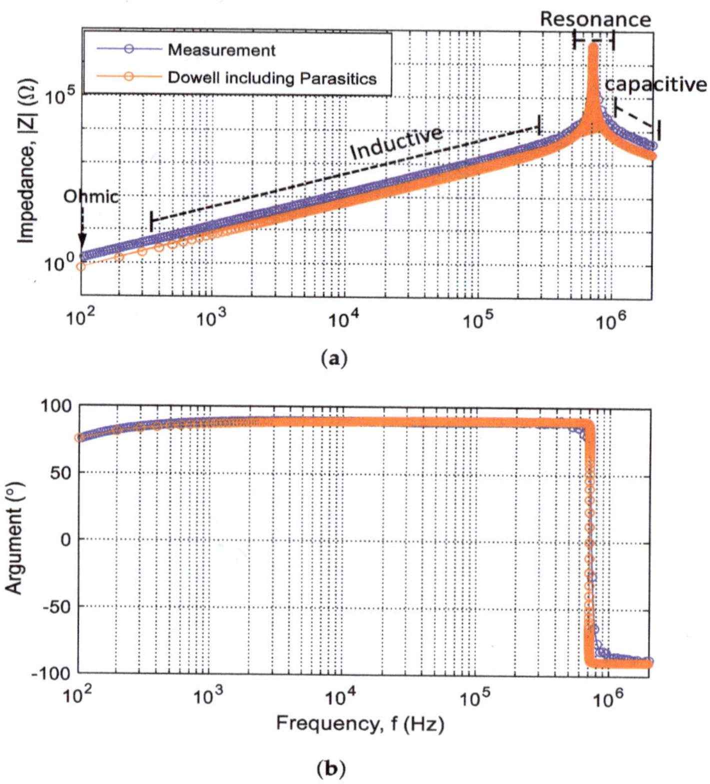

Figure 11. Total winding impedance as a function of frequency. (**a**) Magnitude, and (**b**) argument of modeled impedance compared to experimental measurements.

5. Applications of Loss Models to Simulate the Variable Inductor

After the verification of the loss models, the study is applied to model the double E-core variable inductor shown in Figure 1. The full model is implemented based on the reluctance circuit concept previously stated. However, the two issues associated with the previous reluctance model have been tackled. To avoid algebraic loops due to model causality, the magnetic core has been partitioned according to its operation, as mentioned previously, in Section 2, thus defining each partition in terms of electrical inputs and outputs as explained below.

The middle arm has the main winding, and to model this arm, the input will be the excitation voltage, and the output should be the main inductor current. On the other hand, the lateral arms have the control windings, so for these coils, the input will be the control current, and the output should be the induced voltages in the control windings.

Figure 12a illustrates the magnetic system, which is represented by the reluctance equivalent circuit of the device. The reluctance equivalent circuit is composed of three branches. The left and right branches represent the magnetic circuits of the control arms of the device. The voltage source $(N_b \cdot I_b)$

models the magnetomotive force created by each control winding. The voltage sources $(\phi_R \cdot \Re_R)$ and $(\phi_L \cdot \Re_L)$ model the variable reluctance of the magnetic path of the right and left arms, respectively. Using the control current as the input quantity, the values of the variable voltage sources are calculated based on the JA hysteresis model. The middle branch represents the magnetic circuit of the main arm, the variable reluctance of the magnetic path is similarly represented by the voltage source $(\phi_C \cdot \Re_C)$, while in this case, the magnetomotive force is the output quantity and is represented by a current source. The current in the main winding (I_c) can, thus, be calculated by measuring the voltage across this current source and dividing by the number of turns of the winding (N_c).

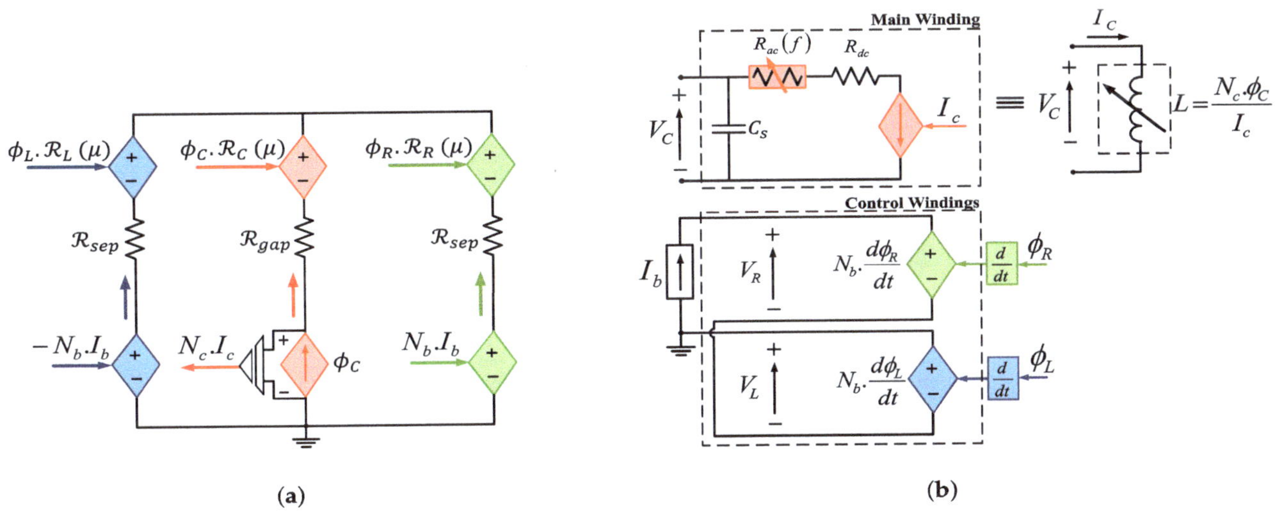

Figure 12. Schematic of the variable inductor model based on the reluctance circuit. (**a**) Magnetic circuit, and (**b**) electric circuit.

On the other hand, Figure 12b illustrates the electrical system, which includes the electrical model of the winding, as well as the input voltage and the output current source. The electrical part of the three windings is represented by Equation (21).

$$V_w = N_w \frac{d\phi_w}{dt}, \qquad w \in L, R, C \tag{21}$$

where V_w is the voltage at winding w, ϕ_w is the magnetic flux created by the winding w, and N_w is the number of turns of the winding w. The winding w refers to the left (L), right (R), or center (C) windings. In order to account for the main winding DC losses, a constant resistor, R_{dc}, is added to the electrical circuit of the main winding in series with the current-controlled current source, which represents the main winding current. Also, to represent the AC winding losses, a variable frequency-dependent resistor, $R_{ac}(f)$, is added in series to R_{dc}.

6. Model Validation Using Simulations and Experimental Results

In order to compare the initial lossless equivalent circuit with the proposed model, which includes core and winding losses, detailed simulations have been carried out. Furthermore, those two simulation models have been compared against experimental measurements obtained from the variable inductor prototype shown in Figure 13. The device has been developed based on the double E-core structure with the design specifications indicated in Table 5, and the models have been adjusted correspondingly.

Figure 13. Variable inductor prototype based on double the E-core structure.

Table 5. Specifications of the test setup developed to validate the full VI model.

Magnetic Core	
Core material	N87
Core shape and size	ETD49/25/16
Main winding no. of turns	23 turns
Control winding no. of turns	55 turns
Power Stage Ratings	
Power level	1 kW
Input voltage	200 V
Peak current	5 A
Frequency (f)	50 kHz

Figure 14a shows a circuit diagram of the developed test platform. It consists of a SiC full-bridge DC-AC converter to apply a square waveform excitation voltage on the inductor main winding. As mentioned previously, the square waveform voltage allows testing the device model under non-sinusoidal conditions, thus assure the validation of the loss study under a general condition of excitation voltage. The converter is controlled using a TMS320F28335 Texas Instruments peripheral board. Additionally, a variable DC voltage source is connected in series with a resistor to provide a DC control current of maximum 1 A to the control winding of the variable inductor. The constructed test platform is illustrated in Figure 14b.

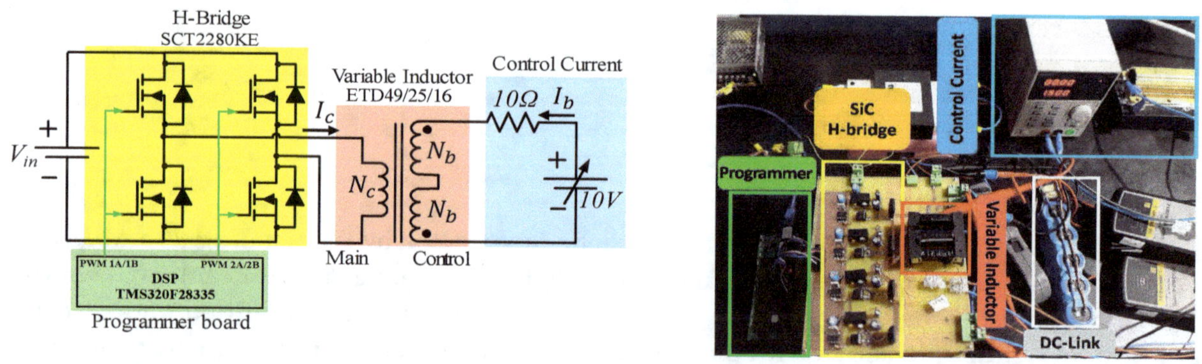

(a) (b)

Figure 14. Experimental setup used to test the variable inductor under small and large-signal analyses. (**a**) Circuit diagram, and (**b**) test platform.

6.1. Small-Signal Analysis

To validate the proposed model under small-signal analysis, the control current was increased from 0 to 1 A in steps of 0.05 A, while keeping zero excitation voltage on the main winding. The equivalent inductance seen from the main winding was measured using the impedance analyzer. Figure 15a illustrates the measurement of the equivalent inductance as a function of the bias control current. Also, the figure illustrates the simulated inductance obtained from the developed models, the initial lossless equivalent circuit, as well as the proposed model, which includes losses. Figure 15b shows the error of each model compared to the experimental results, as calculated by Equation (22). It can be observed that the proposed model that includes losses predicts the inductance within an acceptable error (<6%). On the other hand, the inductance predicted by the initial lossless model shows a clear deviation from the experimental one as the control current increases. It reaches an error of 30% at maximum control current.

$$Error(\%) = \frac{(L_{Experiment} - L_{Model})}{L_{Experiment}} \times 100\,. \tag{22}$$

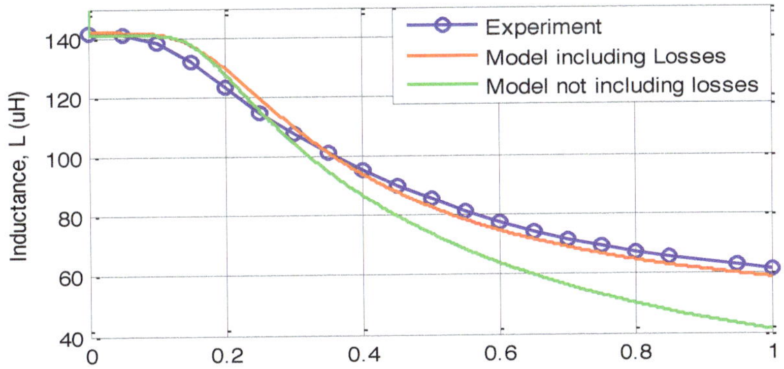

(a)

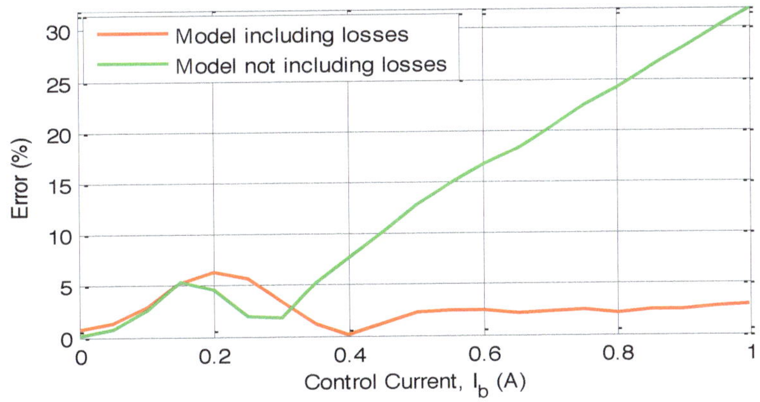

(b)

Figure 15. Small-signal characterization of the variable inductor prototype comparing simulation models with experimental results. (**a**) Inductance value, and (**b**) percentage error as a function of control current.

6.2. *Large-Signal Analysis*

The prototype has also been characterized under large-signal analysis. Similar to the small-signal analysis, a DC control current is applied to the control windings and varied from 0 to 1 A. However, in this case, a square waveform voltage of 30 V is applied to the main winding of the inductor. The inductance is calculated by two different methods using the experimental measurements of the voltage and current through the main winding. The first method calculates the inductance using the RMS values of the waveforms over each cycle at a steady state. On the other hand, the second method uses the instantaneous values of the waveforms. The two methods were then compared to verify the accuracy of the measured inductance value, as explained hereafter.

6.2.1. Impedance Calculation

A simplification applied by considering the RMS value of the first harmonic component of the voltage and current measurements, so the resulting inductance is calculated by Equation (23).

$$L = \frac{X_L}{\omega} = \frac{V_C}{I_C} \cdot \frac{\pi/4}{2\pi f}, \tag{23}$$

where V_C is the voltage applied to the main winding of the variable inductor, and I_C is the current flowing through it.

6.2.2. System Identification Tools

The System Identification Toolbox from Matlab® is used, which requires a set of data that represent the input and output variables of a system; in this case, V_C is the input to the system, and I_C is the output. Using these variables, the tool defines the transfer function of the system based on the user selection of the number of poles and zeros of the system. In the case of an inductor, the system is defined as a 1st order system (1 pole, no zeros), and the transfer function is stated by Equation (24).

$$\frac{I_C(s)}{V_C(s)} = \frac{1}{sL + R_{dc}}. \tag{24}$$

Figure 16 compares both methods of inductance calculation to verify the accuracy of the measured inductance and then uses it to assess the proposed model. It shows that the inductance calculated based on RMS values is in very close agreement with the inductance obtained by the Matlab® identification tool. This conclusion justifies the use of RMS measurements for calculating the inductance in the large-signal analysis in order to simplify the computations.

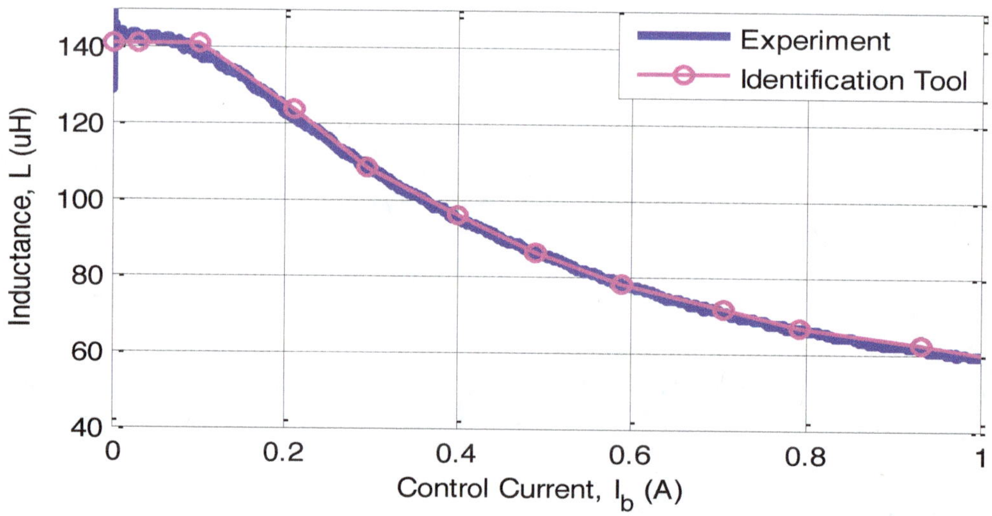

Figure 16. Inductance calculation from experimental measurement.

Similar to the previous small-signal analysis, Figure 17a illustrates the equivalent inductance as a function of the bias control current for the large-signal analysis case. Figure 17b clarifies the trend of deviation or convergence of the models as a function of the control current, compared to the measured results. The results are quite consistent with the previous small-signal analysis; the inductance calculated by the proposed model still approaches the experimental measurements, while the one calculated by the lossless model shows a clear dispersion.

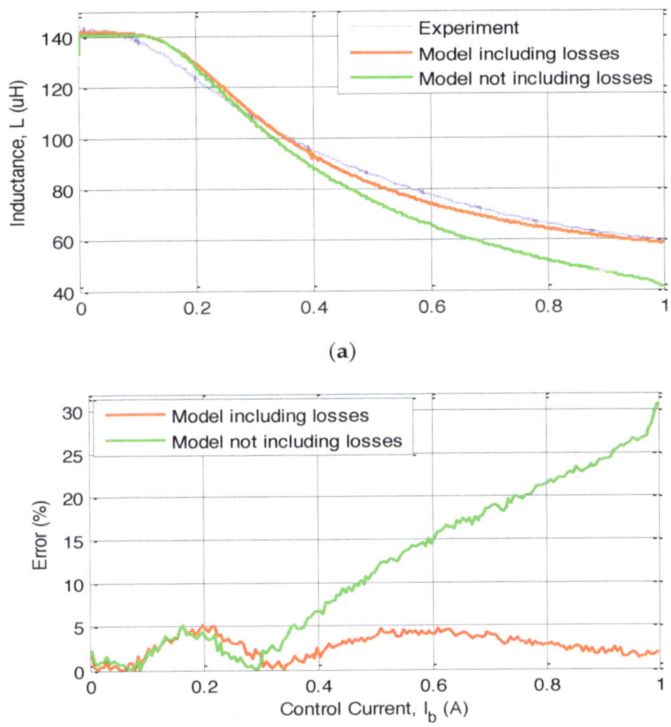

(b)

Figure 17. Large-signal characterization of the variable inductor prototype comparing simulation models with experimental results. (**a**) Inductance value, and (**b**) percentage error as a function of control current.

7. Conclusions and Future Developments

In this paper, a full accurate circuit-based time-domain model for a variable magnetic device has been developed, demonstrated, and experimentally validated. The model can predict the inductance variation as a function of the control current in a variable magnetic element. The main contribution of this work is the development of a model that can be used in several simulation platforms, and, moreover, the inclusion of core losses as well as winding eddy current losses in the magnetic device. Additionally, it solves the causality issues, which are present in previous approaches.

A double E-core variable inductor prototype has been characterized under small-signal as well as large-signal analyses in order to assess the accuracy of the model. The proposed approach is considered an autonomous tool that can analyze any given set of data (simulated or experimental) for a magnetic core, detect the operation frequency, and correspondingly, adjust the magnetic core model with the core and winding loss parameters, and finally predict the inductance as a function of the control current, along with other electric and magnetic quantities that characterize the magnetic core operation. Consequently, only one simulator environment is used for the design and simulation of the electromagnetic system.

The test results validate the model under operation frequencies of hundreds of kHz (<200 kHz). Accordingly, the future developments of this work include the extension of the model's validity to frequency ranges of 1–1.5 MHz. Also, the employment of the developed electromagnetic model to study the behavior of the variable inductor in a power electronic converter, specifically a DAB converter. The proposed model will allow for several studies using time-domain simulations, such as the control of the converter power transfer using the variable inductor, the possibility of the linearization of the system transfer function, and finally, the boost of the efficiency over critical operation ranges, for example, light load and heavy load operation ranges.

Author Contributions: S.S. and J.G. conceived the research and designed and performed the experiments; R.G. contributed to the reviewing and editing; all the authors analyzed the data and contributed in the discussion and conclusions. All authors have read and agreed to the published version of the manuscript.

Abbreviations

The following abbreviations are used in this manuscript:

PEC	Power Electronic Converter
EMI	Electro-Magnetic Interference
DAB	Dual-Active-Bridge
FEA	Finite Elements Analysis
JA	Jiles-Atherton
FFT	Fast Fourier Transform

References

1. Fan, H.; Li, H. High-Frequency Transformer Isolated Bidirectional DC–DC Converter Modules With High Efficiency Over Wide Load Range for 20 kVA Solid-State Transformer. *IEEE Trans. Power Electron.* **2011**, *26*, 3599–3608.
2. Burgio, A.; Menniti, D.; Motta, M.; Pinnarelli, A.; Sorrentino, N.; Vizza, P. A laboratory model of a dual active bridge DC-DC converter for a smart user network. In Proceedings of the 2015 IEEE 15th International Conference on Environment and Electrical Engineering (EEEIC), Rome, Italy, 10–13 June 2015.
3. Saeed, S.; Garcia, J. Extended Operational Range of Dual-Active-Bridge Converters by using Variable Magnetic Devices. In Proceedings of the 2019 IEEE Applied Power Electronics Conference and Exposition (APEC), Anaheim, CA, USA, 17–21 March 2019; pp. 1629–1634.
4. Takach, M.; Lauritzen, P. Survey of magnetic core models. In Proceedings of the 1995 IEEE Applied Power Electronics Conference and Exposition, Dallas, TX, USA, 5–9 March 1995.
5. Chen, Q.; Xu, L.; Ruan, X.; Wong, S.C.; Tse, C.K. Research and Development on New Control Techniques for Electronic Ballasts Based on Magnetic Regulators. Ph.D. Thesis, Dept. Elect. Comput. Eng., Univ. Coimbra, Coimbra, Portugal, 2012.
6. Chen, Q.; Xu, L.; Ruan, X.; Wong, S.C.; Tse, C.K. Gyrator-Capacitor Simulation Model of Nonlinear Magnetic Core. In Proceedings of the 2009 Twenty-Fourth Annual IEEE Applied Power Electronics Conference and Exposition, Washington, DC, USA, 15–19 February 2009.
7. Ludwig, G.; El-Hamamsy, S.A. Coupled inductance and reluctance models of magnetic components. *IEEE Trans. Power Electron.* **1991**, *6*, 240–250.
8. Almaguer, J.; Cárdenas, V.; Espinoza, J.; Aganza-Torres, A.; González, M. Performance and Control Strategy of Real-Time Simulation of a Three-Phase Solid-State Transformer. *Appl. Sci.* **2019**, *9*, 789.
9. Alonso, J.M.; Martinez, G.; Perdigao, M.; Cosetin, M.; do Prado, R.N. Modeling magnetic devices using SPICE: Application to variable inductors. In Proceedings of the 2016 IEEE Applied Power Electronics Conference and Exposition (APEC), Long Beach, CA, USA, 20–24 March 2016.
10. Mandache, L.; Topan, D.; Sirbu, I.G. Accurate Time-Domain Simulation of Nonlinear Inductors Including Hysteresis and Eddy-Current Effects. In Proceedings of the World Congress on Engineering, London, UK, 6–8 July 2011; Volume 2.
11. LTspice® Design Center. Available online: https://www.analog.com/en/design-center/design-tools-and-calculators/ltspice-simulator.html (accessed on 10 April 2020).
12. MathWorks Simulink® Documentation. Available online: https://es.mathworks.com/help/simulink/index.html?s_cid=doc_ftr (accessed on 10 April 2020).
13. PSIM User's Guide. Available online: https://www.myway.co.jp/products/psim/dlfiles/pdf/PSIM_User_Manual_V9.0.2.pdf (accessed on 10 April 2020).
14. Medini, D.; Ben-Yaakov, S. A current-controlled variable-inductor for high frequency resonant power circuits. In Proceedings of the 1994 IEEE Applied Power Electronics Conference and Exposition, Orlando, FL, USA, 13–17 February 1994.
15. Sun, J. Recent Development in Ferrite Material for High Power Application [Passive Components]. *IEEE Power Electron. Mag.* **2018**, *5*, 21–25.

16. Brauer, J. Simple equations for the magnetization and reluctivity curves of steel. *IEEE Trans. Magn.* **1975**. *11*, 81–81.
17. Johansson, P.; Andersson, B. Comparison of Simulation Programs for Supercapacitor Modelling. Master of Science Thesis, Chalmers University of Technology, Gothenburg, Sweden, 2008. Available online: http://webfiles.portal.chalmers.se/et/MSc/AnderssonJohanssonMSc.pdf (accessed on 10 April 2020).
18. Saeed, S.; Garcia, J.; Georgious, R. Modeling of variable magnetic elements including hysteresis and Eddy current losses. In Proceedings of the 2018 IEEE Applied Power Electronics Conference and Exposition (APEC), San Antonio, TX, USA, 4–8 March 2018.
19. Cardelli, E.; Fiorucci, L.; Torre, E.D. Estimation of MnZn ferrite core losses in magnetic components at high frequency. *IEEE Trans. Magn.* **2001**, *37*, 2366–2368.
20. Reinert, J.; Brockmeyer, A.; Doncker, R.D. Calculation of losses in ferro- and ferrimagnetic materials based on the modified Steinmetz equation. *IEEE Trans. Ind. Appl.* **2001**, *37*, 1055–1061.
21. Jiles, D.; Atherton, D. Ferromagnetic hysteresis. *IEEE Trans. Magn.* **1983**, *19*, 2183–2185.
22. Wilson, P.R. Modelling and Simulation of Magnetic Components in Electric Circuits. Ph.D. Thesis, School of Electronics and Computer Science, University of Southampton, Southampton, UK, 2001. Available online: https://eprints.soton.ac.uk/368484/ (accessed on 10 April 2020).
23. Ramesh, A.; Jiles, D.; Roderick, J. A model of anisotropic anhysteretic magnetization. *IEEE Trans. Magn.* **1996**, *32*, 4234–4236.
24. Pop, N.; Caltun, O. Jiles-Atherton Magnetic Hysteresis Parameters Identification. *Acta Phys. Pol. A* **2011**, *120*. Available online: http://przyrbwn.icm.edu.pl/APP/PDF/120/a120z3p22.pdf (accessed on 10 April 2020).
25. Xiao, C.; Chen, G.; Odendaal, W.G.H. Overview of Power Loss Measurement Techniques in Power Electronics Systems. *IEEE Trans. Ind. Appl.* **2007**, *43*, 657–664.
26. Thottuvelil, V.; Wilson, T.; Owen, H. High-frequency measurement techniques for magnetic cores. *IEEE Trans. Power Electron.* **1990**, *5*, 41–53.
27. Mu, M.; Li, Q.; Gilham, D.J.; Lee, F.C.; Ngo, K.D.T. New Core Loss Measurement Method for High-Frequency Magnetic Materials. *IEEE Trans. Power Electron.* **2014**, *29*, 4374–4381.
28. Dowell, P. Effects of eddy currents in transformer windings. *Proc. Inst. Electr. Eng.* **1966**, *113*, 1387.
29. Yin, Y.; Li, L. Improved method to calculate the high-frequency eddy currents distribution and loss in windings composed of round conductors. *IET Power Electron.* **2017**, *10*, 1494–1503.
30. Holguin, F.A.; Asensi, R.; Prieto, R.; Cobos, J.A. Simple analytical approach for the calculation of winding resistance in gapped magnetic components. In Proceedings of the 2014 IEEE Applied Power Electronics Conference and Exposition, Fort Worth, TX, USA, 16–20 March 2014.
31. Massarini, A.; Kazimierczuk, M. Self-capacitance of inductors. *IEEE Trans. Power Electron.* **1997**, *12*, 671–676.

Decoupled Current Controller based on Reduced Order Generalized Integrator for Three-Phase Grid-Connected VSCs in Distributed System

Sen Zhang [1], Jianfeng Zhao [1,*], Zhihong Zhao [2], Kangli Liu [1], Pengyu Wang [1] and Bin Yang [1]

[1] School of Electrical Engineering, Southeast University, Nanjing 210096, China; sen_zhang@seu.edu.cn (S.Z.); 101101727@seu.edu.cn (K.L.); py_wang@seu.edu.cn (P.W.); 230149488@seu.edu.cn (B.Y.)

[2] Department of Electrical Engineering, School of Automation, Nanjing University of Science and Technology, Nanjing 210094, China; zhaozhihong@seu.edu.cn

* Correspondence:jianfeng_zhao@seu.edu.cn

Abstract: Grid-connected voltage source converters (GC-VSCs) are used for interfacing the distributed power generation system (DPGS) to the utility grid. Performance of the current loop is a critical issue for these GC-VSCs. Recently, reduced order generalized integrator (ROGI)-based current controller is proposed, such that AC reference signal of positive or negative sequences can be separately tracked without steady-state error, which has the advantage of less computational burden. However, the cross-coupling within the ROGI-based current controller would deteriorate the transient response of the current loop. In this paper, a ROGI-based decoupled current controller is proposed to eliminate the coupling between α-axis and β-axis. Thus, the faster dynamic response performance can be achieved while maintaining the merits of ROGI-based current controller. An optimal gain parameter design method for the proposed current controller is presented to improve the stability and dynamic response speed of current loop. Simulation and experiments were performed in MATLAB/Simulink and TMS320C28346 DSP-based laboratory prototype respectively, which validated the proposed theoretical approach.

Keywords: decoupling; reduced order generalized integrator (ROGI); optimal gain; distributed power generation system (DPGS); grid-connected voltage source converters (GC-VSCs)

---

1. Introduction

In recent years, distributed power generation system (DPGS) has attracted increasing attention with the aggravation of global energy shortage and environmental issues [1–5]. A typical DPGS is shown in Figure 1; due to the intermittence of renewable energy sources (RESs) such as PV systems and wind turbines, energy storage systems (ESSs) based on battery and supercapacitor are used to suppress power fluctuations, which improves the reliability and quality of power supply. Moreover, integrating the electric energy generated by RESs into the power grid is an important way to improve power generation efficiency. As the typical power electronic interfaces between DPGS and power grid, GC-VSCs have been intensively studied [6,7]. In the grid-connected mode, the voltage and frequency at the point of common coupling are dictated by strong power grid, the GC-VSC is controlled to behave as current-controlled converter. Active and reactive power regulation are performed by changing the grid-connected output current [8,9]. Therefore, accurate and fast control of the grid-connected current is one of the most critical technologies, which directly affects the power quality [8–13].

Various research has focused on current control of GC-VSCs [4,6–12,14–25], among which the most widely studied control method is PI controller [17,19,23,26], which effectively guarantees excellent tracking of AC reference signals and fast dynamic response speed, since AC signals become DC signals through Park transformation and PI controller could produce infinite gain for DC signals [16,17,26]. Nevertheless, d-axis and q-axis current cross-coupling, resulting from the Park transformation, is proportional to the interested control frequency. Actually, the modulation and sampling delay tend to aggravate the coupling and even deteriorate stability of the current loop [17,26]. In addition, resonant controllers are widely used in current loop regulation of GC-VSCs. Proportional-resonant (PR) and vector proportional-integral (VPI) controllers are, respectively, equivalent to conventional PI and complex-vectors PI (cPI) regulators, which are implemented in a positive and negative sequence SRF simultaneously [17]. It assures perfectly tracking AC reference signals without steady-state errors for both sequences [5–7,9,12,14,15,18]. However, the negative-sequence term (i.e., $-j\omega_e$) of conjugated poles introduces two times control frequency fluctuation, bringing slower dynamic response and more severe cross-coupling [16,26]. Moreover, computational burden increases when multiple order harmonics need to be tracked simultaneously [21,22].

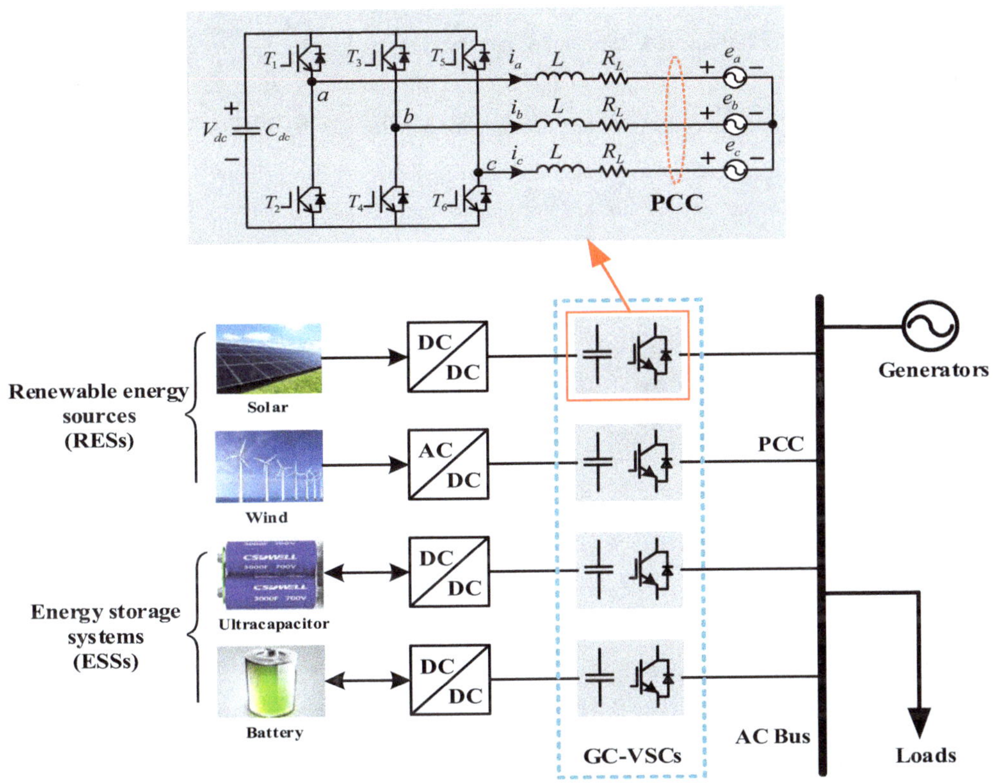

Figure 1. The configuration of a distributed power generation system.

To overcome the disadvantage of the above controllers, some improved methods are proposed. The method of state-feedback decoupling is proposed in [16], which effectively reduces the axes cross-coupling and broadens the bandwidth of current loop. Nevertheless, the decoupling effectiveness could be affected by the time delay and the accuracy of parameter estimation. The alternative decoupling method based on cPI controller is less influenced by these two factors [17]. Thus, a better stability and a faster dynamic response speed of the current loop are achieved. However, the implementation of the aforementioned control algorithms is in SRF and the rotating frame transformation increases the computational burden and complexity. Simultaneously, the error of transformation calculation would emerge if the PLL is not sufficiently accurate [7,9,26]. Overall, a regulator with zero steady-state error in the stationary frame would have some certain advantages in terms of implementation [9–12]. ROGI-based current controller, which is named as PCI controller in [11,12], is proposed with the advantage of less computational burden and improving the bandwidth of harmonics control [21,25]. However, the cross-coupling within the ROGI-based current controller would deteriorate the transient response of the current loop. In addition, the stability of the current loop will decrease when the control frequency is relatively high [23].

In this paper, a ROGI-based decoupled current controller is proposed, which is capable of suppressing the cross-coupling and improving the dynamic response speed, while possessing the advantage of low computation burden and being convenient to implement. An optimal gain method for parameter tuning is also presented to maximize the stability of the current loop.

The paper is organized as follows. Coupling of ROGI-based current controller is analyzed in Section 2. Section 3 introduces the proposed ROGI-based decoupled current controller. Specially, the performance of the proposed controller is analyzed through closed-loop frequency response and mathematical derivation, and the optimal parameter design using the root locus is presented. The simulation and experimental results are provided to validate the theoretical approach in Section 4. Finally, Section 5 summarizes the work.

2. Coupling Analysis of ROGI-Based Current Controller

The ROGI is proposed in [21], its transfer function in the s-domain can be expressed as

$$G_{ROGI}(s) = \frac{1}{s \mp j\omega_e} \tag{1}$$

where ω_e is the fundamental frequency and "$-/+$" denotes the positive/negative sequence. The implementation principle of ROGI is shown in Figure 2a.

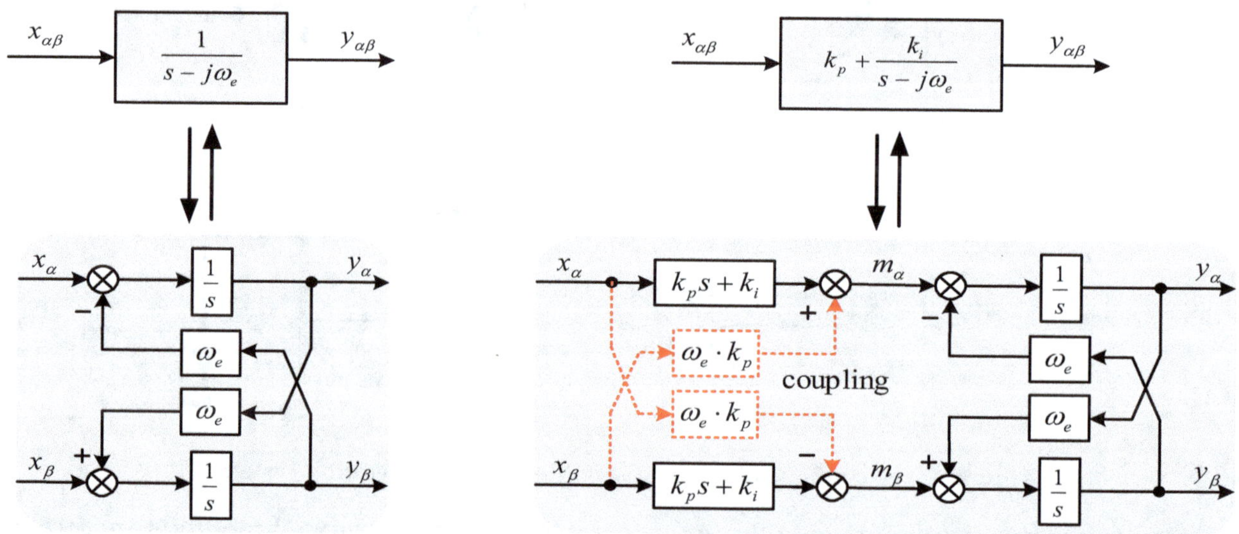

(**a**) ROGI. (**b**) ROGI-based current controller (PCI).

Figure 2. Implementation principle of ROGI and ROGI-based current controller.

A novel proportional complex integral (PCI) is derived from the principle that zero steady-state error can be achieved if the open-loop gain is infinity at the control frequency [10]. Thus, it can be written by,

$$G_{PCI}(s) = k_p + \frac{k_i}{s - j\omega_e} \tag{2}$$

where k_p is the proportional gain and k_i is the integral gain. It is known from Equation (2) that the PCI controller is equivalent to ROGI-based current controller in [10,21]. For convenience, the PCI controller represents the ROGI-based current controller in this paper. The implementation principle of PCI is shown in Figure 2b, where it can be seen that PCI controller includes two $k_p \cdot \omega_e$ coupling branches, which will deteriorate the transient response of the current loop. Besides, the coupling aggravates as the control frequency increases.

Figure 3 shows the diagram of PCI-based current closed-loop for GC-VSCs. $G_d^s(s) = e^{-sT_d}$ represents the digital control delay, which consists of one and a half sampling time delay [19], $T_d = 1.5T_s$ (T_s is the sampling cycle). For a more intuitive sense of the coupling, α-axis and β-axis current response with PCI controller is presented in Figure 4 when a step change occurs in reference current. The reference current $i_{\alpha\beta}^*$ decreases from $10 \cdot e^{j\omega_e t}$ to 0 at $t = 0.04s$, where $i_\alpha^* = \mathrm{Re}\{i_{\alpha\beta}^*\} = 10{\cdot}\cos(\omega_e t)$ and $i_\beta^* = \mathrm{Im}\{i_{\alpha\beta}^*\} = 10{\cdot}\sin(\omega_e t)$. In Figure 4a,b, it can be seen that reference current changes in α-axis affects β-axis current, and vice versa. In addition, the coupling will increase with the rising of the control frequency.

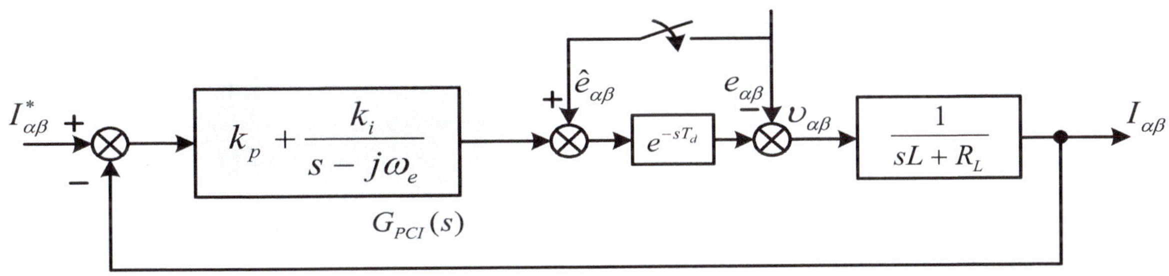

Figure 3. The diagram of PCI-based current closed-loop for GC-VSCs.

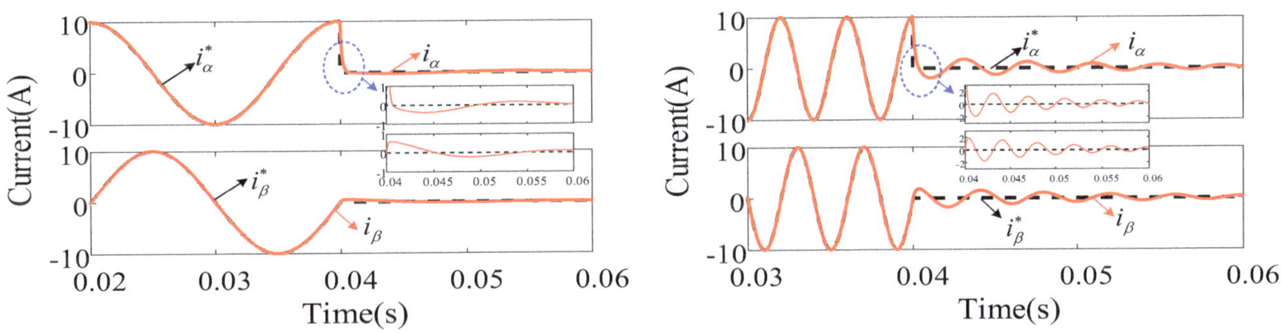

(**a**) $\omega_e = 2{\cdot}50{\cdot}\pi$ rad/s (**b**) $\omega_e = 2{\cdot}250{\cdot}\pi$ rad/s.

Figure 4. α-axis and β-axis current response with PCI controller.

3. The Proposed ROGI-Based Decoupled Current Controller

To suppress the axes-coupling in the PCI-based current loop. an effective decoupling method with ROGI-based decoupled current controller is proposed, which improves dynamic response performance of the current loop. For convenience, the proposed ROGI-based decoupled current controller is referred to as the D-PCI controller in the following section.

3.1. Structure of the D-PCI Controller

As shown in Figure 2b and the analysis of coupling, the key to effective decoupling is to remove the cross-coupling term $k_p \cdot \omega_e$. Thus, the structure of the proposed D-PCI current controller is shown in Figure 5. Thus, the transfer function of D-PCI can be expressed as

$$G_{D-PCI}(s) = k_p + \frac{k_i + j\omega_e k_p}{s - j\omega_e} \tag{3}$$

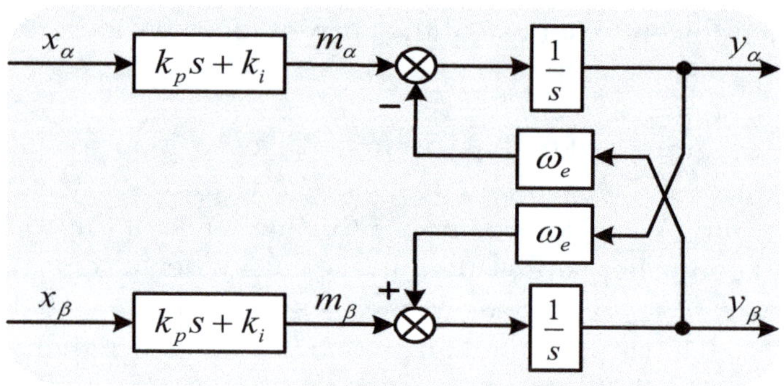

Figure 5. Structure of the proposed D-PCI current controller.

Based on Equations (2) and (3), the Bode plot of PCI and D-PCI controllers tuned at $\omega_e = 2{\cdot}\pi{\cdot}50$ rad/s (i.e., fundamental excitation frequency is 50 Hz) is shown in Figure 6. It can be seen that both PCI and D-PCI controllers achieve infinite gain at fundamental excitation frequency, which means that precise AC reference signal tracking can be realized. Moreover, D-PCI controller has superior performance in suppressing DC reference signal. The magnitude for D-PCI controller is −6.5 dB, while PCI controller is 27 dB.

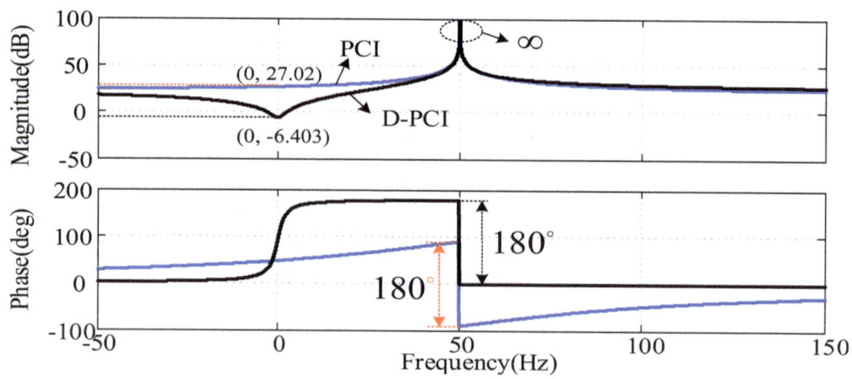

Figure 6. Bode plot of PCI and D-PCI controllers tuned at $\omega_e = 2 \cdot \pi \cdot 50$ rad/s.

Replacing the PCI with D-PCI in Figure 3, α-axis and β-axis current response with D-PCI controller is presented in Figure 7 when a step change occurs in reference current. In Figure 7a,b, it can be seen that, regardless of the control frequency, there is almost no coupling between the current of α-axis and β-axis.

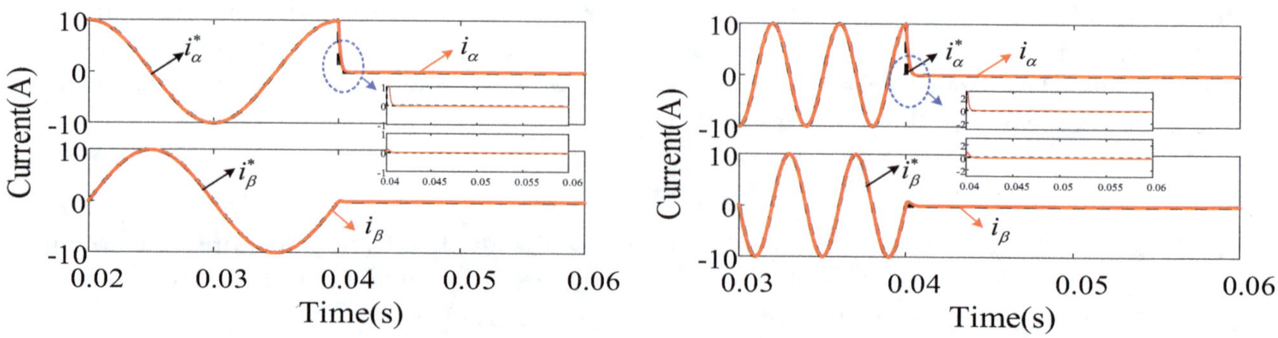

(a) $\omega_e = 2 \cdot 50 \cdot \pi$ rad/s **(b)** $\omega_e = 2{\cdot}250{\cdot}\pi$ rad/s.

Figure 7. α-axis and β-axes current response with D-PCI controller.

3.2. *Performance Analysis of the D-PCI Controller*

From Figure 3 and Equation (3), the output current $I_{\alpha\beta}$ can be derived as

$$I_{\alpha\beta}(s) = \frac{e^{-sT_d}\cdot(k_p s+k_i)}{(sL+R_L)(s-j\omega_e)+e^{-sT_d}\cdot(k_p s+k_i)} I^*_{\alpha\beta}(s) - \frac{s-j\omega_e}{(sL+R_L)(s-j\omega_e)+e^{-sT_d}\cdot(k_p s+k_i)} e_{\alpha\beta}(s) \quad (4)$$

where

$$G^s_{CL}(s) = \frac{(k_p s+k_i)\cdot e^{-sT_d}}{(s-j\omega_e)(sL+R_L)+(k_p s+k_i)\cdot e^{-sT_d}} \quad (5)$$

is the transfer function between $I_{\alpha\beta}(s)$ and $I^*_{\alpha\beta}(s)$, and

$$G^s_{DL}(s) = -\frac{s-j\omega_e}{(sL+R_L)(s-j\omega_e)+e^{-sT_d}\cdot(k_p s+k_i)} \quad (6)$$

is the transfer function between $I_{\alpha\beta}(s)$ and $e_{\alpha\beta}(s)$.

If the frequency of $I^*_{\alpha\beta}(s)$ (reference input signal) and $e_{\alpha\beta}(s)$ (disturbance signal) is equal to ω_e, i.e., $s = j\omega_e$, then $G^s_{CL}(j\omega_e) = 1$ and $G^s_{DL}(j\omega_e) = 0$. It means that D-PCI controller can realize tracking of AC reference signals and suppress AC disturbance signals. Both closed-loop frequency response and mathematical derivation would be used to evaluate the steady-state tracking performance.

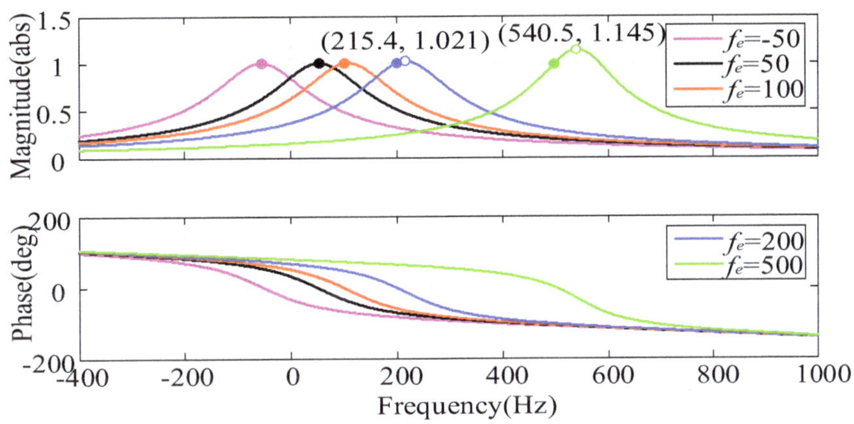

(**a**) Reference tracking $I_{\alpha\beta}(s)$ / $I^*_{\alpha\beta}(s)$.

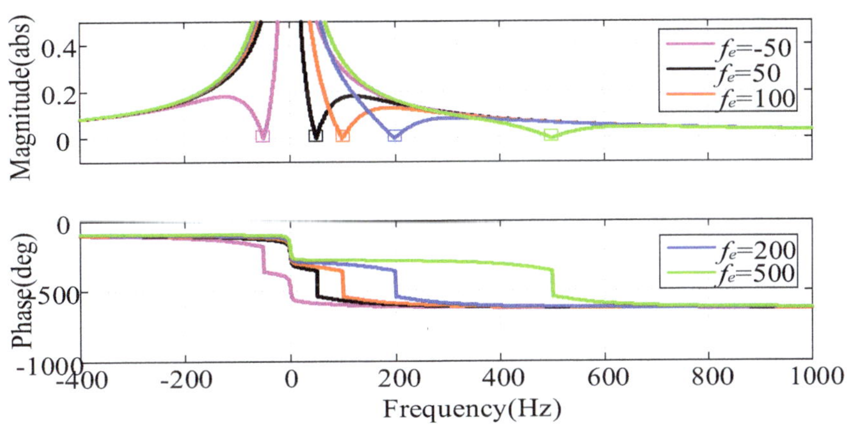

(**b**) Disturbance rejection $I_{\alpha\beta}(s)$ / $I^*_{\alpha\beta}(s)$.

Figure 8. Bode plot of closed-loop current control based on D-PCI controller.

According to Equations (5) and (6), the closed-loop frequency response curves at different control frequencies are shown in Figure 8. The following conclusions can be drawn:

(1) Since the controller provides infinite gain at the interested control frequency (−50 Hz, 50 Hz, 100 Hz, 200 Hz and 500 Hz), unity gain and 0° phase lag output current can be achieved, i.e., $I_{\alpha\beta}(s)/I^*_{\alpha\beta}(s) = 1\angle 0°$, the steady-state error is zero, as shown by "•" in Figure 8a.
(2) As shown by "∘" in Figure 8a, a closed-loop anomalous peak (amplification phenomenon of output current) appears near the control frequency, and, as the control frequency increases, the peak value becomes larger, e.g., no obvious amplification appears at 50 Hz or 100 Hz, while the peak value is 1.02 (1.145) times of the unity gain at 215 Hz (540 Hz). It means that the closed-loop anomalous peak will aggravate the transient oscillation and increase the adjustment time and overshoot with the abrupt change of reference signal. Besides, if phase lock angle is inaccurate, the steady-state output current would be amplified.
(3) As shown in Figure 8b, the disturbance signal at the interested control frequency is completely suppressed, i.e., the magnitude is zero (as shown by "□").

The delay compensation method is used to suppress the closed-loop anomalous peak, and Figure 9 shows the closed-loop frequency response after delay compensation. It can be seen that the unity gain and 0° phase lag output current are achieved at the interested control frequencies, and the closed-loop anomalous peak is completely excluded, as shown by "•" in Figure 9.

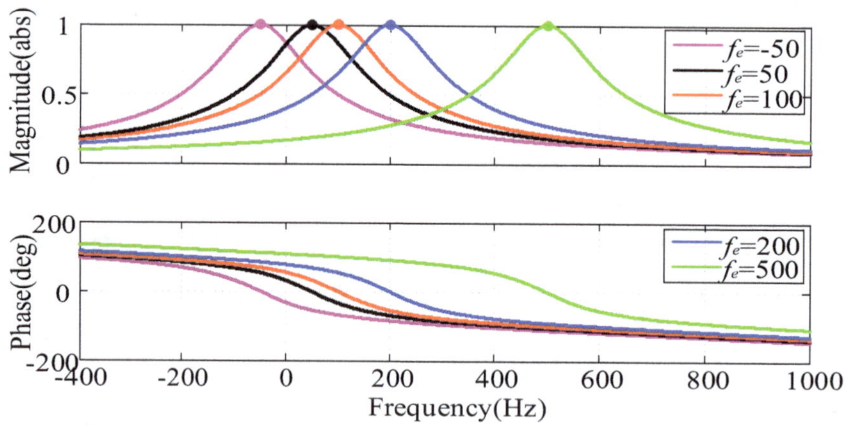

Figure 9. Bode plot of current closed-loop with delay compensation with D-PCI controller.

From the analysis of delay compensation above, the influence of time delay can be ignored. According to the internal model control (IMC) method [13], to cancel the pole of $G^s_{PL}(s)$ by a matching zero in D-PCI controller, $k_i/k_p = R_L/L$ should be satisfied. Hence, Equation (5) can be simplified into Equation (7)

$$G^s_{CL}(s) = \frac{K}{s - j\omega_e + K} \tag{7}$$

where $K = k_p/L$ is the only degree of freedom of the D-PCI controller. Thus, the reference current $i^*_{\alpha\beta}(t)$ can be expressed by

$$i^*_{\alpha\beta}(t) = I_m \cdot \cos(\omega_e t) + jI_m \cdot \sin(\omega_e t) \tag{8}$$

the output current $i_{\alpha\beta}(s)$ is equal to

$$I_{\alpha\beta}(s) = G^s_{CL}(s) \cdot I^*_{\alpha\beta}(s) = \frac{I_m}{s - j\omega_e} - \frac{I_m}{s - j\omega_e + K} \tag{9}$$

in the time domain, Equation (9) can be written as

$$i_{\alpha\beta}(t) = I_m \cdot (\cos\omega_e t - e^{-Kt}) + jI_m \cdot (\sin\omega_e t - e^{-Kt}) \tag{10}$$

if $t \to \infty$ in (10), then

$$\lim_{t\to\infty} i_{\alpha\beta}(t) = I_m \cdot \cos\omega_e t + jI_m \cdot \sin\omega_e t = i^*_{\alpha\beta}(t) \tag{11}$$

It can be seen that D-PCI controller can be used to track ac reference signal without steady-state error, and, with the increase of open-loop gain K, the dynamic response becomes faster. The design of the parameter K is introduced in the next section.

3.3. Parameter Tuning for the D-PCI Controller

In this section, a parameter design method based on root locus is presented. By comprehensive analysis of the IMC method, and as shown in Figure 3, the open-loop transfer function is simplified to

$$G_{OL}^{s}(s) = \frac{(k_p s + k_i) \cdot G_d^s(s)}{(s - j\omega_e)(sL + R_L)} = \frac{K \cdot G_d^s(s)}{(s - j\omega_e)} \tag{12}$$

Taking the following parameters as an example, both the switching and sampling frequency are set to 10 kHz: $L = 5$ mH, $R_L = 0.5\ \Omega$ and $\omega_e = 100\pi$ rad/s. Figure 10 shows the root locus of current loop based on D-PCI controller and Figure 10b is a closer view of the dotted line frame in Figure 10a.

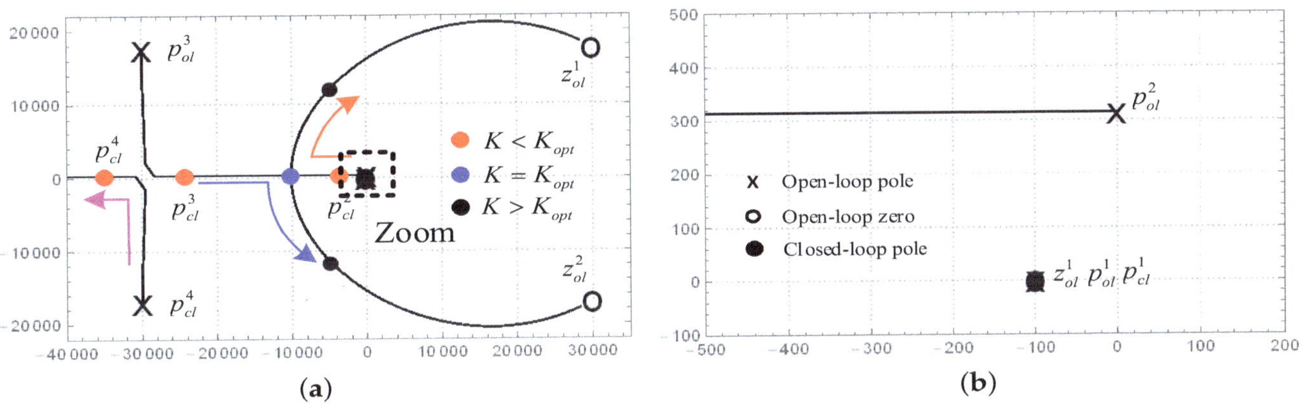

Figure 10. Root locus of current loop based on D-PCI controller: (**a**) general view; and (**b**) closer view of the dotted line frame.

The decay rate σ = Re(p_{cl}) and the damping ratio $\xi = \sigma / |p_{cl}|$ are two important indicators to evaluate the system performance. As shown by the closer view in Figure 10b, the open-loop zero z_{ol}^1 and closed-loop pole p_{ol}^1 overlap with each other, while the effect of p_{ol}^1 on system stability and dynamic response is negligible. When K is low ($K < K_{opt}$), p_{cl}^3 and p_{cl}^4 are far away from imaginary axis (Re(p_{cl}^2) $\ll$ Re $\{(p_{cl}^3),(p_{cl}^4)\}$), p_{cl}^2 becomes the dominant closed-loop pole, the system is in an overdamped state. By increasing K until p_{cl}^2 and p_{cl}^3 overlap (this gain value of K is defined as K_{opt}), the system switches to a state of critical damping ($\xi \approx 1$), as shown by "•" in Figure 10a, and the maximal stability and the fastest dynamic response are obtained simultaneously. If K is increased further, the system becomes underdamped and overshoot occurs in the transient response. At the same time, the stability decreases because p_{cl}^2 and p_{cl}^3 are closer to the imaginary axis. Consequently, K_{opt} is an optimal choice in terms of stability and dynamic response.

Figure 11 shows the time-domain simulation of output current with different gains of K when an abrupt change happens in the reference current. It can be observed that simulation results are consistent with theoretical analysis based on root locus. Overshoot occurs in the current transient response (when $K > K_{opt}$) due to the underdamped characteristics. If $K < K_{opt}$, the excessive damping ratio limits the dynamic performance (although the overshoot is suppressed). When $K = K_{opt}$, there is no overshoot during the transient response. Simulation results are summarized in Table 1, where it can be seen that $K = K_{opt}$ is a best choice if overshot and transient response are considered simultaneously.

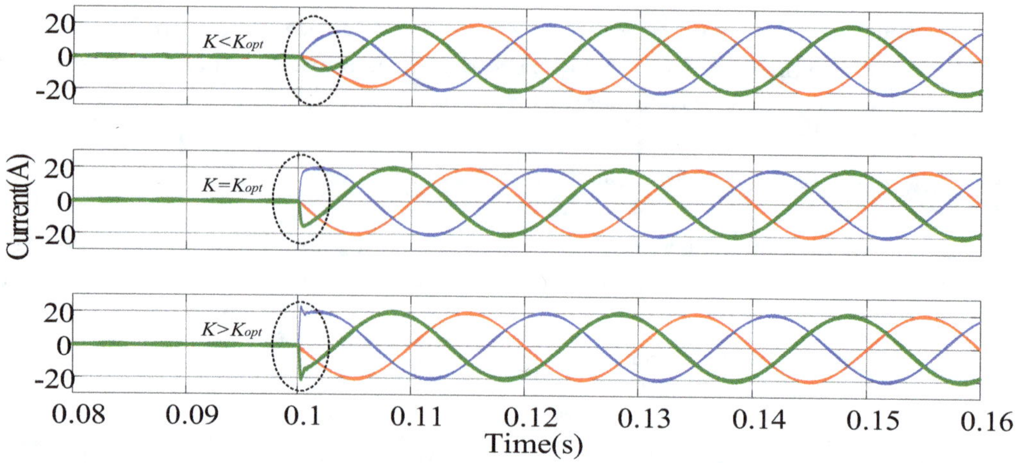

Figure 11. Output current response with different gains of K when an abrupt change happens in the reference current.

Table 1. Evaluation for different value of K

Gain (K)	Damping Ratio ξ	Overshoot	Response Speed
$K < K_{opt}$	$\xi > 1$	no	slow
$K = K_{opt}$	$\xi = 1$	no	fast
$K > K_{opt}$	$\xi < 1$	have	fast

4. Simulation and Experimental Results

To verify the control performance of the proposed D-PCI controller, simulation and experimental results are presented in this section. The main parameters for simulation and experiment are summarized in Table 2.

Table 2. Simulation and experimental setup parameters.

Symbol	Parameters	Value	Unit
V_s	Phase-to-phase voltage	380	V
f	Grid frequency	50	Hz
U_{dc}	DC-link voltage	700	V
L	Inductance of the L filter	5	mH
R_L	Equivalent resistance of the L filter	0.05	Ω
C_{dc}	Capacitor of DC-link	4000	uF
$R(P)$	Active power Load	50(10)	Ω(kW)
I_q^*(Q)	Reactive power Load	21.5(10)	A(kvar)
f_{sw}	Switching frequency	10	kHz
f_s	Sampling frequency	10	kHz
BW	Bandwidth of the current loop	600	Hz
k_{ip}	Proportional gain of the current loop	12.3	/
k_{ii}	Integral gain of the current loop	123	/
k_{vp}	Proportional gain of the voltage loop	0.5	/
k_{vi}	Integral gain of the voltage loop	29.87	/

Figure 12 shows the double closed-loop control scheme of GC-VSC. The outer loop, which has slower dynamics, regulates the DC-link voltage. The inner loop is used to track the reference current, which is the output of the outer loop. In this study, the DC-link voltage was set to 700 V. Comparisons between the proposed and PCI controller were performed in terms of steady-state error and dynamic response.

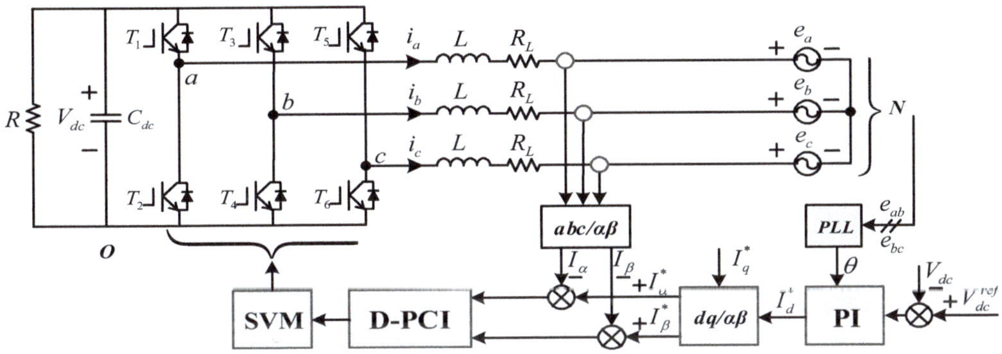

Figure 12. The double closed-loop control scheme of GC-VSC.

4.1. Simulation Results

Figure 13 shows the simulation results of the proposed control strategy. As shown in Figure 13a, the simulation process consisted of several critical time points with load changes, which were used to test the steady-state and dynamic response of the proposed control strategy. Figure 13b–d shows the closer views of Zoom1, Zoom2 and Zoom3 in Figure 13a.

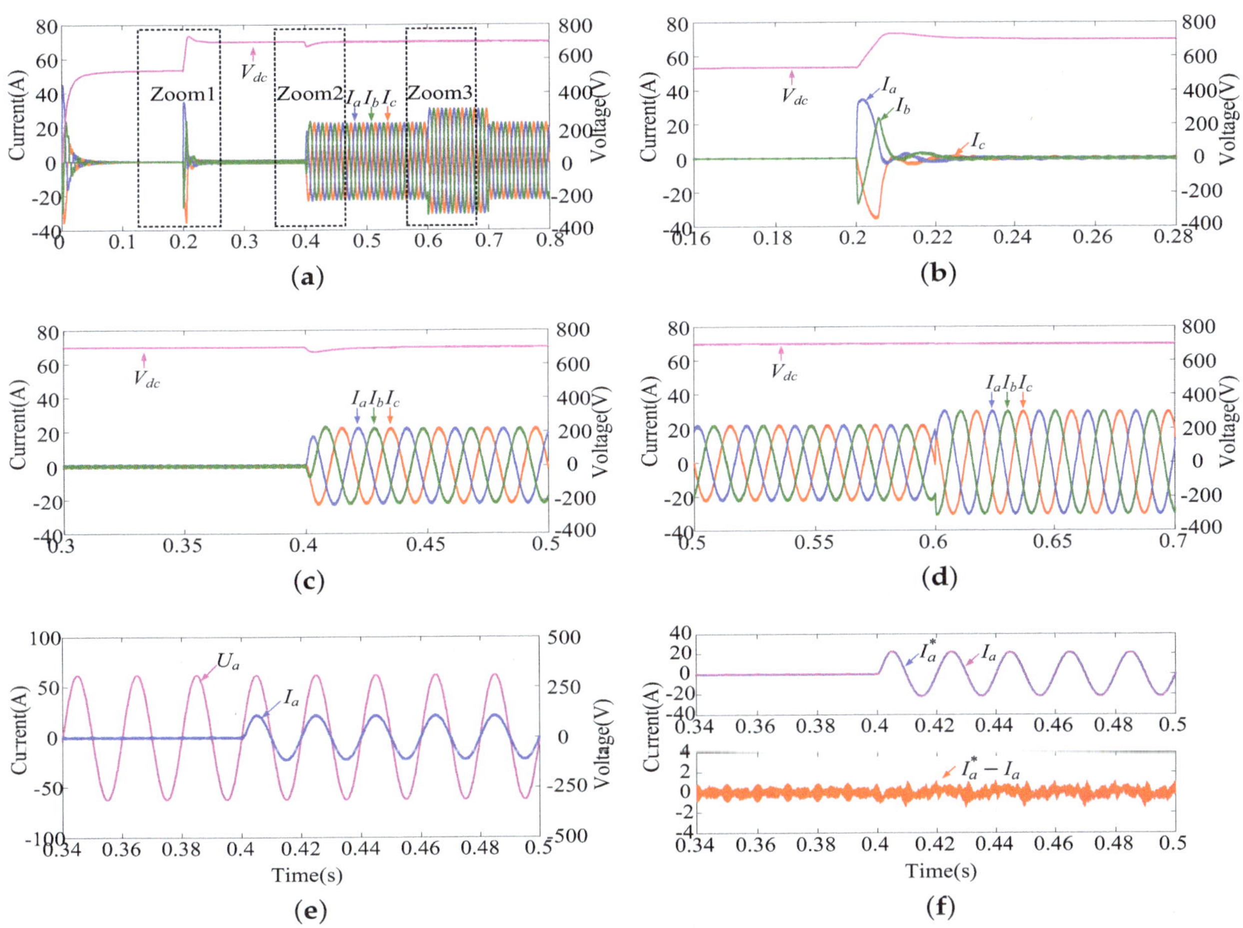

Figure 13. Simulation results of the proposed control strategy: (**a**) general view of the DC-link capacitor voltage and three-phase grid current; (**b**) three-phase grid current when the GC-VSC operates in the state of PWM rectifier; (**c**) three-phase grid current when 10 kW active power is applied; (**d**) three-phase grid current when 10 kvar reactive power is added; (**e**) grid voltage and current of phase A when unity power control is implemented; and (**f**) reference current, actual current and current error of phase A when load changes.

In Figure 13b, it can be seen that the DC-link voltage is regulated to 700 V when the GC-VSC operates in the state of PWM rectifier, which can be replaced by a DC power supply.

Figure 13c shows the transient response of three-phase grid current when 10 kW active power load is applied. It can be seen that the stability of the grid current is achieved after 10 ms. Actually, the dynamic regulation time is less than 10 ms, since the DC-link voltage drop extends the settling time. The excellent dynamic response performance can be further validated by the results in Figure 13d, when 10 kvar reactive power load is added. There is almost no transient regulation process for grid current. In fact, since the reactive power is not consumed, the DC-link voltage does not drop significantly. Moreover, as shown in Figure 13c,d, negligible overshoot is obtained when using the parameter design method in Section 3.

Figure 13e represents the grid voltage and current of phase A with the unit power factor control method. The phase is exactly the same in steady state, which reflects the tracking ability of the D-PCI controller. As shown in Figure 13f, the curves (reference current I_a^* and actual I_a) come closest to coinciding in shape even if a sudden change occurs at 0.4 s, and the current error fluctuates around zero. It shows that the D-PCI controller has the ability to track AC reference signals with zero steady-state error and fast dynamic response.

Figure 14 represents the simulation results with PCI controller. The parameters are consistent with the D-PCI-based control strategy. In Figures 13c and 14a, it can be seen that the dynamic response of D-PCI is faster than PCI, and the excellent performance can be further observed in Figures 13f and 14b. The PCI-based grid current tracks its reference with zero steady-state error after about 30 ms while it takes about 15 ms for the proposed control strategy.

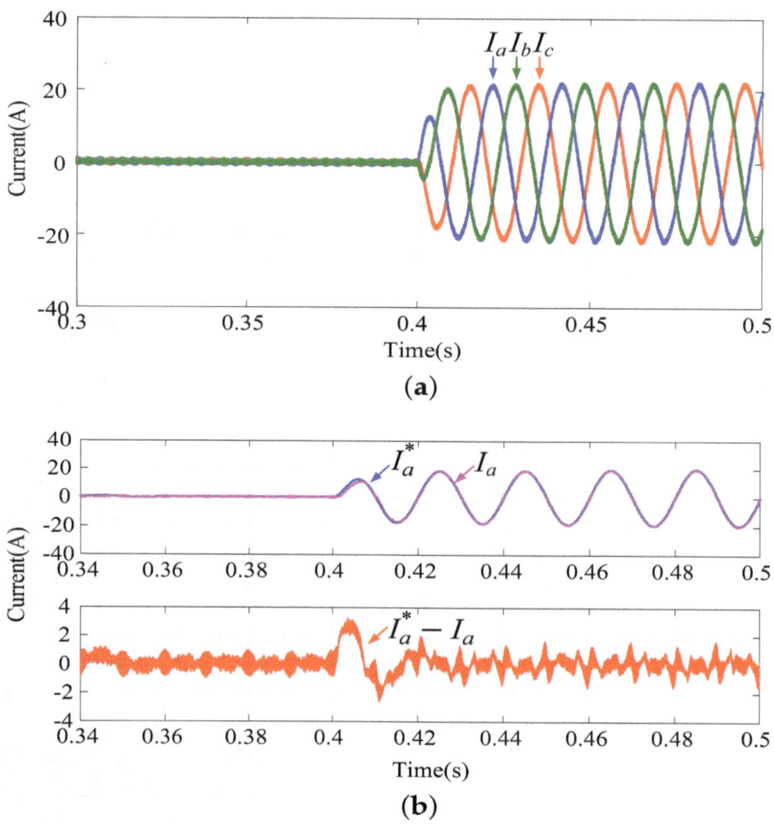

Figure 14. Simulation results of the PCI-based control strategy: (**a**) three-phase grid current when 10 kW active power is applied; and (**b**) reference current, actual current and current error of phase A when load changes.

4.2. Experimental Results

To further support the theoretical analysis and simulation results, the experimental setup of GC-VSC was built in the laboratory, as shown in Figure 15. The parameters in experimental setup were the same as in simulation ones, which are listed in Table 1. Active power load consists of two parallel 100 Ω/10 kW resistors. The real-time algorithm was implemented in the hardware controller, based on TMS320C28346 DSP and EP3C16Q240 FPGA, whose output PWM signals are connected to converter by optical fiber. Experimental waveforms were captured by Yokogawa DL850 oscilloscope. Specifically, the hardware controller lacked of digital-to-analog converter and, since the reference current was a digital variable, it could not be directly measured by the oscilloscope. Therefore, real-time data were acquired by saving memory in Code Composer Studio (Ver.5.5.0) and the waveforms (Figures 16f and 17b) were plotted in MATLAB.

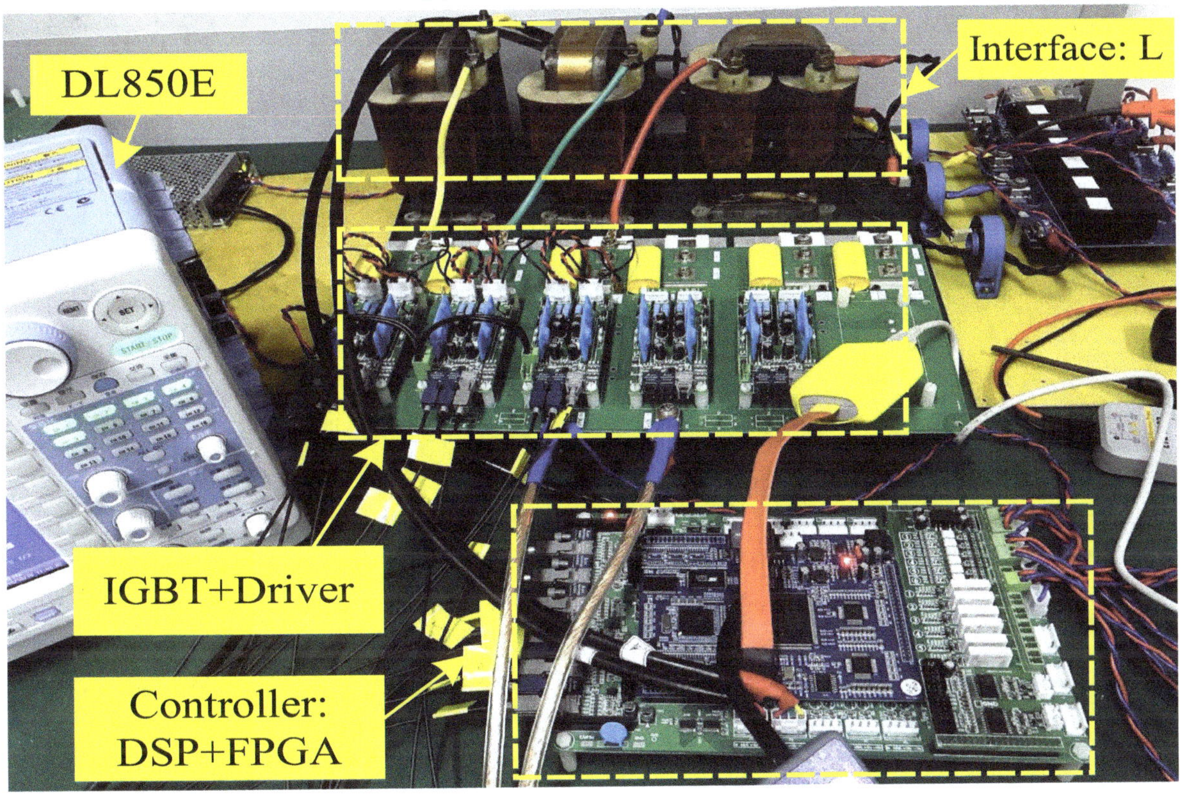

Figure 15. Experimental setup of the GC-VSC.

Figures 16 and 17 show the experimental results of the proposed and PCI-based control strategy, which correspond to the simulation results in Figures 13 and 14, respectively. It can be observed that the experimental and simulation results match to a great extent. The slight difference is the distortion of grid current, which can be mainly attributed to harmonics contained in the grid voltage.

Figure 16 shows that the excellent stability and fast dynamic response of the proposed D-PCI-based control strategy can be obtained even though the GC-VSC experiences different load variation. The superiority can be further verified by the comparative experiments. From the results in Figures 16f and 17b, the following can be concluded: (1) the current error of both strategies tends to zero, which means that tracking of ac reference signal with zero steady-state error can be realized from both methods; and (2) the proposed strategy has faster dynamic response, as the error current of the PCI-based strategy restores to zero after about 30 ms while there is almost no dynamic process for the proposed control strategy.

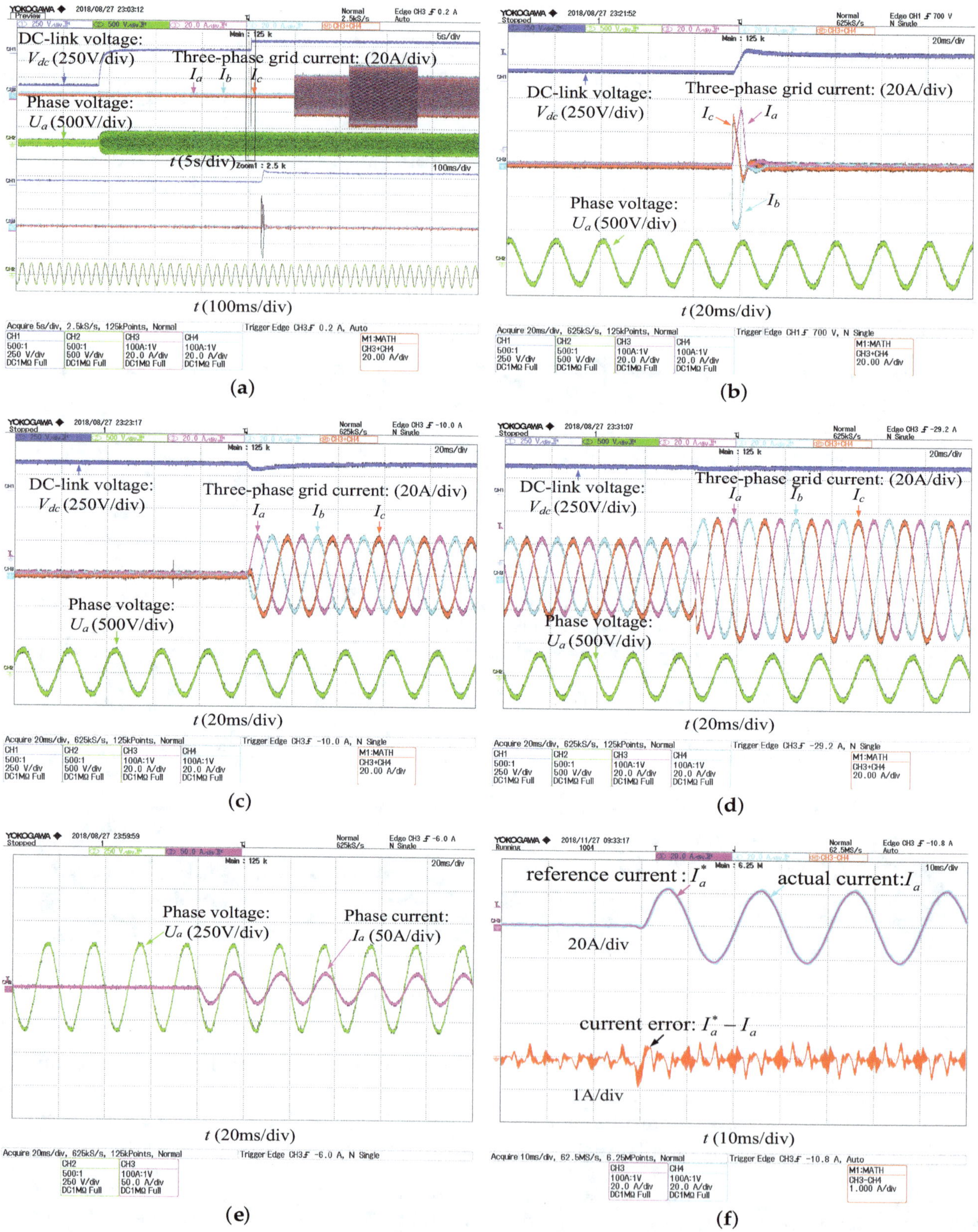

Figure 16. Experimental results of the proposed control strategy: (**a**) general view of the DC-link capacitor voltage and three-phase grid current; (**b**) three-phase grid current when the GC-VSC operates in the state of PWM rectifier; (**c**) three-phase grid current when 10 kW active power is applied; (**d**) three-phase grid current when 10 kvar reactive power is added; (**e**) grid voltage and current of phase A when unity power control is implemented; and (**f**) reference current, actual current and current error of phase A when load changes.

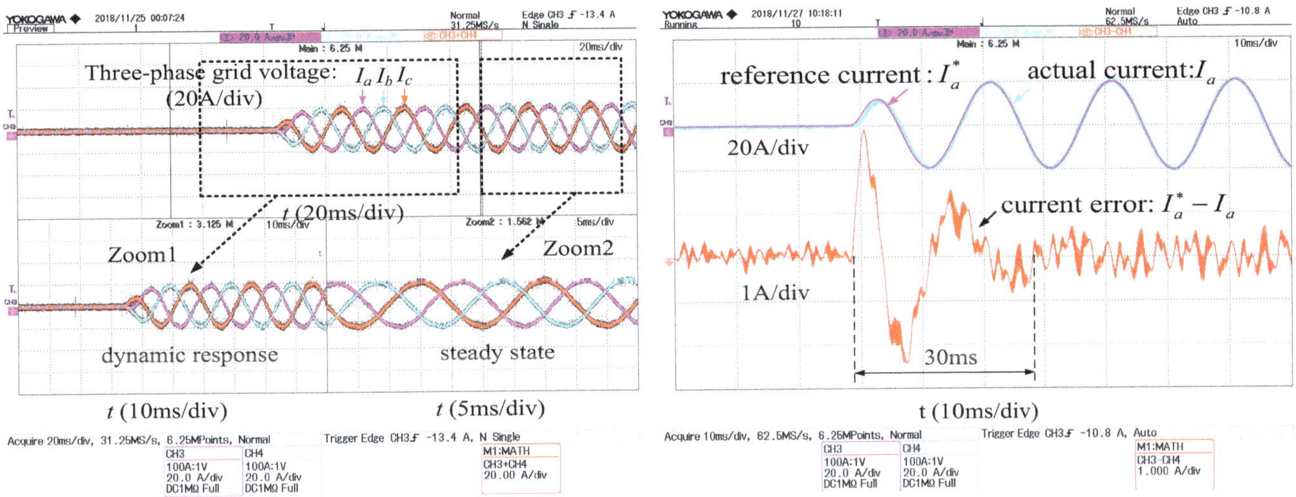

(a) (b)

Figure 17. Experimental results of the PCI-based control strategy: (**a**) three-phase grid current when 10 kvar reactive power is applied; and (**b**) reference current, actual current and current error of phase A when load changes.

5. Conclusions

To attenuate the impact of axes cross-coupling caused by PCI controller, a novel D-PCI controller-based decoupling method in the stationary frame is proposed to track sinusoidal reference signals with zero steady-state error and achieve fast dynamic response. Moreover, an optimal gain method for parameter tuning is presented for enhancing the stability of the current loop. Comparing with PCI, the proposed D-PCI controller can obtain faster dynamic response, as well as better stability performance. Comparative simulations were performed and experimental results between the proposed and PCI controller were compared, which validated the superiority of the proposed D-PCI controller.

Author Contributions: Conceptualization, S.Z. and J.Z.; methodology, S.Z.; software, S.Z.; validation, S.Z., P.W. and B.Y.; data curation, K.L.; writing—original draft preparation, S.Z.; writing—review and editing, Z.Z.; project administration, J.Z. and K.L.; and funding acquisition, J.Z.

References

1. Bose, B. Global Energy Scenario and Impact of Power Electronics in 21st Century. *IEEE Trans. Ind. Electron.* **2013**, *60*, 2638–2651. [CrossRef]
2. Arbab-Zavar, B.; Palacios-Garcia, E.; Vasquez, J.; Guerrero, J. Smart Inverters for Microgrid Applications: A Review. *Energies* **2019**, *12*, 840. [CrossRef]
3. Molina, M. Energy Storage and Power Electronics Technologies: A Strong Combination to Empower the Transformation to the Smart Grid. *Proc. IEEE.* **2017**, *105*, 2191–2219. [CrossRef]
4. Bose, B. Power Electronics, Smart Grid, and Renewable Energy Systems. *Proc. IEEE.* **2017**, *105*, 2011–2018. [CrossRef]
5. Pang, B.; Nian, H. Improved Operation Strategy with Alternative Control Targets for Voltage Source Converter under Harmonically Distorted Grid Considering Inter-Harmonics. *Energies* **2019**, *12*, 1236. [CrossRef]
6. Xie, C.; Zhao, X.; Li, K.; Zou, J.; Guerrero, J.M. A New Tuning Method of Multi-Resonant Current Controllers for Grid-Connected Voltage Source Converters. *IEEE J. Emerg. Sel. Top. Power Electron.* **2019**, *7*, 458–466. [CrossRef]
7. Busada, C.A.; Jorge, S.G.; Solsona, J. A. Resonant Current Controller With Enhanced Transient Response for Grid-Tied Inverters. *IEEE Trans. Ind. Electron.* **2018**, *65*, 2935–2944. [CrossRef]
8. Nie, C.; Wang, Y.; Lei, W.; Li, T.; Yin, S. Modeling and Enhanced Error-Free Current Control Strategy for Inverter with Virtual Resistor Damping. *Energies* **2018**, *11*, 2499. [CrossRef]

9. Holmes, D.G.; Lipo, T.A.; McGrath, B.P.; Kong, W.Y. Optimized Design of Stationary Frame Three Phase AC Current Regulators. *IEEE Trans. Power Electron.* **2009**, *24*, 2417–2426. [CrossRef]
10. Guo, X.; Guerrero, J.M. General Unified Integral Controller With Zero Steady-State Error for Single-Phase Grid-Connected Inverters. *IEEE Trans. Smart Grid.* **2016**, *7*, 74–83. [CrossRef]
11. Guo, X.Q.; Wu, W.Y. Improved current regulation of three-phase grid-connected voltage-source inverters for distributed generation systems. *IET Renew. Power Gener.* **2010**, *4*, 101–115. [CrossRef]
12. Yuan, X.M.; Merk, W.; Stemmler, H.; Allmeling, J. Stationary-frame generalized integrators for current control of active power filters with zero steady-state error for current harmonics of concern under unbalanced and distorted operating conditions. *IEEE Trans. Ind. Appl.* **2002**, *38*, 523–532. [CrossRef]
13. Harnefors, L.; Nee, H. Model-based current control of AC machines using the internal model control method. *IEEE Trans. Ind. Appl.* **1998**, *34*, 133–141. [CrossRef]
14. Lascu, C.; Asiminoaei, L.; Boldea, I.; Blaabjerg, F. Frequency Response Analysis of Current Controllers for Selective Harmonic Compensation in Active Power Filters. *IEEE Trans. Ind. Electron.* **2009**, *56*, 337–347. [CrossRef]
15. Lascu, C.; Asiminoaei, L.; Boldea, I.; Blaabjerg, F. High Performance Current Controller for Selective Harmonic Compensation in Active Power Filters. *IEEE Trans. Power Electron.* **2007**, *22*, 1826–1835. [CrossRef]
16. Vidal, A.; Freijedo, F.D.; Yepes, A.G. Assessment and Optimization of the Transient Response of Proportional-Resonant Current Controllers for Distributed Power Generation Systems. *IEEE Trans. Ind. Electron.* **2013**, *60*, 1367–1383. [CrossRef]
17. Briz, F.; Degner, M.W.; Lorenz, R.D. Analysis and design of current regulators using complex vectors. *IEEE Trans. Ind. Appl.* **2000**, *36*, 817–825. [CrossRef]
18. Zmood, D.N.; Holmes, D.G. Stationary frame current regulation of PWM inverters with zero steady-state error. *IEEE Trans. Power Electron.* **2003**, *18*, 814–822. [CrossRef]
19. Bae, B.; Sul, S.A compensation method for time delay of full-digital synchronous frame current regulator of PWM AC drives. *IEEE Trans. Ind. Appl.* **2003**, *39*, 802–810. [CrossRef]
20. Pérez-Estévez, D.; Doval-Gandoy, J.; Yepes, A.G.; López, Ó. Positive- and Negative-Sequence Current Controller With Direct Discrete-Time Pole Placement for Grid-Tied Converters With LCL Filter. *IEEE Trans. Power Electron.* **2017**, *32*, 7207–7221. [CrossRef]
21. Busada, C.A.; Jorge, S.G.; Leon, A.E.; Solsona, J.A. Current Controller Based on Reduced Order Generalized Integrators for Distributed Generation Systems. *IEEE Trans. Ind. Electron.* **2012**, *59*, 2898–2909. [CrossRef]
22. Xie, C.; Zhao, X.; Li, K.; Liu, D.; Guerrero, J.M; Vasquez, J.C. Phase Compensated Reduced Order Generalized Integrators for Grid-Tied VSCs With Harmonics Compensation Capability. *IEEE Trans. Ind. Appl.* **2018**, *54*, 2568–2578. [CrossRef]
23. Hwang, J.G.; Lehn, P.W.; Winkelnkemper, M. A Generalized Class of Stationary Frame-Current Controllers for Grid-Connected AC–DC Converters. *IEEE Trans. Power Del.* **2010**, *25*, 2742–2751. [CrossRef]
24. Jorge, S.G.; Busada, C.A.; Solsona, J.A. Frequency-Adaptive Current Controller for Three-Phase Grid-Connected Converters. *IEEE Trans. Ind. Electron.* **2013**, *60*, 4169–4177. [CrossRef]
25. Jorge, S.G.; Busada, C.A.; Solsona, J. Reduced order generalised integrator-based current controller applied to shunt active power filters. *IET Power Electron.* **2013**, *7*, 1083–1091. [CrossRef]
26. Yepes, A.G.; Vidal, A.; Malvar, J.; López, O.; Doval-Gandoy, J. Tuning Method Aimed at Optimized Settling Time and Overshoot for Synchronous Proportional-Integral Current Control in Electric Machines. *IEEE Trans. Power Electron.* **2014**, *29*, 3041–3054. [CrossRef]

Permissions

All chapters in this book were first published in MDPI; hereby published with permission under the Creative Commons Attribution License or equivalent. Every chapter published in this book has been scrutinized by our experts. Their significance has been extensively debated. The topics covered herein carry significant findings which will fuel the growth of the discipline. They may even be implemented as practical applications or may be referred to as a beginning point for another development.

The contributors of this book come from diverse backgrounds, making this book a truly international effort. This book will bring forth new frontiers with its revolutionizing research information and detailed analysis of the nascent developments around the world.

We would like to thank all the contributing authors for lending their expertise to make the book truly unique. They have played a crucial role in the development of this book. Without their invaluable contributions this book wouldn't have been possible. They have made vital efforts to compile up to date information on the varied aspects of this subject to make this book a valuable addition to the collection of many professionals and students.

This book was conceptualized with the vision of imparting up-to-date information and advanced data in this field. To ensure the same, a matchless editorial board was set up. Every individual on the board went through rigorous rounds of assessment to prove their worth. After which they invested a large part of their time researching and compiling the most relevant data for our readers.

The editorial board has been involved in producing this book since its inception. They have spent rigorous hours researching and exploring the diverse topics which have resulted in the successful publishing of this book. They have passed on their knowledge of decades through this book. To expedite this challenging task, the publisher supported the team at every step. A small team of assistant editors was also appointed to further simplify the editing procedure and attain best results for the readers.

Apart from the editorial board, the designing team has also invested a significant amount of their time in understanding the subject and creating the most relevant covers. They scrutinized every image to scout for the most suitable representation of the subject and create an appropriate cover for the book.

The publishing team has been an ardent support to the editorial, designing and production team. Their endless efforts to recruit the best for this project, has resulted in the accomplishment of this book. They are a veteran in the field of academics and their pool of knowledge is as vast as their experience in printing. Their expertise and guidance has proved useful at every step. Their uncompromising quality standards have made this book an exceptional effort. Their encouragement from time to time has been an inspiration for everyone.

The publisher and the editorial board hope that this book will prove to be a valuable piece of knowledge for researchers, students, practitioners and scholars across the globe.

List of Contributors

Robert Baždarić, Danjel Vončina and Igor Škrjanc
Faculty of Electrical Engineering, University of Ljubljana, Tržaška 25, Ljubljana 1000, Slovenia

Javier Calvente, Abdelali El Aroudi, Roberto Giral, Angel Cid-Pastor, Enric Vidal-Idiarte and Luis Martínez-Salamero
Departament d'Enginyeria Electrònica, Elèctrica i Automàtica, Universitat Rovira i Virgili, 43007 Tarragona, Spain

Mauricio Dalla Vecchia , Giel Van den Broeck , Simon Ravyts and Johan Driesen
Departement Elektrotechniek, KU Leuven, Kasteelpark Arenberg 10, 3001 Leuven, Belgium EnergyVille, Thor Park 8310, 3600 Genk, Belgium

Mariam Saeed , María R. Rogina, Alberto Rodríguez, Manuel Arias and Fernando Briz
Department of Electrical Engineering, University of Oviedo, 33204 Asturias, Spain

Pablo Zumel, Cristina Fernández, Marlon A. Granda, Antonio Lázaro and Andrés Barrado
Grupo de Sistemas Electrónicos de Potencia, Departamento de Tecnología Electrónica, Escuela Politécnica Superior, Universidad Carlos III de Madrid, 28911 Leganes, Madrid, Spain

Daliang Yang and Li Yin
College of Electrical Engineering, Guangxi University, Nanning 530004, China

Shengguang Xu
Liuzhou Power Supply Bureau, Gxuangxi Power Grid Corporation, Liuzhou 545005, China

Ning Wu
Electric Power Research Institute, Guangxi Power Grid Corporation, Nanning 530023, China

Huipin Lin, Xiao Zhou and Zhengyu Lu
College of Electrical Engineering, Zhejiang University, Hangzhou 310027, China

Lujun Wang
Hubei Key Laboratory for High-Efficiency Utilization of Solar Energy and Operation Control of Energy Storage System, Hubei University of Technology, Wuhan 430068, China

Ahmed H. Okilly and Jeihoon Baek
Electrical & Electronics and Communication Engineering Department, Koreatech University, Cheonan-si 31253, Korea

Namhun Kim
Research Center, ESTRA Automotive, Daegu-si 42981, Korea

Woo-Young Choi and Min-Kwon Yang
Division of Electronic Engineering, Chonbuk National University, Jeonju 561-756, Korea

Leopoldo Gil-Antonio
Faculty of Engineering, Autonomous University of the State of Mexico, Instituto Literario No. 100 Oriente, Toluca 50130, Estado de México, Mexico
Tecnológico de Estudios Superiores de Jocotitlán Carretera Toluca-Atlacomulco km. 44.8, Jocotitlán 50700, Estado de México, Mexico

Belem Saldivar
Faculty of Engineering, Autonomous University of the State of Mexico, Instituto Literario No. 100 Oriente, Toluca 50130, Estado de México, Mexico
Cátedras CONACYT, Av. Insurgentes Sur 1582, Col. Crédito Constructor, Alcaldía Benito Juárez, Ciudad de México 03940, Mexico

Otniel Portillo-Rodríguez, Juan Carlos Ávila-Vilchis and Rigoberto Martínez-Méndez
Faculty of Engineering, Autonomous University of the State of Mexico, Instituto Literario No. 100 Oriente, Toluca 50130, Estado de México, Mexico

Pánfilo Raymundo Martínez-Rodríguez
School of Sciencies, UASLP, San Luis Potosi 78290, SLP, Mexico

Seung-Hwan Lee
School of Electrical and Computer Engineering, University of Seoul, Seoul 02504, Korea

Kyung-Pyo Yi and Myung-Yong Kim
Smart Electrical & Signaling Division, Korea Railroad Research Institute, Uiwang 16105, Korea

Sarah Saeed, Ramy Georgious and Jorge Garcia
LEMUR Research Group, Department of Electrical, Electronics, Computers and Systems Engineering, University of Oviedo, 33204 Gijon, Spain

Sen Zhang, Jianfeng Zhao, Kangli Liu, Pengyu Wang and Bin Yang
School of Electrical Engineering, Southeast University, Nanjing 210096, China

Zhihong Zhao
Department of Electrical Engineering, School of Automation, Nanjing University of Science and Technology, Nanjing 210094, China

Index

www.ingramcontent.com/pod-product-compliance
Lightning Source LLC
LaVergne TN
LVHW080954210926
848485LV00004B/836